Stochastic Methods in Quantum Mechanics

STANLEY P. GUDDER
John Evans Professor of Mathematics
University of Denver

DOVER PUBLICATIONS, INC.
Mineola, New York

Bibliographical Note

This Dover edition, first published in 2005, is an unabridged republication of the work first published by Elsevier North Holland, Inc., New York, in 1979. The figure on page 16 has been revised for this edition.

Library of Congress Cataloging-in-Publication Data

Gudder, Stanley.
   Stochastic methods in quantum mechanics / Stanley P. Gudder.
       p. cm.
   Originally published: New York : Elsevier North Holland, 1979, in series: North Holland series in probability and applied mathematics.
   Includes bibliographical references and index.
   ISBN 0-486-44532-1 (pbk.)
   1. Stochastic processes. 2. Quantum theory. I. Title.

QC174.17.S76G82 2005
530.12—dc22

                                                      2005053024

Manufactured in the United States of America
Dover Publications, Inc., 31 East 2nd Street, Mineola, N.Y. 11501

This book is dedicated to my wife, Paula.
She typed the entire manuscript and provided me
with constant encouragement.

Contents

Preface

Although quantum mechanics has long been recognized as a stochastic subject, only recently have some of the deeper techniques and results of probability theory been substantially utilized. In the last ten years probability theory has not only been applied to clarifying the foundations of quantum mechanics, it has been used in the frontiers of constructive quantum field theory, and applied to the practical problems of communication engineering and laser technology.

In this introductory text we shall survey some of the stochastic methods and techniques in quantum mechanics. It is our intention to strive more for breadth than for depth. We shall try to give the reader the flavor of different areas in this subject, and bring him to a level of understanding so that he can profitably read the current literature. We include an extensive bibliography so that the reader can continue on his own in the direction of his individual interests. Although the bibliography does not approach completeness (we would need over 1,000 references for this) we hope it is representative.

A basic knowledge of functional analysis (about the level of a first-year graduate course) is assumed. Some experience with probability theory and quantum mechanics would be helpful, but the necessary ideas and results from these disciplines will be developed as needed and no prior knowledge is required. Each chapter ends with a selection of exercises. Many of the exercises involve a straightforward verification of results stated in the text, although a few will challenge even the most attentive reader. The proofs of most of the theorems are included. Exceptions are routine proofs that are left as exercises and long or highly technical proofs for which explicit references are given.

The following paraphrase of a statement by B. Simon [250] stresses a point of extreme importance.

In the early days of constructive field theory (1964–1970) probabilistic ideas played an important role in the developments, but a group of workers in the field, of whom I was one of the more vocal, tried hard to eliminate the use of probability theory. I was partly motivated by a not atypical functional analyst's suspicion of probability theory as nothing more than a subset of functional analysis with strange names whose purpose is to confuse those who have not bothered to join the club. There is a sense in which this attitude is correct; any statement in probability theory has a translation into functional analysis. But this somehow misses the point. Like any other foreign language, probability theory is structured around its own natural thought patterns and so is critical to a mode of thinking; put more prosaically, certain exceedingly natural constructions of probability theory look ad hoc and unnatural when viewed as functional analytic constuctions. I confess to have seen the error of my ways and. . . .

Stochastic Methods
in Quantum Mechanics

1

The History of Quantum Mechanics

This chapter has a twofold purpose. It gives the reader a historical perspective, and introduces some of the important concepts of quantum mechanics. As the reader will notice, probability theory did not enter into quantum mechanics at the outset. The pioneers of the field had no reason or intention to make quantum mechanics a probabilistic subject. The stochastic nature of quantum mechanics was reluctantly accepted later when it proved to be an intrinsic, inescapable part of the field.

1.1. Early History

The idea of a quantum of energy was first introduced by Max Planck on December 14, 1900 at a meeting of the German Physical Society. He invented this concept to explain an empirical formula for the energy per unit volume $E(\nu)$ of radiation of frequency ν emitted by a body heated to a temperature T.

In the late 1800s, Rayleigh and Jeans derived a formula for $E(\nu)$ using classical mechanics. Suppose heat radiation is emitted through a small door of a cubical furnace of length L which is heated to a temperature T. The energy of the radiation $e(\nu)$ at frequency ν is found by determining the number of waves $n(\nu)$ with frequency ν and multiplying $n(\nu)$ by the energy associated with these waves. According to a time-honored principle of physics, the equipartition principle, each wave should carry KT units of energy on the average, where K is Boltzmann's constant. Hence $e(\nu) = n(\nu)KT$. To find $n(\nu)$, the waves are represented mathematically by plane

1

waves $\exp[2\pi i(\mathbf{k}\cdot\mathbf{r} - \nu t)]$ where $\mathbf{k}$ is a vector of length $k = \nu/c$ (c is the speed of light) in the direction of wave propagation with components $(n_1/L, n_2/L, n_3/L)$, n_1, n_2, n_3 being integers. The integers appear since we must have a whole number of wavelengths in the furnace cavity. We now find the number of waves $n(\nu)\Delta\nu$ with frequency between ν and $\nu + \Delta\nu$. Draw a three-dimensional lattice with dots at each point with integer coordinates and draw a spherical shell whose inner and outer radii are k and $k + \Delta k$, respectively. Then the number of waves whose wavenumber is between k and $k + \Delta k$ is the number of dots in this shell. This number is approximately the volume of the shell divided by the volume surrounding an individual dot and thus equals

$$\left[(k+\Delta k)^3 - k^3\right]4\pi L^3/3 \approx 4\pi L^3 k^2 \Delta k. \tag{1.1}$$

Hence

$$n(\nu)\Delta\nu = 4\pi L^3 \nu^2 \Delta\nu/c^3$$

and

$$E(\nu) = 4\pi K T\nu^2/c^3. \tag{1.2}$$

Formula (1.2), called the Rayleigh–Jeans law, agrees with experiment for low frequencies ν. However, for high frequencies, (1.2) not only deviates substantially from experiment, it gives a contradiction called the ultraviolet catastrophy. According to Eq. (1.2), as the frequency increases so does the corresponding energy. The ultraviolet catastrophy concludes that the total energy per unit volume due to all frequencies is infinite.

The ultraviolet catastrophy was one of the first instances in which classical mechanics was found to be inadequate. Planck's energy quantum, invented to derive the correct formula for $E(\nu)$ and thus avoid this catastrophy, ushered in a new era, the era of quantum mechanics. Planck's idea was so unusual and seemed so grotesque at the time, that he himself could hardly believe it. Planck suggested that instead of allowing the energy of radiation waves to have arbitrary values, we make the following postulate.

The energy of electromagnetic waves can exist only in the form of discrete packages, or quanta, the energy content of each package being directly proportional to its frequency.

Using this postulate, Planck derived his radiation law. A wave with frequency ν can have only a countable number of energy values $e(n) = nh\nu$, $n = 0, 1, 2, \ldots$, where h is a constant. According to classical statistical mechanics, these energy values are distributed according to the Gibbs probability $p_n = c_0 \exp[-e(n)/KT]$, where c_0 is a normalization constant.

Since $\Sigma p_n = 1$ we obtain

$$c_0 = 1 - \exp(-h\nu/KT). \tag{1.3}$$

The average energy $e_\nu = \Sigma e(n)p_n$ of a wave with frequency ν becomes

$$e_\nu = h\nu \exp(-h\nu/KT)[1 - \exp(-h\nu/KT)]^{-1}. \tag{1.4}$$

Using our previous formula for $n(\nu)$ the energy per unit volume $E(\nu) = n(\nu)e_\nu/L^3$ becomes

$$E(\nu) = 4\pi h\nu^3 c^{-3}[1 - \exp(-h\nu/KT)]^{-1} \exp(-h\nu/KT). \tag{1.5}$$

Planck's radiation law (1.5) agrees with the Rayleigh–Jeans law (1.2) at low frequencies. Moreover, (1.5) agrees remarkably with experiment for all frequencies if we let $h = 6.62 \times 10^{-27}$ erg sec.

Planck's basic postulate was also substantiated later by other phenomena such as Einstein's photoelectric effect and the Compton effect. However, one of the most important applications appeared in Bohr's quantum orbits. In 1911 Ernest Rutherford, using experiments involving the scattering of α particles through gold foil, discovered the model for the atom which is essentially accepted today. Rutherford concluded from his experiments that an atom must consist of a very small positively charged nucleus surrounded at a relatively large distance by enough negatively charged electrons to balance the nuclear charge. These findings again contradicted classical theory. The electrons could not be stationary since then they would be attracted to the nucleus, causing the atom to collapse, so the electrons must be orbiting the nucleus. But by classical electrodynamics, a charged particle which is accelerating must radiate energy, and one can calculate, using classical theory, that the electrons should radiate away all of their kinetic energy within 10^{-8} sec. Having lost all their kinetic energy, atomic electrons must fall into the nucleus and the atom ceases to exist!

Another phenomenon concerning atoms which classical physics could not explain was the light spectrum emitted by heated elements. If an element is heated to a high temperature it will emit light consisting of a sequence of sharp spectral lines which show up on a photographic plate after the light is refracted by a prism. In the 1800s spectroscopists had amassed huge quantities of data concerning the frequencies of these spectral lines. They found, for example, that the spectral lines of light emitted by heated hydrogen had frequencies $\nu_{m,n}$ labeled by two positive integers $n < m$ and given by

$$\nu_{m,n} = R(n^{-2} - m^{-2}), \tag{1.6}$$

where $R = 3.289 \times 10^{15}$ sec^{-1}.

In 1913, Niels Bohr, using Planck's postulate, gave a theoretical explanation not only for why the atom does not collapse but for the formula (1.6). Bohr reasoned that the light emitted by an excited hydrogen atom was due to a loss of energy by its single orbital electron. It follows that if no light is emitted then the electron's energy remains constant. Moreover, since the energy carried by the light radiation must, according to Planck's postulate, have only discrete values, the energy of the orbital electron must also have discrete energy values $E_1, E_2, \ldots$. When the electron jumps from an energy level E_m to a lower energy level E_n, the energy of the emitted light is $E_m - E_n$. Again, by Planck's postulate, this energy must be proportional to the frequency of the emitted light. Hence, the frequency may be labeled by two positive integers $n < m$ and is given by $h\nu_{m,n} = E_m - E_n$ or

$$\nu_{m,n} = (E_m - E_n)/h. \tag{1.7}$$

Guided by the above reasoning, Bohr announced the following postulates.

1. The hydrogen electron moves in a circular orbit and the electron's energy remains constant so long as it is in the same orbit.
2. Only a discrete number of orbits are allowed. In the allowed orbits, the electron must have angular momentum equal to a multiple of $h/2\pi = \hbar$.
3. When an electron shifts from one possible orbit to another, the energy difference $E_m - E_n$ is transformed into radiation of frequency $(E_m - E_n)/h$.

We now show that the spectroscopic data in Eq. (1.6) follows from Bohr's three postulates. Suppose the electron is in the nth possible orbit. This orbit must be circular by Postulate 1. Let r_n be the radius of the orbit and let μ and e be the known mass and charge, respectively, of the electron. If v_n is the speed of the electron (which is a constant) then it is well known that the acceleration has magnitude v_n^2/r_n. Since the electron experiences a Coulomb attractive force of magnitude e^2/r_n^2, Newton's second law gives $\mu v_n^2/r_n = e^2/r_n^2$. Hence $v_n = e(\mu r_n)^{-1/2}$. Since the angular momentum has magnitude $\mu v_n r_n$, Postulate 2 implies that $\mu v_n r_n = nh$. Solving these last two equations for r_n we obtain $r_n = n^2\hbar^2/\mu e^2$. (In particular $r_1 = \hbar^2/\mu e^2 \approx 10^{-8}$ cm, which agrees with experiment.) Since the total energy E_n is the sum of the kinetic energy K_n and the potential energy V_n we have

$$E_n = K_n + V_n = \mu v_n^2/2 - e^2/r_n = -2\pi^2 e^4 \mu/n^2 h^2. \tag{1.8}$$

Hence, by Postulate 3

$$\nu_{n,m} = (E_m - E_n)/h = 2\pi^2 e^4 \mu h^{-3}(n^{-2} - m^{-2}) \tag{1.9}$$

From the known values of e, μ, and h one finds that $2\pi^2 e^4 \mu h^{-3} = 3.289 \times 10^{15}$ sec^{-1} and so Eq. (1.9) reduces to Eq. (1.6).

Electromagnetic radiation is naturally associated with a wave-type motion. However, in certain phenomena such as the photoelectric effect and the Compton effect, radiation acts like a stream of particles or photons. In 1925, de Broglie postulated his wave–particle duality principle, which asserts that particles also play this dual role. In particular, electrons should have wavelike properties. He asserted that the hydrogen orbital electron in its nth quantum orbit should have associated with it a wave which "fits" in its orbit. Thus the wavelength of the wave must be $\lambda_n = 2\pi r_n / n$. Substituting v_n for r_n according to our previous formula gives $\lambda_n = h / \mu v_n$. De Broglie then concluded that an electron should have associated with it a wave of wavelength $h / \mu v$. This was substantiated experimentally in 1927 when Davisson and Germer showed that electron beams gave diffraction patterns similar to light waves and that these patterns correspond to waves of wavelength $h / \mu v$.

In 1926, Schrödinger reasoned that since electromagnetic waves in a vacuum satisfy a wave equation $c^{-2}(\partial^2 E / \partial t^2) = \nabla^2 E$ [∇^2 is the Laplacian $(\partial / \partial x)^2 + (\partial / \partial y)^2 + (\partial / \partial z)^2$ and E is the electric field] then by wave–particle duality, electron waves (or de Broglie waves) should also satisfy a wave equation. He asserted that the equation which governs de Broglie waves in vacuum should be

$$i\hbar \frac{\partial \psi}{\partial t} = -\frac{\hbar^2}{2\mu} \nabla^2 \psi. \tag{1.10}$$

For the hydrogen atom an additional Coulomb potential term is added to give the equation

$$i\hbar \frac{\partial \psi}{\partial t} = -\frac{\hbar^2}{2\mu} \nabla^2 \psi - \frac{e^2}{r} \psi, \tag{1.11}$$

where $r = (x^2 + y^2 + z^2)^{1/2}$.

This equation was later generalized by Dirac to include relativistic effects. Moreover, Schrödinger postulated that the eigenvalues E of the eigenvalue equation

$$-\frac{\hbar^2}{2\mu} \nabla^2 \psi - \frac{e^2}{r} \psi = E\psi \tag{1.12}$$

with suitable boundary conditions give the allowed values for the electron's energy. These eigenvalues turn out to be the values given in Eq. (1.8) and thus Schrödinger's postulate leads to the same result as Bohr's postulates.

In 1927 Heisenberg announced his famous uncertainty principle. This principle states that the position and momentum of a physical object cannot be measured with arbitrary precision at the same time. More precisely, if Δp and Δq are the errors made in a momentum and coordinate measurement at the same time, then $\Delta p \, \Delta q \geqslant h$. This principle was substantiated by experiment. As a crude example, suppose an electron moves along the x axis and we want to measure its position and momentum. To find the electron's position we must "see" it and hence bounce a light ray off of it. If the light ray has wavelength λ, then we cannot locate the electron more precisely than λ, so $\Delta x \geq \lambda$. According to Planck's postulate the light ray must carry an energy of at least $h\nu$, where ν is the light ray's frequency. Since the speed of the light ray is c, the light ray has an associated momentum of at least $h\nu / c$. Since $\lambda\nu = c$, this becomes h/λ. Hence the light ray will impart a momentum h/λ to the electron, so $\Delta p \geq h/\lambda$ and $\Delta x \, \Delta p \geq h$. This gives an example of "interfering" measurements. Since h is very small, this phenomenon is not noticeable for objects that are roughly human size or larger.

1.2. Enter Probability

The situation in 1926–1927 was confusing. Although quantum mechanics could explain phenomena that classical mechanics could not, the quantum mechanics of that time consisted of a collection of vaguely related assertions. The assertions were stated in terms of certain *ad hoc* postulates such as Planck's energy quanta, Bohr's quantization rules, de Broglie–Schrödinger waves, and Heisenberg's uncertainty principle. The justification was that these postulates led to results which agreed with experiment. However, they applied only to specific phenomena, could not be easily generalized to include more complicated situations, and were not held together by a systematic unified theory. Moreover, there were some puzzling unanswered questions that needed resolving. What is the significance of the function ψ in Eqs. (1.10), (1.11), and (1.12)? Why do the eigenvalues of Eq. (1.12) give the allowed energy values of a hydrogen electron?

It turned out that the unifying, underlying theory was a quantum probability theory. This was first realized by Born during the period 1926–1929. It is interesting that during this same period Kolmogorov developed the foundations of classical probability theory [160]. The relationships between these two types of probability theory were not systematically exploited until much later [172, 173]. The Heisenberg uncertainty principle indicated to Born that quantum mechanics was a stochastic theory. In classical mechanics one can, in principle, simultaneously measure the position and momentum of a particle to arbitrary precision. The fact that one cannot do this for quantum mechanical particles indicates that these quantities have an intrinsic dispersion due to some kind of statistical fluctuations. When the position and momentum are measured one is really determining an average of these quantities and the uncertainty principle shows that there is a lower bound to the product of the dispersion of these quantities from their average values. Born proposed that the de Broglie waves were not waves in a physical sense, but were "probability waves" in the sense that the de Broglie–Schrödinger wave function $\psi(t,x,y,z)$ is a complex-valued function which determines the probability that the electron is at the point (x,y,z) at time t. More precisely, the probability that the electron is in a set $\Delta \subseteq \mathbb{R}^3$ at time t is

$$\int_\Delta |\psi(t,x,y,z)|^2 \, dx\, dy\, dz \qquad (1.13)$$

assuming that

$$\int_{\mathbb{R}^3} |\psi(t,x,y,z)|^2 \, dx\, dy\, dz = 1. \qquad (1.14)$$

Born's ideas were developed into a systematic theory by Dirac and von Neumann in the early 1930s. Although their theories were similar and Dirac's approach was succinct and elegant, we shall follow von Neumann since his methods are more rigorous mathematically. Von Neumann recognized that the two important concepts in quantum mechanics are states and observables. The states correspond to a theoretically complete description of the system and the observables correspond to the measurable quantities such as position, momentum, and energy. He decided that the wave function $\psi(t,x,y,z)$ describes the state of the system at time t. Equation (1.14) shows that for fixed t,ψ is a unit vector in the complex Hilbert space $L^2(\mathbb{R}^3)$. Since the Schrödinger equation (1.12) is an operator eigenvalue equation $H\psi = E\psi$ and the eigenvalues E are energy values, von Neumann and others reasoned that the operator H corresponds to the energy observable. On a suitable domain, H is self-adjoint, which is physically reasonable since the energy must have real values. Moreover, just as position in a certain direction is conjugate to momentum in that direction, energy is conjugate to time. Hence the energy operator H can be used to describe the time evolution of the system. This is further substantiated by the time-dependent Schrödinger equation (1.11) which can be written $i\hbar(\partial\psi/\partial t) = H\psi$. Putting these ideas together, von Neumann proposed the following axioms.

A1. The states of a quantum system are unit vectors in a complex Hilbert space $\mathcal{H}$.

A2. The observables are self-adjoint operators on $\mathcal{H}$.

A3. The probability that an observable A has a value in a Borel set $\Delta \subseteq \mathbb{R}$ when the system is in the state ψ is $\langle P^A(\Delta)\psi, \psi \rangle$, where $P^A(\cdot)$ is the resolution of the identity for A.

A4. If the state at time $t=0$ is ψ, then the state at time t is $\psi_t = \exp(-itH/\hbar)\psi$, where H is the energy observable.

To be precise, von Neumann did not say that the states are unit vectors and the observables are self-adjoint operators, but that the states are described by unit vectors and the observables are described by self-adjoint operators. Also, it is only the *pure* states that are unit vectors; there are other types of states called mixed states which we shall discuss later.

Axioms A1, A2, and A4 are descriptive. They tell how the physical concepts can be described by mathematical constructs and how the evolution of the system can be described mathematically. Axiom A3 is extremely important since it gives the contact between the mathematical structure and reality. It is this axiom which gives the distribution of values for an observable. This distribution is precisely what is measured in the laboratory, thus enabling one to test the theory and to make predictions about the behavior of the physical system. Since A3 is stochastic in nature we obtain a "quantum probability theory." Not only are the above postulates elegant and general, but as we shall see, they contain the ideas of Planck, Bohr, de Broglie, Schrödinger, Heisenberg, Born, and others as special cases.

For generality, A2 does not prescribe which self-adjoint operators represent which observables. This must be decided upon according to the specific situation being considered. In the case of the hydrogen electron, Eq. (1.12) indicates that the energy observable is represented by the operator

$$H = -(\hbar^2/2\mu)\nabla^2 - e^2/r. \qquad (1.15)$$

In classical mechanics the energy H_c has the form $H_c = p^2/2\mu + V(x,y,z)$, where $p^2 = p_x^2 + p_y^2 + p_z^2$ is the kinetic energy and $V(x,y,z)$ is the potential energy. Using this analogy one is tempted to postulate that the x momentum observable is represented by the operator $-i\hbar(\partial/\partial x)$ and the x coordinate observable is represented by the operator $f(x,y,z) \mapsto xf(x,y,z)$. The y,z momenta and coordinate operators are defined analogously. This correspondence does indeed give the desired results.

Differentiating A4 with respect to t gives

$$\frac{\partial \psi_t}{\partial t} = -\frac{iHe^{-itH/\hbar}}{\hbar}\psi = -\frac{iH}{\hbar}\psi_t. \qquad (1.16)$$

Equation (1.16) is the general Schrödinger equation and it reduces to Eq. (1.11) for the H defined in Eq. (1.15). The resolution of the identity of the x coordinate operator is $f(x,y,z) \mapsto (\chi_\Delta f)(x,y,z)$, $\Delta \subseteq \mathbb{R}$, where χ_Δ is the indicator (or characteristic) function of Δ. Applying A3, the probability that the x coordinate of the electron is in Δ when the system is in state ψ becomes

$$\langle \chi_\Delta \psi, \psi \rangle = \int_{\Delta \times \mathbb{R} \times \mathbb{R}} |\psi(x,y,z)|^2 \, dx \, dy \, dz.$$

If we form similar expressions for the y, z coordinates, we obtain Born's equation (1.13). Now suppose we want to find the allowed energy values for the hydrogen electron. If H has the value E with certainty, then A3 and Schwarz's inequality imply that

$$1 = |\langle P^H(\{E\})\psi, \psi \rangle| \leq \|P^H(\{E\})\psi\|.$$

It follows that $P^H(\{E\})\psi = \psi$ and hence

$$H\psi = E\psi \tag{1.17}$$

This is Schrödinger's eigenvalue equation (1.12), which shows that the allowed energy values are the eigenvalues of H.

Applying A3 and the spectral theorem, we can find the expectation $E[A]$ of the observable A (if it exists) in the state ψ:

$$E[A] = \int \lambda \langle P^A(d\lambda)\psi, \psi \rangle = \left\langle \int \lambda P^A(d\lambda)\psi, \psi \right\rangle = \langle A\psi, \psi \rangle.$$

The *variance* of A in the state ψ is defined by $\text{Var}(A) = E[(A - E[A])^2]$. The *standard deviation* $\Delta A = [\text{Var}(A)]^{1/2}$ gives a measure of the dispersion of the values of A about its expectation. Suppose A, B are self-adjoint operators and suppose $\psi, A\psi, B\psi$ are in the domains of A and B.

If $\langle A\psi, \psi \rangle = \langle B\psi, \psi \rangle = 0$ we have

$$\begin{aligned}
|\langle (AB - BA)\psi, \psi \rangle| &= |\langle A\psi, B\psi \rangle - \langle B\psi, A\psi \rangle| \\
&= 2|\text{Im}\langle A\psi, B\psi \rangle| \leq 2\|A\psi\| \, \|B\psi\| \\
&= 2\langle \psi, A^2\psi \rangle^{1/2} \langle \psi, B^2\psi \rangle^{1/2} = 2\Delta A \, \Delta B.
\end{aligned}$$

If $\langle A\psi, \psi \rangle, \langle B\psi, \psi \rangle \neq 0$ replace A by $A - \langle A\psi, \psi \rangle$ and B by $B - \langle B\psi, \psi \rangle$ to again obtain

$$\Delta A \, \Delta B \geq \tfrac{1}{2}|\langle (AB - BA)\psi, \psi \rangle|. \tag{1.18}$$

Now let $A = p_x = -i\hbar(\partial/\partial x)$ and $B = q_x = $ multiplication by x be the x momentum and x coordinate observables, respectively. The operators p_x, q_x satisfy the Heisenberg commutation relation

$$p_x q_x - q_x p_x = -i\hbar I. \tag{1.19}$$

(To make (1.19) rigorous one must specify the domains of the operators. We shall do this in a more satisfactory way later using the so-called Weyl form of the relation.) Substituting (1.19) into (1.18) gives

$$\Delta p_x \Delta q_x \geq \tfrac{1}{2}\hbar.$$

This shows that (1.18) is a general form of the Heisenberg uncertainty relations.

9

1.3. Some Comparisons

Von Neumann's axioms have stood the test of time and are essentially unchanged today. They form the basis of nonrelativistic quantum mechanics with a finite number of degrees of freedom used by most contemporary physicists. Even in relativistic quantum field theory with its infinite number of degrees of freedom much of von Neumann's theory remains intact.

Axiom A3 establishes quantum mechanics as a stochastic theory, but one which is quite different from the classical Kolmogorov theory. In order to compare the two, let us briefly review classical probability theory. In this theory, a basic role is played by a triple (Ω, Σ, μ), where (Ω, Σ) is a measurable space and μ is a nonnegative measure on the σ-algebra Σ satisfying $\mu(\Omega) = 1$. We call (Ω, Σ, μ) a *probability space* and μ a *probability measure*. The elements of Ω correspond to the possible outcomes of a random experiment, the sets in Σ correspond to random events, and the measure $\mu(A)$ for $A \in \Sigma$ gives the probability that the event A occurs.

A *random variable* is a measurable function $f: \Omega \to \mathbb{R}$. Random variables correspond to measurable quantities for the random experiment. Denote the class of Borel subsets of $\mathbb{R}$ by $\mathcal{B}(\mathbb{R})$. For $B \in \mathcal{B}(\mathbb{R})$ and random variable f, $f^{-1}(B)$ is the event that f has a value in B and $\mu[f^{-1}(B)]$ is the probability of that event. The probability measure μ_f on $\mathcal{B}(\mathbb{R})$ defined by $\mu_f(B) = \mu[f^{-1}(B)]$ is called the *distribution* of f. The *expectation* of f (if it exists) is $E[f] = \int f d\mu$ and it is easily seen that

$$E[f] = \int_{\mathbb{R}} \lambda \, d\mu_f(\lambda). \tag{1.20}$$

Moreover, if $u: \mathbb{R} \to \mathbb{R}$ is a Borel function and f is a random variable, then $u(f) = u \circ f$ is a random variable and

$$E[u(f)] = \int u(f) \, d\mu = \int_{\mathbb{R}} u(\lambda) \, d\mu_f(\lambda). \tag{1.21}$$

If f, g are random variables, it is important to know the probability of the simultaneous occurrence of events such as $f^{-1}(A) \cap g^{-1}(B)$, $A, B \in \mathcal{B}(\mathbb{R})$. The *joint distribution* of f, g is defined as the probability measure $\mu_{f,g}$ on $\mathcal{B}(\mathbb{R}^2)$ satisfying

$$\mu_{f,g}(A \times B) = \mu[f^{-1}(A) \cap g^{-1}(B)] \tag{1.22}$$

for all $A, B \in \mathcal{B}(\mathbb{R})$. It is easily shown that $\mu_{f,g}$ always exists and satisfies the consistency conditions $\mu_{f,g}(A \times \mathbb{R}) = \mu_f(A)$, $\mu_{f,g}(\mathbb{R} \times B) = \mu_g(B)$. Thus $\mu_{f,g}$ determines the distributions μ_f and μ_g but one can give examples which show that μ_f, μ_g do not determine $\mu_{f,g}$. Two random variables f and g are *independent* if

$$\mu\left[f^{-1}(A)\cap g^{-1}(B)\right]=\mu\left[f^{-1}(A)\right]\mu\left[g^{-1}(B)\right] \qquad (1.23)$$

for all $A,B\in\mathcal{B}(\mathbb{R})$. Using (1.22) we see that f and g are independent if and only if $\mu_{f,g}=\mu_f\times\mu_g$. Intuitively, f and g are independent if and only if the probability that an event $f^{-1}(A)$ occurs is unchanged if it is known that an event $g^{-1}(B)$ occurs for all $A,B\in\mathcal{B}(\mathbb{R})$. Notice that in the case of independent random variables μ_f and μ_g do determine $\mu_{f,g}$.

There are certain elementary comparisons that one can make between classical probability theory and quantum probability theory. Axiom A3 tells us that the probability that the observable A has a value in the Borel set B is $\langle P^A(B)\psi,\psi\rangle$. Thus observables correspond to random variables and the projection operators $P^A(B)$ correspond to events. In fact, we interpret $P^A(B)$ as the event that the observable A has a value in the set B. In general, we interpret the set $\mathcal{P}$ of orthogonal projections on the Hilbert space $\mathcal{K}$ as the set of quantum mechanical events. Since there is a natural one-to-one correspondence between orthogonal projections and closed subspaces of $\mathcal{K}$ we can use the two concepts interchangeably and also consider the set of closed subspaces of $\mathcal{K}$ as the set of quantum mechanical events.

We have seen that quantum probability theory replaces the measurable space (Ω,Σ) of classical probability theory with the pair $(\mathcal{K},\mathcal{P})$, where $\mathcal{P}$ is the set of orthogonal projections or, equivalently, the set of closed subspaces of $\mathcal{K}$. The usual operations of containment $A\subseteq B$, union $A\cup B$, intersection $A\cap B$, and complementation A^c in Σ have their natural counterparts in $\mathcal{P}$. In subspace language these are $M\subseteq N$, $\overline{\operatorname{span}}M\cup N$ (where $\overline{\operatorname{span}}$ means the closed span), $M\cap N$, and orthogonal complement $M^\perp$. In projection language these are $P\leq Q$ $(PQ=QP=P)$, $P\vee Q$ (the orthogonal projection onto the closed subspace spanned by the ranges of P and Q), $P\wedge Q$ (the orthogonal projection onto the intersection of the ranges of P and Q), and $I-P$.

We have seen that in quantum probability theory the observables play the role of random variables. Moreover, A3 tells us that the probability measure μ is replaced by the map $P\mapsto\langle P\psi,\psi\rangle$, $P\in\mathcal{P}$, where ψ is a state. The map $P\mapsto\langle P\psi,\psi\rangle$ is a probability measure in the sense that

$$0\leq\langle P\psi,\psi\rangle\leq1 \qquad\text{for all}\quad P\in\mathcal{P} \qquad (1.24\text{a})$$

$$\langle I\psi,\psi\rangle=1 \qquad (1.24\text{b})$$

$$\left\langle\sum_{i=1}^{\infty}P_i\psi,\psi\right\rangle=\sum_{i=1}^{\infty}\langle P_i\psi,\psi\rangle,\quad P_i\perp P_j,\quad i\neq j, \qquad (1.24\text{c})$$

11

where $\sum_{i=1}^{\infty} P_i$ converges in the weak-operator topology. By analogy with classical probability theory, we call the probability measure $B \mapsto \langle P^A(B)\psi, \psi \rangle$ the *distribution* of the observable A in the state ψ.

If ψ is a unit vector in $\mathcal{H}$ we call the map $m_\psi(P) = \langle P\psi, \psi \rangle$, $P \in \mathcal{P}$, a *pure state*. This is an abuse of terminology but it should cause no confusion. A *mixed state* is a map $m: \mathcal{P} \rightarrow [0, 1]$ of the form $m = \sum_{i=1}^{\infty} \lambda_i m_{\psi_i}$, where ψ_i are unit vectors and $\lambda_i \geq 0$, $\sum \lambda_i = 1$. It can be shown that the ψ_i can always be chosen to be orthogonal. If we let W be the positive trace-class operator of trace one given by $W = \sum \lambda_i P_i$ where P_i is the orthogonal projection onto the span of ψ_i, then for all $P \in \mathcal{P}$ we have

$$m(P) = \sum \lambda_i \langle P\psi_i, \psi_i \rangle = \sum \lambda_i \operatorname{tr}(P_i P)$$
$$= \operatorname{tr}\left(\sum \lambda_i P_i P \right) = \operatorname{tr}(WP). \tag{1.25}$$

It follows that there exists a bijection $m \mapsto W_m$ between the set of states and the set of positive trace-class operators of trace one satisfying $m(P) = \operatorname{tr}(W_m P)$ for all $P \in \mathcal{P}$. Mixed states correspond to the fact that a convex combination of probability measures is again a probability measure. We call positive trace class operators of trace one *density operators*.

1.4. Notes and References

Our history of quantum mechanics was condensed and simplified. To keep the presentation concise, we omitted many important developments and contributions. For more details the reader is referred to the references [30, 81, 128, 137, 235, 275]. Good introductions to quantum mechanics may be found in [24, 186, 208, 219]. We refer the reader to [270] for the work of von Neumann and [63] to the work of Dirac. Some of the standard references on classical probability theory are [31, 65, 160, 164, 166]. Two of the earlier papers on probability theory and quantum mechanics are [190, 280].

1.5. Exercises

1. Supply the details for deriving Eq. (1.1).

2. Derive Eqs. (1.3) and (1.4).

3. Show that Eq. (1.5) approaches Eq. (1.2) as ν approaches 0.

4. Supply the details for deriving Eq. (1.8).

5. Show that the Heisenberg uncertainty principle is not detectable for objects the size of a bullet.

6. Supply the details for deriving Eq. (1.17).

7. Show that $\text{Var}(A) = 0$ in the state ψ if and only if ψ is an eigenvector of A with eigenvalue $E[A]$.

8. Show that Eq. (1.19) holds on a dense subspace of $L^2(\mathbb{R}^3)$.

9. Prove Eqs. (1.20) and (1.21).

10. Show that Eqs. (1.24a), (1.24b), and (1.24c) hold.

11. If A is an observable and ψ is a state, show that $B \mapsto \langle P^A(B)\psi, \psi \rangle$ is a probability measure on $\mathcal{B}(\mathbb{R})$.

12. Show that the joint distribution of two random variables always exist.

13. Show that the distributions μ_f and μ_g do not necessarily determine $\mu_{f,g}$.

2

A Different Kind of Probability

In this chapter we show that although there are similarities between classical and quantum probability theory, there are also important differences. Many of these differences stem from the fact that quantum mechanics includes interference effects as exemplified by the Heisenberg uncertainty principle. In quantum probability theory, operators play the roles of random variables and the interference effects are frequently exhibited mathematically by the noncommutativity of certain operators. For this reason, quantum probability theory is sometimes called a noncommutative probability theory.

2.1. Interference Effects

In Section 1.3 we began comparing the two probability theories. We now consider some of their important differences. We first illustrate the concept of probability interference. Although such interference does not occur in classical probability theory it frequently does in the quantum theory.

The standard example of a quantum interference effect is the double-slit experiment. In this experiment, a stream of separated noninteracting particles (say, electrons) issues through the single slit on the left, passes through the double-slit screen in the middle, and hits the detection screen on the right, where its position is recorded (see Fig. 1).

If there were no quantum interference, then the electrons would act like true particles and the distribution of arrivals would appear somewhat as in Fig. 2.

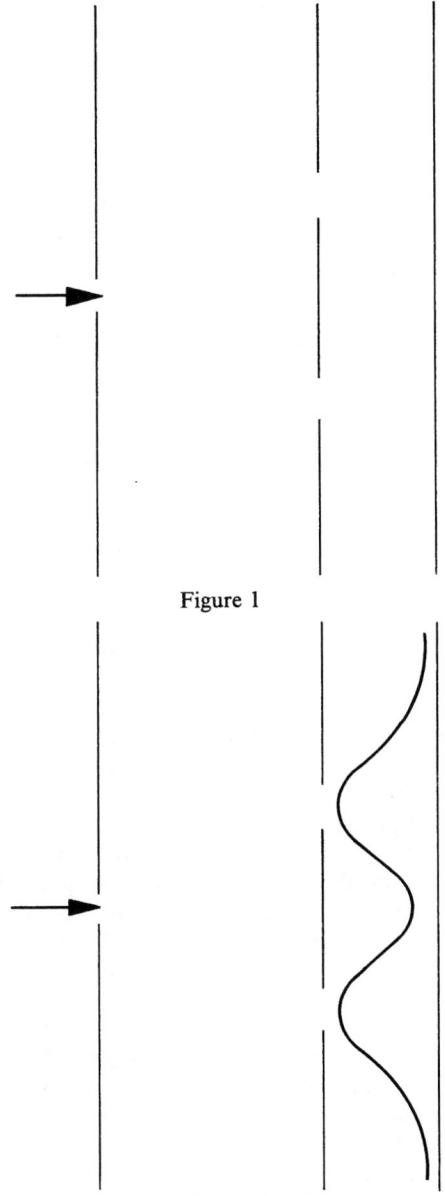

Figure 1

Figure 2

2. A Different Kind of Probability

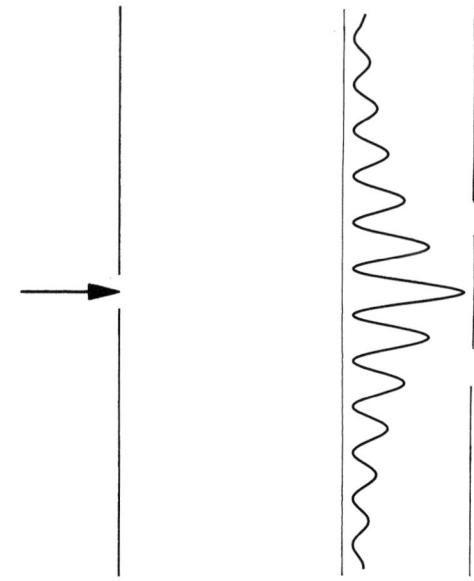

Figure 3

In reality one finds a diffraction pattern typical of wave interference such as in Fig. 3.

If one of the double slits is closed, the diffraction pattern is not observed. Moreover, if an observation is made (using strong light of short wavelength) to see which of the two slits a particle goes through, there is again no diffraction pattern.

This interference phenomenon may be explained as follows. According to Axiom A3 (Section 1.2), the probability that an observable A has a value in $\Delta \in \mathfrak{B}(\mathbb{R})$ when the system is in the state ψ is $\langle P^A(\Delta)\psi, \psi \rangle$. Now let $\phi \in \mathcal{H}$ be another state. Then the orthogonal projection Q onto the one-dimensional subspace spanned by ϕ is an observable. Also, Q can be considered to be an event. The observable Q has only the two values 0 and 1, where $P^Q(\{1\}) = Q$, $P^Q(\{0\}) = 0$. If the observable Q has the value 1 then the event Q occurs and if Q has the value 0 then the event Q does not occur. Thus Q may be thought of as the observable which determines whether or not the system is in the state ϕ. By Axiom A3, if the system is in the state ψ, then the probability that the system is in state ϕ (upon measuring the state determining observable Q) is

$$\langle P^Q(\{1\})\psi, \psi \rangle = \langle Q\psi, \psi \rangle = |\langle \psi, \phi \rangle|^2.$$

We call $|\langle \psi, \phi \rangle|^2$ the *transition probability* from state ψ to state ϕ. Notice that the function $T(\psi, \phi) = |\langle \psi, \phi \rangle|^2$ has the properties that one might desire for a transition probability, namely,

$$0 \le T(\psi, \phi) \le 1, \tag{2.1a}$$

$$T(\psi, \phi) = T(\phi, \psi), \tag{2.1b}$$

$$T(\psi, \phi) = 1 \quad \text{if and only if} \quad \psi = c\phi, \quad |c| = 1. \tag{2.1c}$$

Now let ϕ_1 be the state for a particle passing through the upper of the two slits and ϕ_2 the state for a particle passing through the lower slit. Since a particle cannot pass through the two slits simultaneously we must have $T(\phi_1, \phi_2) = 0$ and hence $\phi_1 \perp \phi_2$. If m is the state of the system at the detection screen, then m will determine the observed arrival distribution on this screen. Suppose we make an observation to determine whether a particle passes through the upper or lower slit and find that a particle passes through the upper slit with probability λ (and hence through the lower slit with probability $1 - \lambda$). Then the state of the system at the detection screen is the mixed state $m = \lambda m_{\phi_1} + (1 - \lambda) m_{\phi_2}$. This state would give a distribution as in Fig. 2 for a classical particle.

Now suppose that no such observation has been made. If a particle begins at the left in a pure state and no subsequent observation is made, then by Axiom A4 the particle will remain in a pure state. (If an observation is made, then a measuring apparatus must be introduced and the system is changed to the combined system of particle plus apparatus, so A4 is no longer valid for the original system.) Thus the state of the system at the detection screen is a pure state ψ. Now expand ψ in an orthonormal basis $\psi = \sum \langle \psi, \phi_i \rangle \phi_i$ including ϕ_1 and ϕ_2. The transition probabilities $|\langle \psi, \phi_1 \rangle|^2$, $|\langle \psi, \phi_2 \rangle|^2$ give the probabilities that a particle comes through the upper or lower slit, respectively, and arrives at the detection screen in the state ψ. Since a particle must pass through either the upper or lower slit we have $|\langle \psi, \phi_1 \rangle|^2 + |\langle \psi, \phi_2 \rangle|^2 = 1$. It follows from Parseval's equality that $|\langle \psi, \phi_i \rangle|^2 = 0$ for $i \ne 1, 2$. Hence $\psi = \langle \psi, \phi_1 \rangle \phi_1 + \langle \psi, \phi_2 \rangle \phi_2$. We say that ψ is a *superposition* of ϕ_1 and ϕ_2. Such superpositions give interference effects between the two wave functions ϕ_1 and ϕ_2 and cause diffraction patterns such as in Fig. 3. Superpositions are a quantum mechanical effect that does not occur in classical probability theory.

The double-slit experiment gives an example of interference of probabilities. Another way that interference occurs in quantum probability but not in classical probability is through the interference of measurements. The fact that the measurements of two observables can interfere is typified by the Heisenberg uncertainty principle. Notice, however, that in the uncertainty principle if two observables A and B commute, then there is no positive lower bound for the product of their variances. This indicates that commuting observables do not interfere (to be precise, we say that two observables *commute* if their resolutions of identity commute). This fact is demonstrated succinctly by a theorem due to von Neumann. This theorem states that two observables A and B commute if and only if there exists a third observable C and two Borel functions u, v such that $A = u(C), B = v(C)$. (We shall prove a generalization of this theorem later.) Thus, if A and B commute, they can be measured simultaneously by measuring C and hence do not interfere. In this way commuting observables act like random variables and their stochastic properties can be found using classical probability theory. However, for noncommuting observables this is far from the case. One way of seeing this is the fact that noncommuting observables do not possess joint probability distributions in the sense discussed below.

We have already mentioned that if f and g are random variables, then their joint distribution $\mu_{f,g}$ exists but is not determined by the distributions μ_f and μ_g. However, the distributions of $x_1 f + x_2 g$ for all $x_1, x_2 \in \mathbb{R}$ do determine $\mu_{f,g}$. In fact, it can be shown that $\mu_{f,g}$ is the unique measure on $\mathfrak{B}(\mathbb{R}^2)$ that satisfies

$$\mu_{f,g}\{(y_1, y_2) : x_1 y_1 + x_2 y_2 \in E\} = \mu\{\omega \in \Omega : x_1 f(\omega) + x_2 g(\omega) \in E\} \quad (2.2)$$

for every $E \in \mathfrak{B}(\mathbb{R})$, $x_1, x_2 \in \mathbb{R}$. Now let A_1 and A_2 be observables such that $x \cdot A = x_1 A_1 + x_2 A_2$ are essentially self-adjoint for every $x = (x_1, x_2) \in \mathbb{R}^2$. Motivated by (2.2) we say that A_1 and A_2 *have a joint distribution* in the state ψ if there exists a measure μ_{A_1, A_2} on $\mathfrak{B}(\mathbb{R}^2)$ such that

$$\mu_{A_1, A_2}\{y \in \mathbb{R}^2 : x \cdot y \in E\} = \langle P^{x \cdot A}(E)\psi, \psi \rangle \quad (2.3)$$

for every $E \in \mathfrak{B}(\mathbb{R})$, $x \in \mathbb{R}^2$. If A_1 and A_2 have a joint distribution, it is easily seen that there exist random variables f and g such that

$$\mu_{f,g}\{y \in \mathbb{R}^2 : x \cdot y \in E\} = \langle P^{x \cdot A}(E)\psi, \psi \rangle \quad (2.4)$$

for all $E \in \mathfrak{B}(\mathbb{R})$ and $x \in \mathbb{R}^2$.

Theorem 2.1 (Nelson [198]). *Let A_1 and A_2 be self-adjoint operators on a Hilbert space $\mathcal{H}$ such that $x \cdot A$ is essentially self-adjoint for every $x \in \mathbb{R}^2$. Then A_1 and A_2 commute if and only if they have a joint distribution in every state.*

PROOF. Necessity is easily proved using the theorem of von Neumann mentioned above. For sufficiency, suppose that for every unit vector $\psi \in \mathcal{H}$ there exists a measure $\mu_\psi = \mu_{A_1, A_2}$ satisfying (2.3). If we integrate first over the hyperplane orthogonal to $x \in \mathbb{R}^2$ we find that

$$\int_{\mathbb{R}^2} e^{ix\cdot y}\, d\mu_\psi(y) = \int_{-\infty}^{\infty} e^{i\lambda}\, d\mu_\psi\{y : x\cdot y \le \lambda\}$$

$$= \int_{-\infty}^{\infty} e^{i\lambda}\langle dP^{x\cdot A}(-\infty,\lambda])\psi,\psi\rangle = \langle e^{ix\cdot A}\psi,\psi\rangle.$$

Hence, the measure μ_ψ is the Fourier transform of the function $\langle e^{ix\cdot A}\psi,\psi\rangle$. By the polarization identity, if ϕ and ψ are vectors in $\mathcal{H}$, there exists a complex measure $\mu_{\phi\psi}$ such that $\mu_{\phi\psi}$ is the Fourier transform of $\langle e^{ix\cdot A}\phi,\psi\rangle$ and $\mu_{\psi\psi} = \mu_\psi$. Since $(\phi,\psi)\mapsto\mu_{\phi\psi}(\Delta)$, $\Delta \in \mathcal{B}(\mathbb{R}^2)$, is a bounded sesquilinear form, there exists a unique bounded linear operator $S(\Delta)$ such that $\langle S(\Delta)\phi,\psi\rangle = \mu_{\phi\psi}(\Delta)$. Hence,

$$\int_{\mathbb{R}^2} e^{ix\cdot y}\langle S(dy)\phi,\psi\rangle = \langle e^{ix\cdot A}\phi,\psi\rangle.$$

The operator $S(\Delta)$ is positive since μ_ψ is a positive measure. Therefore, for any finite set of vectors $\psi_j \in \mathcal{H}$ and corresponding points $x_j \in \mathbb{R}^2$ we have

$$\sum_{j,k}\left\langle e^{i(x_j - x_k)\cdot A}\psi_j,\psi_k\right\rangle = \sum_{j,k}\int_{\mathbb{R}^2} e^{i(x_j - x_k)\cdot y}\left\langle S(dy)\psi_j,\psi_k\right\rangle$$

$$= \int_{\mathbb{R}^2}\left\langle S(dy)\psi(y),\psi(y)\right\rangle \ge 0,$$

where $\psi(y) = \sum_j e^{ix_j\cdot y}\psi_j$. Moreover, $e^{i0\cdot A} = 1$ and $e^{i(-x)\cdot A} = (e^{ix\cdot A})^*$. Under these conditions, the theorem on unitary dilations of Nagy [227, Appendix, p. 21] implies that there exists a Hilbert space $\mathcal{\tilde{K}}$ containing $\mathcal{H}$ and a unitary representation $x\mapsto U(x)$ of $\mathbb{R}^2$ on $\mathcal{\tilde{K}}$ such that if P is the orthogonal projection of $\mathcal{\tilde{K}}$ onto $\mathcal{H}$, then $PU(x)\psi = e^{ix\cdot A}\psi$ for every $x \in \mathbb{R}^2$ and $\psi \in \mathcal{H}$. Since $e^{ix\cdot A}$ is already unitary,

$$\|U(x)\psi\| = \|e^{ix\cdot A}\psi\| = \|\psi\|,$$

so that $\|PU(x)\psi\| = \|U(x)\psi\|$. Hence, $PU(x)\psi = U(x)\psi$ and each $U(x)$ maps $\mathcal{H}$ into itself so that $U(x)\psi = e^{ix\cdot A}\psi$ for every $\psi \in \mathcal{H}$. Since $x\to U(x)$ is a unitary representation of the commutative group $\mathbb{R}^2$, the $e^{ix\cdot A}$ all commute. It follows that A_1 and A_2 commute. $\qquad\square$

In essence, the above theorem says that a pair (or by the same proof, any finite number) of observables can be treated as random variables if and only if they commute. Related results of varying degrees of generality may be found in [42, 76, 97, 209, 246, 265, 267]. Although noncommuting observables do not possess joint distributions in the above sense, we shall see in Chapter 4 that there is a more general way of formulating this concept which is meaningful for noncommuting observables.

2.2. Classical and Quantum Probability

We have seen that the quantum probability theory of von Neumann gives different results than the classical Kolmogorov probability theory. We now show that quantum probability theory contains the classical theory as a special case and hence the difference lies in the greater generality of the former over the latter.

If $\mathcal{H}$ is a complex Hilbert space, $\mathcal{P}$ the set of orthogonal projections on $\mathcal{H}$, and $\psi \in \mathcal{H}$ is a unit vector, then the triple $(\mathcal{H}, \mathcal{P}, \psi)$ is the quantum probability counterpart to the probability space (Ω, Σ, μ). (We could replace ψ by a mixed state, but we do not need that generality for the present.) A *σ-isomorphism* $h: \Sigma \rightarrow \mathcal{P}$ is an injection that satisfies $h(\Omega) = I$, if $A, B \in \Sigma$ and $A \cap B = \varnothing$ then $h(A) \perp h(B)$ and $h(\cup_{i=1}^{\infty} A_i) = \Sigma h(A_i)$ whenever $A_i \cap A_j = \varnothing$, $i \neq j$. It can be shown that a σ-isomorphism preserves all the set theoretic operations in Σ.

Theorem 2.2. *If (Ω, Σ) is a measurable space, there exists a Hilbert space $\mathcal{H}$ and a σ-isomorphism h from Σ to the set $\mathcal{P}$ of orthogonal projections on $\mathcal{H}$. For any probability measure μ on (Ω, Σ) there exists a pure state ϕ_μ on $\mathcal{P}$ such that $\mu(\Delta) = \langle h(\Delta)\phi_\mu, \phi_\mu \rangle$ for every $\Delta \in \Sigma$. For any class of random variables $\{f_\alpha\}$ on (Ω, Δ) there exists a class of observables $\{A_\alpha\}$ such that $P^{A_\alpha}(E) = h[f_\alpha^{-1}(E)]$ for every $E \in \mathcal{B}(\mathbb{R})$.*

PROOF. Let $\{\mu_\delta : \delta \in D\}$ be the collection of all probability measures on (Ω, Σ) and let $H_\delta = L^2(\Omega, \Sigma, \mu_\delta)$. If $\Delta \in \Sigma$, let $h_\delta(\Delta)$ be the projection onto the closure of the subspace $\{f \in H_\delta : f(\Delta^c) = 0\}$. It is straightforward to show that h_δ is a σ-homomorphism from Σ into the set of orthogonal projections of H_δ. Now $\mathcal{H} = \Sigma \{ H_\delta : \delta \in D \}$, the direct sum of the Hilbert spaces H_δ is a Hilbert space. For $\Delta \in \Sigma$, define $h(\Delta) = \Sigma \{ h_\delta(\Delta) : \delta \in D \}$. It is routine to show that h is a σ-homomorphism from Σ into the set $\mathcal{P}$ of orthogonal projections of $\mathcal{H}$. To show that h is injective, suppose $h(\Delta_1) = h(\Delta_2)$. Then by definition, $h_\delta(\Delta_1) = h_\delta(\Delta_2)$ for all $\delta \in D$, and it follows that $\mu_\delta[(\Delta_1 \cap \Delta_2^c) \cup (\Delta_1^c \cap \Delta_2)] = 0$. But then $\Delta_1 \cap \Delta_2^c = \varnothing$ since if $\omega \in \Delta_1 \cap \Delta_2^c$ and μ is the probability measure concentrated at ω, then $\mu(\Delta_1 \cap \Delta_2^c) = 1$, a contradiction. Similarly $\Delta_1^c \cap \Delta_2 = \varnothing$ and hence $\Delta_1 = \Delta_2$. Now if μ_β is a probability measure, let $\phi \in \mathcal{H}$ be defined by $\phi_\beta \equiv 1$, $\phi_\alpha \equiv 0$, $\alpha \in D - \{\beta\}$. Now it is easily seen that $(h(\Delta)\psi)_\delta = \chi_\Delta \psi_\delta$, $\delta \in D$, for all $\psi \in \mathcal{H}$. We therefore have

$$\langle h(\Delta)\phi, \phi \rangle = \sum \{ \langle (h(\Delta)\phi)_\delta, \phi_\delta \rangle : \delta \in D \}$$

$$= \sum \{ \langle \chi_\Delta \phi_\delta, \phi_\delta \rangle : \delta \in D \} = \langle \chi_\Delta \phi_\beta, \phi_\beta \rangle = \int_\Delta d\mu_\beta = \mu_\beta(\Delta).$$

The last statement of the theorem is straightforward. $\qquad \square$

Due to the greater generality of quantum probability theory, some of the basic results of the classical theory do not hold there. For example, it is well known that a probability measure μ is subadditive; that is, $\mu(\Delta_1 \cup \Delta_2) \leq \mu(\Delta_1) + \mu(\Delta_2)$ for all Δ_1, Δ_2. The analogous result $m(P_1 \vee P_2) \leq m(P_1) + m(P_2)$ need not hold for a state m. For a counterexample, let $\phi = (1, 0)$, $\phi_1 = (0, 1)$, and $\phi_2 = 2^{-1/2}(1, 1)$ be unit vectors in the two-dimensional Hilbert space C^2. Let m be the pure state corresponding to ϕ (i.e., $m(P) = \langle P\phi, \phi \rangle$) and let P_1, P_2 be the projections onto the one-dimensional subspaces spanned by ϕ_1, ϕ_2, respectively. We then have $m(P_1 \vee P_2) = 1$ and

$$m(P_1) + m(P_2) = \langle P_1\phi, \phi \rangle + \langle P_2\phi, \phi \rangle = 0 + \tfrac{1}{2} = \tfrac{1}{2}.$$

Hence $m(P_1 \vee P_2) > m(P_1) + m(P_2)$.

Certain limit theorems of classical probability hold and others do not in quantum probability theory. To illustrate this, let $\Delta_i \in \Sigma$ be a sequence of events. In the classical theory we define

$$\limsup \Delta_i = \bigcap_{k=1}^{\infty} \bigcup_{j=k}^{\infty} \Delta_j, \qquad \liminf \Delta_i = \bigcup_{k=1}^{\infty} \bigcap_{j=k}^{\infty} \Delta_j.$$

The $\limsup \Delta_i$ is the event which occurs if and only if infinitely many of the Δ_i occur and $\liminf \Delta_i$ the event which occurs if and only if all but finitely many of the Δ_i occur. If $\limsup \Delta_i = \liminf \Delta_i = \Delta$ we write $\Delta = \lim \Delta_i$. Basic results in the classical theory are:

$$\liminf \Delta_i \subseteq \limsup \Delta_i; \tag{2.5}$$

$$\mu(\liminf \Delta_i) = \lim_{i \to \infty} \mu\left(\bigcap_{k=i}^{\infty} \Delta_k \right); \tag{2.6}$$

$$\mu(\limsup \Delta_i) = \lim_{i \to \infty} \mu\left(\bigcup_{k=i}^{\infty} \Delta_k \right); \tag{2.7}$$

$$\text{if } \lim \Delta_i = \Delta, \qquad \text{then } \mu(\Delta) = \lim \mu(\Delta_i) \quad \forall \mu, \tag{2.8}$$

where μ is a probability measure.

If P_i is a sequence of orthogonal projections we have the analogous definitions:

$$\limsup P_i = \bigwedge_{k=1}^{\infty} \bigvee_{j=k}^{\infty} P_j, \qquad \liminf P_i = \bigvee_{k=1}^{\infty} \bigwedge_{j=k}^{\infty} P_j.$$

If $\limsup P_i = \liminf P_i = P$ we write $P = \lim P_i$. Then results analogous to (2.5)–(2.8) hold [94]. The difference is that the converse of (2.8) holds in classical but not in quantum probability theory.

2. A Different Kind of Probability

Lemma 2.3. *If $\lim \mu(\Delta_i)$ exists and equals $\mu(\Delta)$ for every probability measure μ, then $\lim \Delta_i = \Delta$.*

PROOF. If the hypothesis of the lemma holds, then applying (2.6) and (2.7) we have

$$\mu(\liminf \Delta_i) = \lim \mu \left(\bigcap_{k=i}^{\infty} \Delta_k \right) \leq \lim \mu(\Delta_i)$$

$$= \mu(\Delta) \leq \lim \mu \left(\bigcup_{k=i}^{\infty} \Delta_k \right) = \mu(\limsup \Delta_i).$$

If $\liminf \Delta_i \neq \varnothing$, let $\omega \in \liminf \Delta_i$ and let μ be the probability measure concentrated at ω. Then $\mu(\liminf \Delta_i) = 1$ and hence $\mu(\Delta) = 1$. Thus $\omega \in \Delta$ and $\liminf \Delta_i \subseteq \Delta$. Similarly, $\Delta \subseteq \limsup \Delta_i$. Suppose $\liminf \Delta_i \neq \Delta$ and let $\omega \in \Delta$, $\omega \notin \liminf \Delta_i$. Let μ be the probability measure concentrated at ω. Now ω is not in an infinite number of Δ_i's and hence there is a subsequence $\Delta_{i(j)}$ of Δ_i such that $\mu(\Delta_{i(j)}) = 0$. But then

$$1 = \mu(\Delta) = \lim \mu(\Delta_i) = \lim \mu(\Delta_{i(j)}) = 0,$$

which is a contradiction. Hence, $\liminf \Delta_i = \Delta$. Now suppose $\limsup \Delta_i \neq \Delta$ and let $\omega \in \limsup \Delta_i$, $\omega \notin \Delta$. Again, let μ be concentrated at ω. Then ω is in an infinite number of the Δ_i's and hence $\lim \mu(\Delta_i) = 1$. But $\mu(\Delta) = 0$ gives a contradiction, so $\limsup \Delta_i = \Delta$. $\square$

We now show that the analog of Lemma 2.3 does not hold in quantum probability theory. Let ϕ_i be a sequence of distinct unit vectors in C^2 which converge to a vector ϕ_0 and let P_i be the projection onto the one-dimensional subspace spanned by ϕ_i. If ϕ is an arbitrary unit vector and m the corresponding pure state we have

$$m(P_i) = \langle P_i \phi, \phi \rangle = |\langle \phi, \phi_i \rangle|^2 \to |\langle \phi, \phi_0 \rangle|^2 = m(P_0).$$

By taking convex combinations one can easily show that this also holds for mixed states so $m(P_0) = \lim m(P_i)$ for every state m. However,

$$\liminf P_i = 0 < P_0 < I = \limsup P_i. \tag{2.9}$$

In classical probability theory a central role is played by the concept of independence. We have already discussed the independence of two random variables in Section 1.3. A finite set of random variables $f_i, \ldots, f_n$ in a probability space (Ω, Σ, μ) is *independent* if for any sets $B_1, \ldots, B_n \in \mathcal{B}(\mathbb{R})$ we have

$$\mu[f_1^{-1}(B_1) \cap \cdots \cap f_n^{-1}(B_n)] = \mu[f_1^{-1}(B_1)] \cdots \mu[f_n^{-1}(B_n)].$$

An arbitrary collection of random variables is *independent* if any finite subset is independent. Analogously, we say that a finite set of observables $A_1, \ldots, A_n$ in a quantum probability space $(\mathcal{H}, \mathcal{P}, m)$ is *independent* if for any sets $B_1, \ldots, B_n \in \mathcal{B}(\mathbb{R})$ we have

$$m[P^{A_1}(B_1) \wedge \cdots \wedge P^{A_n}(B_n)] = m[P^{A_1}(B_1)] \cdots m[P^{A_n}(B_n)].$$

An arbitrary collection of observables is *independent* if any finite subset is independent. To illustrate this concept we prove the following simple result.

Lemma 2.4. *Two observables A_1 and A_2 on $(\mathcal{H}, \mathcal{P})$ are independent in every state if and only if A_1 or A_2 is a constant (i.e., has the form λI, $\lambda \in \mathbb{R}$).*

PROOF. Sufficiency is straightforward. For necessity, let $B_1, B_2 \in \mathcal{B}(\mathbb{R})$ and let $P_1 = P^{A_1}(B_1)$, $P_2 = P^{A_2}(B_2)$. If $P_1 - P_1 \wedge P_2 \neq 0$ and $P_2 - P_1 \wedge P_2 \neq 0$, then there exist states m_1 and m_2 with $m_1(P_1 - P_1 \wedge P_2) = 1$ and $m_2(P_2 - P_1 \wedge P_2) = 1$. Then $m_1(P_1 \wedge P_2) = m_2(P_1 \wedge P_2) = 0$. If $m = \frac{1}{2}m_1 + \frac{1}{2}m_2$, then $m(P_1)m(P_2) > 0$ and $m(P_1 \wedge P_2) = 0$, which is a contradiction. Thus we have either $P_1 \leq P_2$ or $P_2 \leq P_1$. If $0 < P_1 \leq P_2 < I$, then the inequalities for P_1 and $I - P_2$ would lead to a contradiction. Hence, if $0 < P_1 < I$, then for every $B \in \mathcal{B}(\mathbb{R})$ we obtain $P^{A_2}(B) = 0$ or $P^{A_2}(B) = 1$. It follows that $A_2 = \lambda I$ for some $\lambda \in \mathbb{R}$. $\square$

Cantelli's lemma is an important elementary result in classical probability theory which states that $\mu(\limsup \Delta_i) = 0$ if $\sum \mu(\Delta_i) < \infty$. This result does not hold in quantum probability theory. For example, let $\phi = (0, 1)$, $\phi_n = (1, 2^{-n}) \in C^2$, $n = 1, 2, \ldots$, let P_n be the projection onto the one-dimensional subspace spanned by ϕ_n, and let m be the pure state corresponding to ϕ. Then

$$m(\limsup P_i) = \lim m\left(\bigvee_{k=i}^{\infty} P_k \right) = 1, \qquad \sum m(P_i) < \infty. \qquad (2.10)$$

If the projections P_i are independent, then not only does Cantelli's lemma hold but also the following stronger result due to Borel and Cantelli (for a proof see [94]).

Theorem 2.5. *If P_i, $i = 1, 2, \ldots$, are independent projections in the state m, then $m(\limsup P_i) = 0$ if $\sum m(P_i) < \infty$ and $m(\limsup P_i) = 1$ if $\sum m(P_i) = \infty$.*

Let A, A_i, $i = 1, 2, \ldots$, be observables on $(\mathcal{H}, \mathcal{P}, m)$ such that $A - A_i$ is essentially self-adjoint for all i. We can define various types of conver-

2. A Different Kind of Probability

gence generalizing the usual concepts in the classical theory. For example, A_i *converges in probability to* A if

$$\lim_{i \to \infty} m\left[P^{A-A_i}([-\varepsilon,\varepsilon]) \right] = 1 \qquad \forall \varepsilon > 0. \tag{2.11}$$

A_i *converges almost everywhere to* A if

$$m\left[\limsup\left(P^{A-A_i}([-\varepsilon,\varepsilon]) \right) \right] = 1 \qquad \forall \varepsilon > 0. \tag{2.12}$$

A_i *converges almost uniformly to* A if for any $\delta > 0$ there exists a projection P with $m(P) > 1 - \delta$ such that for every $\varepsilon > 0$ there exists an n for which

$$i \geq n \Rightarrow P^{A-A_i}([-\varepsilon,\varepsilon]) \geq P. \tag{2.13}$$

One can easily show that almost uniform convergence implies almost everywhere convergence implies convergence in probability. However, unlike the classical theory, it is not true that if a sequence converges in probability, there exists a subsequence that converges almost everywhere [191]. Theorems generalizing the dominated convergence theorem and Fatou's lemma hold [191]. However, Egorov's theorem does not hold [191].

It is natural to try to prove results analogous to the famous limit theorems of classical probability theory. This has been done for certain versions and special cases of some of these theorems although much work remains in this area. For a law of large numbers, central limit theorems, and a functional central limit theorem, the reader is referred to [40, 47, 94, 206]. To illustrate these we now consider a central limit theorem due to Cushen and Hudson [47].

Let p,q be a pair of self-adjoint operators on the Hilbert space $\mathcal{H}$ satisfying the Heisenberg commutation relation

$$[p,q] \equiv pq - qp = -i\hbar I. \tag{2.14}$$

For $r,s \in \mathbb{R}$, let $W(r,s)$ be the unitary operator $e^{i(rp+sq)}$. Using (2.14) and formal manipulations we obtain the *Weyl relation*

$$W(r,s)W(r',s') = \omega(r,s,r',s')W(r+r',s+s') \tag{2.15}$$

for every $(r,s) \in \mathbb{R}^2$, where ω is the function on $\mathbb{R}^4$:

$$\omega(r,s,r',s') = \exp\left[\tfrac{1}{2} i\hbar(rs' - sr') \right].$$

If (p,q) satisfies (2.15) we call (p,q) a *canonical pair*. Of course, the usual momentum and position observables form a canonical pair. The von Neumann uniqueness theorem (whose proof we shall indicate later) says that these are the only canonical pairs in a certain sense. To be precise, the von Neumann uniqueness theorem states that if a family of unitary operators $W(r,s)$ satisfies (2.15) then $W(r,s)$ is unitarily equivalent to a direct sum of copies of the family $W_0(r,s) = e^{i(rp_0 + sq_0)}$, where p_0, q_0 are the Schrödinger operators defined in $L^2(\mathbb{R}, dt)$ given by $p_0\psi(t) = -i\hbar(d\psi(t)/dt)$, $q_0\psi(t) = t\psi(t)$.

Expressing the above result slightly differently, there exists a unitary map $T_{p,q} : \mathcal{H} \to L^2(\mathbb{R}, dt) \otimes \mathcal{H}'$, where $\mathcal{H}'$ is some other Hilbert space which satisfies

$$W(r,s) = T_{p,q}^{-1} W_0(r,s) \otimes I_{\mathcal{H}'} T_{p,q} \qquad (2.16)$$

for all $(r,s) \in \mathbb{R}^2$, $I_{\mathcal{H}'}$ being the identity operator on $\mathcal{H}'$ and $\otimes$ being the tensor product. If D is a density operator on $\mathcal{H}$ then there exists a unique density operator $D_{p,q}$ on $L^2(\mathbb{R}, dt)$ such that

$$\operatorname{tr} A D_{p,q} = \operatorname{tr} T_{p,q}^{-1} A \otimes I_{\mathcal{H}'} T_{p,q} D \qquad (2.17)$$

for every bounded linear operator. A on $L^2(\mathbb{R}, dt)$. Knowledge of $D_{p,q}$ determines the expectations of observables which are "functions of p and q" and we therefore call $D_{p,q}$ the *distribution operator* for (p,q) in the state D. This is analogous to the situation where f, g are random variables and the joint distribution $\mu_{f,g}$ is a measure on $\mathcal{B}(\mathbb{R}^2)$ which determines the distributions of any Borel function of f and g. Although, in general, we cannot define a joint distribution for two noncommuting observables, we can for the special observables p and q because of the von Neumann uniqueness theorem.

Now suppose $(p_1, q_1), \ldots, (p_n, q_n)$ are canonical pairs on $\mathcal{H}$ which commute in the sense that the corresponding operators $W_j(r,s) = e^{i(rp_j + sq_j)}$ commute with each other. It follows from the von Neumann uniqueness theorem that there exists a unitary map

$$T_{p_1, q_1, \ldots, p_n, q_n} : \mathcal{H} \to L^2(\mathbb{R}, dt) \otimes \cdots \otimes L^2(\mathbb{R}, dt) \otimes \mathcal{H}'$$

such that

$$W_j(r,s) = T_{p_1, q_1, \ldots, p_n, q_n}^{-1} (I \otimes \cdots \otimes W_0(r,s) \otimes \cdots \otimes I \otimes I_{\mathcal{H}'}) T_{p_1, q_1, \ldots, p_n, q_n},$$

where I is the identity on $L^2(\mathbb{R}, dt)$. Corresponding to a density operator D on $\mathcal{H}$, just as in (2.17) there exists a unique density operator

$$D_{p_1, q_1, \ldots, p_n, q_n} \qquad \text{on} \quad L^2(\mathbb{R}, dt) \otimes \cdots \otimes L^2(\mathbb{R}, dt)$$

such that for every bounded operator A on $L^2(\mathbb{R}, dt) \otimes \cdots \otimes L^2(\mathbb{R}, dt)$

$$\operatorname{tr} A D_{p_1, q_1, \ldots, p_n, q_n} = \operatorname{tr} T_{p_1, q_1, \ldots, p_n, q_n}^{-1} (A \otimes I_{\mathcal{H}'}) T_{p_1, q_1, \ldots, p_n, q_n} D.$$

We call $D_{p_1, q_1, \ldots, p_n, q_n}$ the *joint distribution operator* for $(p_1, q_1), \ldots, (p_n, q_n)$ in the state D.

2. A Different Kind of Probability

Recall that random variables $f_1,\ldots,f_n$ are independent if and only if their joint distribution $\mu_{f_1,\ldots,f_n}$ equals the product measure $\mu_{f_1} \times \cdots \times \mu_{f_n}$. Analogously, we say that the canonical pairs $(p_1,q_1),\ldots,(p_n,q_n)$ are *independent* in the state D if the joint distribution operator can be expressed as

$$D_{p_1,q_1,\ldots,p_n,q_n} = D_{p_1,q_1} \otimes \cdots \otimes D_{p_n,q_n}.$$

Now suppose (p_i,q_i), $i=1,2,\ldots$, is an infinite sequence of commuting canonical pairs on $\mathcal{H}$ and let D be a density operator on $\mathcal{H}$. We say that the sequence (p_i,q_i) is *independent* if every finite subsequence is independent. We say that (p_i,q_i) are *identically distributed* if the distribution operators $D_{p_1,q_1}, D_{p_2,q_2},\ldots$ are identical.

Let D_i, $i=1,2,\ldots$, and D be density operators on $L^2(\mathbb{R},dt)$. We say that D_i *converges weakly* to D if for every bounded operator A on $L^2(\mathbb{R},dt)$ $\operatorname{tr} AD_n \to \operatorname{tr} AD$. We say that a sequence of canonical pairs (p_i,q_i) on $(\mathcal{H},D)$ *converges in distribution* to the limit distribution D_0 if the sequence of distribution operators D_{p_i,q_i} on $L^2(\mathbb{R},dt)$ converges weakly to the density operator D_0 on $L^2(\mathbb{R},dt)$. This definition is analogous to the definition of a sequence of random variables converging in distribution.

Let (p,q) be a canonical pair on $(\mathcal{H},D)$. The *nth moments* of p and q in the state D are defined by

$$\langle p^n \rangle_D = \int \lambda^n \operatorname{tr} P^p(d\lambda)D, \quad \langle q^n \rangle_D = \int \lambda^n \operatorname{tr} P^q(d\lambda)D.$$

If the second moments $\Gamma_{11} = \langle p^2 \rangle_D, \Gamma_{22} = \langle q^2 \rangle_D$ are finite, then the expectations $m_1 = \langle p \rangle_D, m_2 = \langle q \rangle_D$ are finite and these can be expressed as

$$\Gamma_{11} = \sum \langle p\psi_i, Dp\psi_i \rangle, \qquad \Gamma_{22} = \sum \langle q\psi_i, Dq\psi_i \rangle \qquad (2.18)$$

$$m_1 = \sum \langle p\psi_i, D\psi_i \rangle, \qquad m_2 = \sum \langle p\psi_i, D\psi_i \rangle, \qquad (2.19)$$

where ψ_i is an orthonormal basis in the domains of p and q. We define the off-diagonal elements of the 2×2 *covariance matrix* $\Gamma = [\Gamma_{ij}]$ by

$$\Gamma_{12} = \Gamma_{21} = \tfrac{1}{2} \sum (\langle p\psi_i, Dq\psi_i \rangle + \langle q\psi_i, Dp\psi_i \rangle).$$

If it exists, the covariance matrix is always nonsingular and $\det \Gamma \geq \tfrac{1}{4}\hbar^2$, which is essentially the Heisenberg uncertainty principle.

Now let $(p_1,q_1),\ldots,(p_n,q_n)$ be commuting canonical pairs on $\mathcal{H}$. Then the operators $\bar{p}_n, \bar{q}_n$ defined by

$$\bar{p}_n = \frac{1}{\sqrt{n}}(p_1 + \cdots + p_n), \qquad \bar{q}_n = \frac{1}{\sqrt{n}}(q_1 + \cdots + q_n) \qquad (2.20)$$

give a new canonical pair $(\bar{p}_n, \bar{q}_n)$.

We have the necessary ingredients to state a quantum mechanical central limit theorem. Before doing this let us consider the analogous classical central limit theorem. The *normal distribution* with mean zero and variance σ is the probability measure μ_σ on $\mathscr{B}(\mathbb{R})$ defined by

$$\mu_\sigma(A) = \frac{1}{\sqrt{2\pi}\,\sigma} \int_A e^{-t^2/2\sigma^2} dt$$

for all $A \in \mathscr{B}(\mathbb{R})$. A sequence of random variables f_i converges in distribution to μ_σ if

$$\int g\,d\mu_{f_i} \to \int g\,d\mu_\sigma$$

for every bounded continuous function g on $\mathbb{R}$. Now let f_i be a sequence of independent, identically distributed random variables with zero expectation and finite variance σ, and let $\bar{f}_n = (n)^{-1/2}(f_1 + \cdots + f_n)$. The central limit theorem states that $\bar{f}_n$ converges in distribution to μ_σ.

Theorem 2.6 (Cushen and Hudson [47]). *Let (p_i, q_i) be a sequence of commuting canonical pairs on a Hilbert space $\mathcal{H}$ which are independent and identically distributed with finite covariance matrix Γ and zero expectations in the state D. If $\det\Gamma > \frac{1}{4}\hbar^2$, then the sequence of canonical pairs $(\bar{p}_n, \bar{q}_n)$ given by (2.20) converge in distribution to the limit distribution*

$$D_\Gamma = N_\Gamma \exp\left[-\tfrac{1}{2}\eta_\Gamma\left(\gamma_{11}p_0^2 + \gamma_{12}p_0q_0 + \gamma_{21}q_0p_0 + \gamma_{22}q_0^2\right) \right],$$

where

$$N_\Gamma = \left[\operatorname{tr} \exp\left(-\tfrac{1}{2}\eta_\Gamma\left(\gamma_{11}p_0^2 + \gamma_{12}p_0q_0 + \gamma_{21}q_0p_0 + \gamma_{22}q_0^2\right)\right) \right]^{-1},$$

$$[\gamma_{ij}] = \Gamma^{-1}, \qquad \eta_\Gamma = 2\hbar^{-1}\det\Gamma \coth^{-1}(2\hbar^{-1}\det\Gamma).$$

If $\det\Gamma = \frac{1}{4}\hbar^2$, the distribution operators of all the $\bar{p}_n, \bar{q}_n$ are already equal to the limit distribution which is the pure state determined by the eigenvector corresponding to the smallest eigenvalue of the positive self-adjoint operator

$$\gamma_{11}p_0^2 + \gamma_{12}p_0q_0 + \gamma_{21}q_0p_0 + \gamma_{22}q_0^2.$$

The proof is fairly long and will be omitted.

2.3. Conditional Expectation

Conditional expectations play a basic role in classical probability theory. In fact, some of the most important areas of the theory such as Markov processes and martingales [65, 68] rely heavily on this concept. Although

2. A Different Kind of Probability

there has been much discussion [48, 56, 113, 114, 192, 193, 262, 263], until recently conditional expectation has not been satisfactorily generalized to quantum probability.

We first review the concept of conditional expectation in the classical theory. Let (Ω, Σ, μ) be a probability space and let $B \in \Sigma$ with $\mu(B) \neq 0$. For any $A \in \Sigma$ the *conditional probability* $\mu(A|B)$ *of A given B* is defined as $\mu(A|B) = \mu(A \cap B)/\mu(B)$. We interpret $\mu(A|B)$ as the probability that the event A occurs given that the event B has occurred. Notice that the function $A \mapsto \mu(A|B)$ is a probability measure on (Ω, Σ).

Now let $B_1, \ldots, B_n$ be mutually disjoint sets in Σ which satisfy $\cup B_i = \Omega$, $\mu(B_i) \neq 0$, $i = 1, \ldots, n$. Let Σ_0 be the sub-σ-algebra of Σ generated by $B_1, \ldots, B_n$. For any $A \in \Sigma$ define the random variable $\mu(A|\Sigma_0)$ by $\mu(A|\Sigma_0)(\omega) = \mu(A|B_i)$, where $\omega \in B_i$. We call $\mu(A|\Sigma_0)$ the *conditional probability of A given* Σ_0. Notice that $\mu(A|\Sigma_0)$ satisfies the following two properties:

$$\mu(A|\Sigma_0) \quad \text{is measurable with respect to} \quad \Sigma_0; \tag{2.21}$$

$$\int_B \mu(A|\Sigma_0) \, d\mu = \int_B \chi_A \, d\mu \qquad \forall B \in \Sigma_0. \tag{2.22}$$

Property (2.21) is obvious. Property (2.22) clearly holds for B equal to some B_i, and since every $B \in \Sigma_0$ is a union of B_i's the result holds by additivity. Conversely, suppose f is a random variable on (Ω, Σ, μ) which satisfies:

$$f \quad \text{is measurable with respect to} \quad \Sigma_0; \tag{2.23}$$

$$\int_B f \, d\mu = \int_B \chi_A \, d\mu \qquad \forall B \in \Sigma_0. \tag{2.24}$$

Applying (2.23), f is constant on each B_i, and from (2.24) we have for every $\omega \in B_i$

$$f(\omega) = \left[\mu(B_i) \right]^{-1} \int_{B_i} f(\omega) \, d\mu = \mu(A \cap B_i)/\mu(B_i) = \mu(A|B_i).$$

Hence, $f = \mu(A|\Sigma_0)$, so we see that (2.23) and (2.24) characterize $\mu(A|\Sigma_0)$.

Now let f be an arbitrary random variable on (Ω, Σ, μ) and again let $B \in \Sigma$, $\mu(B) \neq 0$. We define the *conditional expectation of f given B* as

$$E(f|B) = \int f \, d\mu(\cdot|B) = \left[\mu(B) \right]^{-1} \int_B f \, d\mu.$$

Let $B_1, \ldots, B_n$ and Σ_0 be defined as in the previous paragraph and define the random variable $E(f|\Sigma_0)(\omega) = E(f|B_i)$, where $\omega \in B_i$. We call $E(f|\Sigma_0)$ the *conditional expectation of f given* Σ_0. As before, $E(f|\Sigma_0)$ satisfies two properties:

$$E(f|\Sigma_0) \quad \text{is measurable with respect to} \quad \Sigma_0; \quad (2.25)$$

$$\int_B E(f|\Sigma_0)\,d\mu = \int_B f\,d\mu \quad \forall B \in \Sigma_0. \quad (2.26)$$

Moreover, as before, $E(f|\Sigma_0)$ is characterized by (2.25) and (2.26). Notice that conditional probability is a special case of conditional expectation since $\mu(A|\Sigma_0) = E(\chi_A|\Sigma_0)$.

More generally, let Σ_0 be an arbitrary sub-σ-algebra of Σ. We now introduce a general concept of conditional expectation which reduces to the previous definitions as special cases. The *conditional expectation* $E(f|\Sigma_0)$ *of* f *given* Σ_0 is a random variable satisfying (2.25) and (2.26). If $E(f)$ exists, then $E(f|\Sigma_0)$ exists and is unique a.e. $[\mu]$. Indeed, if we define the measure $\nu(B) = \int_B f\,d\mu$, $B \in \Sigma_0$, then ν is absolutely continuous relative to μ restricted to Σ_0. By the Radon–Nikodym theorem [67, 123] there exists a unique (a.e. $[\mu]$) function g which is measurable with respect to Σ_0 satisfying $\nu(B) = \int_B g\,d\mu$. Hence g satisfies (2.25) and (2.26) (with $E(f|\Sigma_0)$ replaced by g). In particular, if f is a bounded random variable, then $E(f|\Sigma_0)$ exists and is unique (a.e. $[\mu]$). The *conditional probability* is now defined as $\mu(A|\Sigma_0) = E(\chi_A|\Sigma_0)$ for all $A \in \Sigma$. The following lemma summarizes some of the important properties of conditional expectations. For a probability space (Ω, Σ, μ), $\mathcal{C}(\Sigma)$ denotes the set of essentially bounded complex-valued random variables with norm $\|f\| = \operatorname{ess\,sup}|f(\omega)|$.

Lemma 2.7. *Let* (Ω, Σ, μ) *be a probability space and let* Σ_0 *be a sub-σ-algebra of* Σ. *Then* $f \mapsto E(f|\Sigma_0)$ *is a map from* $\mathcal{C}(\Sigma)$ *onto* $\mathcal{C}(\Sigma_0)$ *which satisfies the following:*

1. $f \mapsto E(f|\Sigma_0)$ *is linear;*
2. $E(f|\Sigma_0) = f \;\; \forall f \in \mathcal{C}(\Sigma_0)$;
3. $\|E(f|\Sigma_0)\| \leq \|f\|$;
4. $E(f^*|\Sigma_0) = E(f|\Sigma_0)^*$, *where* f^* *is the complex conjugate of* f;
5. $E(f^*f|\Sigma_0) \geq 0$;
6. $E(f^*f|\Sigma_0) = 0$ *implies* $f = 0$ *a.e.;*
7. $E(gf|\Sigma_0) = gE(f|\Sigma_0) \;\; \forall g \in \mathcal{C}(\Sigma_0)$;
8. $E(f|\Sigma_0)^* E(f|\Sigma_0) \leq E(f^*f|\Sigma_0)$;
9. *if* f_n *is a bounded increasing sequence of nonnegative functions in* $\mathcal{C}(\Sigma)$, *then* $\lim E(f_n|\Sigma_0) = E(\lim f_n|\Sigma_0)$.

The proof is straightforward and is left as an exercise for the reader.

2. A Different Kind of Probability

How can we generalize the concept of conditional expectation to quantum probability theory? As we have seen, in quantum probability theory the probability space (Ω, Σ, μ) is replaced by a triple $(\mathcal{K}, \mathcal{P}(\mathcal{K}), W)$ where $\mathcal{K}$ is a complex Hilbert space, $\mathcal{P}(\mathcal{K})$ is the set of orthogonal projections on $\mathcal{K}$, and W is a density operator. The random variables are replaced by observables (self-adjoint operators) on $\mathcal{K}$, and the expectation of an observable A in the state W is $E(A) = \operatorname{tr} WA$. It is natural to replace the sub-σ-algebra Σ_0 by a sub-σ-lattice $\mathcal{P}_0$ of $\mathcal{P}(\mathcal{K})$. Precisely, a *sub-σ-lattice* $\mathcal{P}_0$ of $\mathcal{P}(\mathcal{K})$ is a collection of orthogonal projections on $\mathcal{K}$ satisfying

$$I \in \mathcal{P}_0, \tag{2.27}$$

$$\text{if } P \in \mathcal{P}_0, \quad \text{then } I - P \in \mathcal{P}_0, \tag{2.28}$$

$$\text{if } P_i \in \mathcal{P}_0, \quad \text{then } \bigvee P_i \in \mathcal{P}_0. \tag{2.29}$$

We say that an observable B is *measurable* relative to a sub-σ-lattice $\mathcal{P}_0$ if $P^B(\Delta) \in \mathcal{P}_0$ for every $\Delta \in \mathcal{B}(\mathbb{R})$. We could define the conditional expectation of an observable A given a sub-σ-lattice $\mathcal{P}_0$ to be an observable $E(A|\mathcal{P}_0)$ satisfying properties analogous to (2.25) and (2.26). Clearly, we would replace (2.25) by

$$E(A|\mathcal{P}_0) \quad \text{is measurable relative to} \quad \mathcal{P}_0. \tag{2.30}$$

In order to find a condition analogous to (2.26) let us rewrite (2.26) in the following equivalent form:

$$E(\chi_B E(f|\Sigma_0)) = E(\chi_B f) \qquad \forall B \in \Sigma_0. \tag{2.31}$$

For simplicity, let us restrict our attention to a bounded observable A. Since $\chi_B f = \chi_B f \chi_B$, and since PA need not be self-adjoint for $P \in \mathcal{P}_0$, it is natural to replace the random variable $\chi_B f, B \in \Sigma_0$, by the observable $PAP, P \in \mathcal{P}_0$. (Another reason for doing this is given in [113].) We thus replace condition (2.26) by

$$E(PE(A|\mathcal{P}_0)P) = E(PAP) \qquad \forall P \in \mathcal{P}_0, \tag{2.32}$$

that is, $\operatorname{tr}(WPE(A|\mathcal{P}_0)P) = \operatorname{tr}(WPAP)$ for all $P \in \mathcal{P}_0$. Hence a natural generalization of conditional expectation to quantum probability theory would be the following. The *conditional expectation $E(A|\mathcal{P}_0)$ of A given $\mathcal{P}_0$* is an observable satisfying (2.30) and (2.32). Although this definition has some uses [113], it has the unfortunate drawback that $E(A|\mathcal{P}_0)$ may not exist for every bounded observable A except under special circumstances [113]. Moreover, even when it does exist, it does not in general satisfy counterparts of all the desirable conditions (1)–(9) of Lemma 2.7 [114]. We now reformulate the concept of conditional expectation in classical probability theory so that it has a more satisfactory generalization to quantum probability theory.

Let (Ω, Σ, μ) be a probability space and let Σ_0 be a sub-σ-algebra of Σ. It is clear that the Hilbert space $L^2(\Omega, \Sigma_0, \mu)$ is a closed subspace of the Hilbert space $L^2(\Omega, \Sigma, \mu)$. Now let f be a bounded random variable on (Ω, Σ, μ). Define the bounded linear operator $\pi(f): L^2(\Omega, \Sigma, \mu) \to L^2(\Omega, \Sigma, \mu)$ by $[\pi(f)g](\omega) = f(\omega)g(\omega)$. Let $\mathcal{E}(f|\Sigma_0)$ satisfy

$$\mathcal{E}(f|\Sigma_0) \quad \text{is a bounded operator on} \quad L^2(\Omega, \Sigma_0, \mu), \tag{2.33}$$

$$E(g_1 \mathcal{E}(f|\Sigma_0)g_2) = E(g_1 \pi(f)g_2) \quad \forall g_1, g_2 \in L^2(\Omega, \Sigma_0, \mu). \tag{2.34}$$

Notice that in terms of the inner product on $L^2(\Omega, \Sigma_0, \mu)$, (2.34) can be written $\langle g_1, \mathcal{E}(f|\Sigma_0)g_2 \rangle = \langle g_1, \pi(f)g_2 \rangle$ for all $g_1, g_2 \in L^2(\Omega, \Sigma_0, \mu)$. It thus follows that $\mathcal{E}(f|\Sigma_0)$ is unique. Now $E(f|\Sigma_0)$ is a bounded random variable and by (2.25) $\pi[E(f|\Sigma_0)]g \in L^2(\Omega, \Sigma_0, \mu)$, for every $g \in L^2(\Omega, \Sigma_0, \mu)$. Hence, the restriction $\pi_r[E(f|\Sigma_0)]$ of $\pi[E(f|\Sigma_0)]$ to $L^2(\Omega, \Sigma_0, \mu)$ satisfies

$$\pi_r[E(f|\Sigma_0)] \quad \text{is a bounded operator on} \quad L^2(\Omega, \Sigma_0, \mu). \tag{2.35}$$

Moreover, it easily follows from (2.26) that

$$E(g_1 \pi_r[E(f|\Sigma_0)]g_2) = E(g_1 \pi(f)g_2) \quad \forall g_1, g_2 \in L^2(\Omega, \Sigma_0, \mu). \tag{2.36}$$

It follows that $\pi_r[E(f|\Sigma_0)] = \mathcal{E}(f|\Sigma_0)$. In this sense, we can identify $\mathcal{E}(f|\Sigma_0)$ with $E(f|\Sigma_0)$ and (2.33), (2.34) are equivalent formulations of the defining conditions (2.25), (2.26) for $E(f|\Sigma_0)$.

Now consider quantum probability theory on the structure $(\mathcal{H}, \mathcal{P}(\mathcal{H}), W)$. We can generalize (2.33), (2.34) to this situation. As we have seen, in this case a sub-σ-algebra is replaced by a sub-σ-lattice $\mathcal{P}_0$ of $\mathcal{P}(\mathcal{H})$. Let $\mathcal{C}(\mathcal{P}_0)$ be the set of bounded linear operators on $\mathcal{H}$ of the form $A_1 + iA_2$, where A_1 and A_2 are observables which are measurable relative to $\mathcal{P}_0$. It turns out that $\mathcal{C}(\mathcal{P}_0)$ has a well-studied structure called a von Neumann algebra. Precisely, a *von Neumann algebra* $\mathcal{C}$ is an algebra of bounded linear operators on a complex Hilbert space $\mathcal{H}$ satisfying

$$I \in \mathcal{C}, \tag{2.37}$$

$$\text{if } A \in \mathcal{C}, \quad \text{then } A^* \in \mathcal{C}, \tag{2.38}$$

$$\mathcal{C} \quad \text{is closed in the weak operator topology.} \tag{2.39}$$

Condition (2.39) means that if A is a bounded linear operator on $\mathcal{H}$ for which there exists a net $A_\alpha \in \mathcal{C}$ satisfying $\langle A_\alpha x, y \rangle \to \langle Ax, y \rangle$ for every $x, y \in \mathcal{H}$, then $A \in \mathcal{C}$. Of course, the set of all bounded linear operators $\mathcal{L}(\mathcal{H}) = \mathcal{C}(\mathcal{P}(\mathcal{H}))$ is a von Neumann algebra.

2. A Different Kind of Probability

We say that the state W is *faithful* if $\text{tr}(WA^*A) = 0$ implies that $A = 0$ for $A \in \mathcal{L}(\mathcal{K})$. We shall assume in the remainder of this section that W is faithful. We now form a Hilbert space $L^2(\mathcal{K}, \mathcal{P}(\mathcal{K}), W)$ which is analogous to the classical Hilbert space $L^2(\Omega, \Sigma, \mu)$. Define the inner product $\langle A, B \rangle = \text{tr} \, WB^*A$, $A, B \in \mathcal{L}(\mathcal{K})$. Then $L^2(\mathcal{K}, \mathcal{P}(\mathcal{K}), W)$ is the completion of $\mathcal{L}(\mathcal{K})$ under this inner product. For $A \in \mathcal{L}(\mathcal{K})$ define the linear operator $\tilde{\pi}(A) : \mathcal{L}(\mathcal{K}) \to \mathcal{L}(\mathcal{K})$ by $\tilde{\pi}(A)B = AB$. It is easily seen that $\tilde{\pi}(A)$ is bounded and can hence be uniquely extended to a bounded linear operator $\pi(A)$ on $L^2(\mathcal{K}, \mathcal{P}(\mathcal{K}), W)$. It is straightforward to show that π preserves linearity, $*$, and $\|\pi(A)\| \le \|A\|$ (we shall prove these in Chapter 5). In other words, π is a continuous $*$-representation of $\mathcal{L}(\mathcal{K})$. The above construction is a special case of a general method called the Gelfand–Naimark–Segal (GNS) construction which we shall treat in detail in Chapter 5.

The GNS construction can be carried out in the same way for the von Neumann algebra $\mathcal{Q}(\mathcal{P}_0)$ to form the Hilbert space $L^2(\mathcal{K}, \mathcal{P}_0(\mathcal{K}), W)$. Then $L^2(\mathcal{K}, \mathcal{P}_0(\mathcal{K}), W)$ is a closed subspace of $L^2(\mathcal{K}, \mathcal{P}(\mathcal{K}), W)$ and we denote the restriction of π to $L^2(\mathcal{K}, \mathcal{P}_0(\mathcal{K}), W)$ by π_r. We can now formulate the quantum generalization of conditions (2.33), (2.34). For $A \in \mathcal{L}(\mathcal{K})$, the *conditional expectation* $\mathcal{E}(A|\mathcal{P}_0)$ of A given $\mathcal{P}_0$ satisfies

$$\mathcal{E}(A|\mathcal{P}_0) \in \mathcal{L}(L^2(\mathcal{K}, \mathcal{P}_0(\mathcal{K}), W)), \tag{2.40}$$

$$E[\pi(B_1)\mathcal{E}(A|\mathcal{P}_0)B_2] = E(B_1AB_2) \quad \forall B_1, B_2 \in \mathcal{Q}(\mathcal{P}_0). \tag{2.41}$$

By continuity, (2.41) can be written $\langle B_1, \mathcal{E}(A|\mathcal{P}_0)B_2 \rangle = \langle B_1, \pi(A)B_2 \rangle$ for every $B_1, B_2 \in L^2(\mathcal{K}, \mathcal{P}_0(\mathcal{K}), W)$. The next theorem shows that $\mathcal{E}(A|\mathcal{P}_0)$ always exists, is unique, and enjoys many important properties analogous to those satisfied by $E(f|\Sigma_0)$.

Theorem 2.8. *There exists a unique bounded linear operator* $\mathcal{E}(A|\mathcal{P}_0)$ *satisfying* (2.40), (2.41), *and* $\mathcal{E}(A|\mathcal{P}_0) = P\pi(A)|L^2(\mathcal{K}, \mathcal{P}_0(\mathcal{K}), W)$, *where P is the orthogonal projection of* $L^2(\mathcal{K}, \mathcal{P}(\mathcal{K}), W)$ *onto* $L^2(\mathcal{K}, \mathcal{P}_0(\mathcal{K}), W)$. *Moreover, we have:*

1. $A \mapsto \mathcal{E}(A|\mathcal{P}_0)$ *is linear;*
2. $\mathcal{E}(B|\mathcal{P}_0) = \pi_r(B) \quad \forall B \in \mathcal{Q}(\mathcal{P}_0)$;
3. $\|\mathcal{E}(A|\mathcal{P}_0)\| \le \|A\|$;
4. $\mathcal{E}(A^*|\mathcal{P}_0) = \mathcal{E}(A|\mathcal{P}_0)^*$;
5. $\mathcal{E}(A^*A|\mathcal{P}_0) \ge 0$;
6. $\mathcal{E}(A^*A|\mathcal{P}_0) = 0$ *implies* $A = 0$;
7. $\mathcal{E}(B_1AB_2|\mathcal{P}_0) = \pi_r(B_1)\mathcal{E}(A|\mathcal{P}_0)\pi_r(B_2) \quad \forall B_1, B_2 \in \mathcal{Q}(\mathcal{P}_0)$;
8. $\mathcal{E}(A|\mathcal{P}_0)^*\mathcal{E}(A|\mathcal{P}_0) \le \mathcal{E}(A^*A)$;
9. *if A_n is a bounded increasing sequence of positive operators in $\mathcal{L}(\mathcal{K})$, then* $\lim \mathcal{E}(A_n|\mathcal{P}_0) = \mathcal{E}(\lim A_n|\mathcal{P}_0)$.

The proof is straightforward and is left as an exercise for the reader. In Chapter 4 another formulation of conditional expectation will be discussed.

2.4. Notes and References

A good discussion of the double-slit experiment and the interference of probabilities may be found in [198]. The "superposition" of states and the superposition principle is stressed in [63]. Abstract formulations of the superposition principle are discussed in [102,268]. A further discussion of independence and limit laws in quantum probability theory may be found in [94]. The discussion in Section 2.3 is based on the work in [114], where more details may be found. In this paper the relation between conditional expectation and course-graining or statistical inference is discussed. Connections to maximal entropy states are studied in [183,184].

2.5. Exercises

1. Prove Eqs. (2.1a), (2.1b), and (2.1c).

2. Prove that $\mu_{f,g}$ is the unique measure on $\mathcal{B}(\mathbb{R}^2)$ satisfying (2.2) for every $E \in \mathcal{B}(\mathbb{R})$, $x_1, x_2 \in \mathbb{R}$.

3. Show that if A_1 and A_2 have a joint distribution in the state ψ, then there exist random variables f and g satisfying (2.4) for all $E \in \mathcal{B}(\mathbb{R})$, $x \in \mathbb{R}^2$.

4. Prove the necessity of Theorem 2.1.

5. Show that a σ-isomorphism $h: \Sigma \rightarrow \mathcal{P}$ preserves all the usual set theoretic operations in Σ.

6. Supply the details in the proof of Theorem 2.2.

7. Prove that $\limsup \Delta_i$ is the set of elements in infinitely many of the Δ_i and $\liminf \Delta_i$ is the set of elements in all but finitely many of the Δ_i.

8. Prove Eqs. (2.5)–(2.8).

9. Prove that (2.9) holds.

10. Prove the sufficiency of Lemma 2.4.

11. Prove that (2.10) holds.

12. Show that (2.11), (2.12), and (2.13) generalize the usual concepts of convergence in measure, convergence almost everywhere, and convergence almost uniformly, respectively.

13. Show that $(2.13) \Rightarrow (2.12) \Rightarrow (2.11)$.

2. A Different Kind of Probability

14. Use formal manipulations to derive (2.15) from (2.14).

15. Show that (2.16) follows from the von Neumann uniqueness theorem.

16. Prove the statement that includes (2.17).

17. Prove that the nth moments of p and q in the state D equal the corresponding nth moments of the Schrödinger operators p_0 and q_0 in the state $D_{p,q}$.

18. Prove (2.18) and (2.19).

19. Show that if the covariance matrix Γ exists, then it is nonsingular and $\det \Gamma \geq \frac{1}{4}\hbar^2$.

20. Let $(p_1, q_1), \ldots, (p_n, q_n)$ be commuting canonical pairs and define the operators $\bar{p}_n, \bar{q}_n$ as in (2.20). Prove that $(\bar{p}_n, \bar{q}_n)$ is a canonical pair.

21. Let (p_1, q_1) and (p_2, q_2) be commuting canonical pairs which are independent in the state D, have finite covariance matrices $\Gamma^{(1)}, \Gamma^{(2)}$, respectively, and zero expectations. Prove that $\bar{p}_2 = (2)^{-1/2}(p_1 + p_2)$, $\bar{q}_2 = (2)^{-1/2}(q_1 + q_2)$ have finite covariance matrix $\bar{\Gamma}^{(2)} = (2)^{-1}(\Gamma^{(1)} + \Gamma^{(2)})$ and zero expectations.

22. Prove Lemma 2.7.

23. Prove (2.36).

24. Give an example of a faithful state; a nonfaithful state.

25. Prove Theorem 2.8.

3

The Quantum Logic Approach

The von Neumann quantum probability theory served an important role in understanding and unifying the concepts of quantum mechanics. However, von Neumann's axioms have the disadvantage of being *ad hoc* and physically unmotivated. These axioms lead to some puzzling questions. Where does the Hilbert space come from? Why represent pure states by vectors and observables by self-adjoint operators? Why does the dynamical law have the postulated form? Such questions have led investigators to develop an even more general probability theory based on physically motivated axioms. One attempt in this direction is the quantum logic approach. This approach was originated by Birkhoff and von Neumann in 1936 [20] and was refined and developed later by Mackey [172–174], Piron [211, 212], Varadarajan [267, 268], Jauch [138–143], and others [22, 75, 77, 79, 91, 95, 100, 101, 104, 105, 110, 115, 118, 175, 176, 191, 204, 205, 213–216, 222, 232, 284, 285]. Another method, the operational approach, will be discussed in Chapter 4.

The quantum logic approach has little to do with classical logic. The name derives from the similarity between the structure developed and the usual Boolean structure of the propositional calculus of logic. We do not mean to imply that the framework developed here is a new kind of logic governing our thought process (although this viewpoint has been considered [124, 220]).

3.1. States and Observables

As pointed out by von Neumann, Dirac, and others, two of the most basic concepts in quantum mechanics are the concepts of state and observable. In the laboratory, the experimentalist is involved in two main activities, preparation and measurement. In studying a physical system, he first prepares the system so that it is in a specific state or condition. He then subjects the system to a measuring apparatus and measures an observable quantity such as energy, momentum, position, charge, spin, magnetic moment, etc. Suppose we have some fixed physical system which may be prepared in any one of a collection of states $S = \{s, s_1, s_2, \dots\}$ and that in this system one may measure the observables $\mathcal{O} = \{x, y, z, \dots\}$. Now the result of measuring an observable x can usually be formulated as a number; for example, the spin of the particle is $+2$, the energy of the proton is 3 erg. Of course, in practice, one repeats the experiment many times (keeping the state as fixed as possible) and obtains only a statistical distribution for the values of x. Thus, given an observable x and a state s, the experimentalist obtains a probability distribution $E \mapsto p(x, s, E)$. That is, $E \mapsto p(x, s, E)$ is a probability measure on $\mathcal{B}(\mathbb{R})$ where $p(x, s, E)$ is interpreted as the probability that the observable x has a value in E when the system is in the state s. This gives us our first axiom.

Axiom 3.1. There is a map $p: \mathcal{O} \times S \times \mathcal{B}(\mathbb{R}) \to [0, 1]$ where $p(x, s, \cdot)$ is a probability measure on $\mathcal{B}(\mathbb{R})$.

If two observables have the same distribution in every state, then there is no experimental way to distinguish them, so they are identical. Similarly, if two states give the same distributions for all observables, they must be equal. We are thus led to our next axiom.

Axiom 3.2. If $p(x, s, E) = p(y, s, E)$ for every $s \in S$ and $E \in \mathcal{B}(\mathbb{R})$, then $x = y$. If $p(x, s_1, E) = p(x, s_2, E)$ for every $x \in \mathcal{O}$ and $E \in \mathcal{B}(\mathbb{R})$, then $s_1 = s_2$.

In Section 2.1 we used the concept of transition probability to study interference effects. Using only Axioms 3.1 and 3.2, Cantoni [35] has shown that a generalized transition probability can be defined which reduces to the usual concept considered in Section 2.1. This gives a direct physically motivated definition for transition probabilities without introducing any further postulates.

Let $(\mathcal{O}, S, p)$ be a triple satisfying Axioms 3.1 and 3.2. For every $x \in \mathcal{O}$ and $s \in S$ let s_x be the probability measure on $\mathcal{B}(\mathbb{R})$ defined by $s_x(E) = p(x, s, E)$. For $s, t \in S$ we would like to define the transition probability $T(s, t)$ of s to t. If the system is originally in the state s and we measure a t-determining observable, then $T(s, t)$ is the probability that after this measurement the system is in the state t. It thus seems plausible that we can define $T(s,t)$ in terms of the numbers $\{s_x, t_x : x \in \mathcal{O}\}$. For simplicity, suppose that x is an observable with only two values, say 0 and 1. Then s_x and t_x are probability measures concentrated on the set $\{0, 1\}$. Now x determines the pairs of numbers $(s_x(0), s_x(1))$ and $(t_x(0), t_x(1))$. If these pairs are "close," then this indicates that s and t are "close," so $T(s,t)$ would be close to 1. Conversely, if these pairs are quite different we would expect $T(s,t)$ to be close to 0. A simple function which brings out this relationship is given by

$$T_x(s,t) = \left[s_x(0) t_x(0) \right]^{1/2} + \left[s_x(1) t_x(1) \right]^{1/2}.$$

Indeed, if $s_x(0) = t_x(0)$, then since $s_x(0) + s_x(1) = 1$ and $s_y(0) + s_y(1) = 1$ we have $s_x(1) = t_x(1)$ and hence $T_x(s,t) = s_x(0) + s_x(1) = 1$. On the other hand, if s and t are quite different, then we have $s_x(0) \approx 1$, $t_x(0) \approx 0$ or $s_x(0) \approx 0$, $t_x(0) \approx 1$. Suppose, for concreteness, that the former holds. Then

$$T_x(s,t) = \left[s_x(0) t_x(0) \right]^{1/2} + \left[1 - s_x(0) \right]^{1/2} \left[1 - t_x(0) \right]^{1/2}$$

$$\approx t_x(0)^{1/2} + \left[1 - s_x(0) \right]^{1/2} \approx 0.$$

Since we would like to bring out the difference between s and t as much as possible we could define $T(s,t) = \inf\{ T_x(s,t) : x \in \mathcal{O} \}$. It turns out to be more convenient to consider $T_x(s,t)^2$. We now treat the general case where x is an arbitrary observable.

For $s, t \in S$, $x \in \mathcal{O}$ there are many finite measures on $\mathcal{B}(\mathbb{R})$ relative to which s_x and t_x are both absolutely continuous. For example, $s_x + t_x$ is such a measure. Let s_x and t_x be absolutely continuous relative to a finite measure σ and define the number $T_x(s,t)$ by

$$T_x^{1/2}(s,t) = \int_{\mathbb{R}} \left[(ds_x/d\sigma)(dt_x/d\sigma) \right]^{1/2} d\sigma.$$

That this number is independent of σ can be seen as follows. Suppose s_x and t_x are also absolutely continuous relative to σ_1. Now there exists a finite measure σ_2 on $\mathcal{B}(\mathbb{R})$ relative to which both σ and σ_1 are absolutely continuous. We then have

$$\int \left(\frac{ds_x}{d\sigma} \frac{dt_x}{d\sigma} \right)^{1/2} d\sigma = \int \left(\frac{ds_x}{d\sigma} \frac{dt_x}{d\sigma} \right)^{1/2} \frac{d\sigma}{d\sigma_2} d\sigma_2$$

$$= \int \left(\frac{ds_x}{d\sigma_2} \frac{dt_x}{d\sigma_2} \right)^{1/2} d\sigma_2 = \int \left(\frac{ds_x}{d\sigma_1} \frac{dt_x}{d\sigma_1} \right)^{1/2} \frac{d\sigma_1}{d\sigma_2} d\sigma_2$$

$$= \int \left(\frac{ds_x}{d\sigma_1} \frac{dt_x}{d\sigma_1} \right)^{1/2} d\sigma_1.$$

3. The Quantum Logic Approach

We define the *generalized transition probability* for s to t by $T(s,t) = \inf\{T_x(s,t) : x \in \theta\}$. In Section 2.1 we gave three conditions [(2.1a), (2.1b) and (2.1c)] which seemed desirable and which are satisfied by the usual transition probability. The next lemma shows that the generalized transition probability satisfies these conditions.

Lemma 3.1. (a) $T(s,t) = T(t,s)$ for all $s,t \in S$. (b) $0 \le T(s,t) \le 1$ for all $s,t \in S$. (c) $T(s,t) = 1$ if and only if $s = t$.

PROOF. (a) Trivial. (b) Clearly $T(s,t) \ge 0$. Moreover, if for $s,t \in S$, $x \in \theta$ the measure σ is chosen as above, then $(ds_x/d\sigma)^{1/2}$ and $(dt_x/d\sigma)^{1/2}$ are unit vectors in the Hilbert space $L^2(\mathbb{R}, \mathfrak{B}(\mathbb{R}), \sigma)$. It follows from Schwarz's inequality that $T_x^{1/2}(s,t) \le 1$. Hence, $T(s,t) \le 1$. (c) Since $T_x^{1/2}(s,s) = s_x(\mathbb{R}) = 1$, $T(s,s) = 1$. Conversely, if $T(s,t) = 1$, then since $T_x^{1/2}(s,t) \le 1$ we must have $T_x^{1/2}(s,t) = 1$ for every $x \in \theta$. Again, by Schwarz's inequality this implies that $ds_x/d\sigma = dt_x/d\sigma$. Hence $s_x = t_x$ for every $x \in \theta$, and by Axiom 3.2 $s = t$. $\square$

We now show that $T(s,t)$ reduces to the usual transition probability of Section 2.1. The proof of the next theorem is a simplification of the proof given by Cantoni [35].

Theorem 3.2. If ϕ and ψ are unit vectors in a Hilbert space $\mathcal{H}$ which represent pure states s and t, respectively, then $T(s,t) = |\langle \phi, \psi \rangle|^2$.

PROOF. Let A be a fixed self-adjoint operator. By the spectral theorem [67] we can assume that $\mathcal{H} = L^2(\mathbb{R}, \mathfrak{B}(\mathbb{R}), \sigma)$, where σ is a finite measure and $(Af)(\lambda) = \lambda f(\lambda)$ for all $f \in \mathcal{H}$. Then

$$s_A(E) = \langle P^A(E)\phi, \phi \rangle = \int_E |\phi|^2 \, d\sigma$$

and similarly $t_A(E) = \int_E |\psi|^2 \, d\sigma$. Hence, $ds_A/d\sigma = |\phi|^2$ and $dt_A/d\sigma = |\psi|^2$. We then have

$$T_A^{1/2}(s,t) = \int |\phi| \, |\psi| \, d\sigma \ge \left| \int \phi^* \psi \, d\sigma \right| = |\langle \phi, \psi \rangle|.$$

Hence $T_A(s,t) \ge |\langle \phi, \psi \rangle|^2$ and so $T(s,t) \ge |\langle \phi, \psi \rangle|^2$. Next, let $A = P_\phi$, the one-dimensional projection onto the span of ϕ. Then $P^A(\{0\}) = I - P_\phi$ and $P^A(\{1\}) = P_\phi$. Hence, $s_A(\{1\}) = 1$, $t_A(\{0\}) = \langle (I - P_\phi)\psi, \psi \rangle$, and $t_A(\{1\}) = \langle P_\phi \psi, \psi \rangle$. Let σ be the measure concentrated on $\{0,1\}$ which satisfies $\sigma(\{0\}) = 1$, $\sigma(\{1\}) = 1$. Then $ds_A/d\sigma = \chi_{\{1\}}$ and

$$dt_A/d\sigma = t_A(\{0\})\chi_{\{0\}} + t_A(\{1\})\chi_{\{1\}}.$$

Hence,

$$T_A^{1/2}(s,t) = \int \left(\frac{ds_A}{d\sigma} \frac{dt_A}{d\sigma} \right)^{1/2} d\sigma = t_A(\{1\})^{1/2} = |\langle \phi, \psi \rangle|$$

and we have $T(s,t) \le |\langle \phi, \psi \rangle|^2$. $\qquad\qquad\qquad\qquad\qquad\qquad\square$

We now return to our study of states and observables. If we measure an observable x, then it is just as easy to measure the observable x^2; simply take the measured value of x and square it. The probability that x^2 has a value in $E \in \mathcal{B}(\mathbb{R})$ equals the probability that x has a value in $\{\lambda : \lambda^2 \in E\}$. More generally, if $f: \mathbb{R} \to \mathbb{R}$ is a Borel function and x is an observable, then $f(x)$ should also be an observable and the probability that $f(x)$ has a value in $E \in \mathcal{B}(\mathbb{R})$ equals the probability that x has a value in $f^{-1}(E) = \{\lambda : f(\lambda) \in E\}$. We are thus led to the following axiom.

Axiom 3.3. If $x \in \mathcal{O}$ and $f: \mathbb{R} \to \mathbb{R}$ is a Borel function, there exists a $y \in \mathcal{O}$ such that $p(y, s, E) = p(x, s, f^{-1}(E))$ for every $s \in S$, $E \in \mathcal{B}(\mathbb{R})$.

It follows from Axiom 3.2 that the observable in Axiom 3.3 is unique. We denote this observable by $y = f(x)$.

Now there is a particular type of observable which is extremely simple. These are the observables with at most the two possible values 0 and 1. We call such observables *generalized events* (or *events*, for short). For example, a counter is a generalized event since it gives a measurement with only two possible outcomes: unactivated (or 0) and activated (or 1). Thus, a generalized event either does not occur (value 0) or occurs (value 1). To be precise, a *generalized event* is an observable x that satisfies $p(x, s, \{0, 1\}) = 1$ for every $s \in S$; that is, x has the value 0 or 1 with certainty in every state. It is easy to show that $x \in \mathcal{O}$ is a generalized event if and only if $x^2 = x$. There is another convenient way to describe generalized events. It is not hard to show that $x \in \mathcal{O}$ is a generalized event if and only if $x = \chi_E(y)$ for some $y \in \mathcal{O}$, $E \in \mathcal{B}(\mathbb{R})$. The generalized event $\chi_E(y)$ has the value 1 if y has a value in E and the value 0 if y has a value in E^c. Notice that if $\chi_E(x) = \chi_E(y)$ for every $E \in \mathcal{B}(\mathbb{R})$ then

$$p(x, s, E) = p\big(x, s, \chi_E^{-1}(\{1\})\big) = p\big(\chi_E(x), s, \{1\}\big)$$
$$= p\big(\chi_E(y), s, \{1\}\big) = p(y, s, E)$$

for every $s \in S$, $E \in \mathcal{B}(\mathbb{R})$, and hence, by Axiom 3.2, $x = y$. We thus see that not only can we associate with any observable x a collection of generalized events $\{\chi_E(x) : E \in \mathcal{B}(\mathbb{R})\}$ but that this associated collection of generalized events determines x.

Denote the set of generalized events by $\mathcal{E}$. The pair $(\mathcal{E}, S)$ contains all the information given in $(\mathcal{O}, S)$. We now consider the mathematical properties of $(\mathcal{E}, S)$. If $a \in \mathcal{E}$ and $s \in S$ we define $s(a) = p(a, s, \{1\})$. Now $s(a)$ may be interpreted as the probability that a has the value 1 (or that a occurs) in the state s. Notice that for $a, b \in \mathcal{E}$, if $s(a) = s(b)$ for every $s \in S$, then, by Axiom 3.2, $a = b$. It also follows from Axiom 3.2 that if $s_1(a) = s_2(a)$ for every $a \in \mathcal{E}$, then $s_1 = s_2$. Hence there are sufficiently many states to determine an event and vice versa. For $a, b \in \mathcal{E}$, we define $a \leq b$ if $s(a) \leq s(b)$ for every $s \in S$. Thus, $a \leq b$ if a has a smaller probability of occurring than b in every state. It is easy to check that $\leq$ is a partial order relation on $\mathcal{E}$; that is, $a \leq a$ for all $a \in \mathcal{E}$, $a \leq b$ and $b \leq a$ imply that $a = b$, and $a \leq b$ and $b \leq c$ imply that $a \leq c$. Thus $(\mathcal{E}, \leq)$ is a partially ordered set or poset.

Let f be the function $f(\lambda) = 1 - \lambda$. If $a \in \mathcal{E}$, we define the observable a' by $a' = f(a)$. Notice that

$$p(a', s, \{0, 1\}) = p(a, s, f^{-1}(\{0, 1\})) = p(a, s, \{0, 1\}) = 1$$

for every $s \in S$ so $a' \in \mathcal{E}$. Also

$$s(a') = p(a', s, \{1\}) = p(a, s, f^{-1}(\{1\})) = p(a, s, \{0\})$$
$$= 1 - p(a, s, \{1\}) = 1 - s(a).$$

Thus, a' corresponds to the complement of the event a. If f_0 and f_1 are the functions that are identically zero and one, respectively, and $x \in \mathcal{O}$, we define the observables 0 and 1 by $0 = f_0(x)$ and $1 = f_1(x)$. Notice that

$$p(0, s, \{0, 1\}) = p(x, s, f_0^{-1}(\{0, 1\})) = p(x, s, \mathbb{R}) = 1.$$

Hence $0 \in \mathcal{E}$ and also

$$s(0) = p(0, s, \{1\}) = p(x, s, f_0^{-1}(\{1\})) = p(x, s, \varnothing) = 0$$

for every $s \in S$. Similarly, $1 \in \mathcal{E}$ and $s(1) = 1$ for every $s \in S$. Thus, $0 \leq a \leq 1$ for every $a \in \mathcal{E}$. We may interpret 0 and 1 as the events which never occur and always occur, respectively. In the poset $(\mathcal{E}, \leq)$ we say that c is the *least upper bound* (or sup) of $a, b \in \mathcal{E}$ if $a, b \leq c$ and $a, b \leq d$ implies that $c \leq d$. The sup need not exist, but when it does it is unique. We denote the sup of a and b by $a \vee b$ when it exists. We define the *greatest lower bound* (or inf) of a and b dually and denote it by $a \wedge b$ when it exists. The proof of the following lemma is straightforward.

Lemma 3.3. *The operation $a \mapsto a'$ is an orthocomplementation on $(\mathcal{E}, \leq)$; that is, $a'' = a$ for every $a \in \mathcal{E}$, if $a \leq b$, then $b' \leq a'$, and $a \vee a' = 1$ for every $a \in \mathcal{E}$.*

This lemma shows that $(\mathcal{E}, \leq, ')$ is an orthocomplemented poset. Notice that we have generalized the structure of classical probability theory. If (Ω, Σ) is a measurable space, let $\mathcal{O}$ be the set of random variables and let S be the set of probability measures on (Ω, Σ). For $f \in \mathcal{O}$, $\mu \in S$, $E \in \mathcal{B}(\mathbb{R})$, define $p(f, \mu, E) = \mu[f^{-1}(E)]$. Then Axiom 3.1 clearly holds. Moreover, it is easy to show that Axiom 3.2 holds. The generalized events are random variables of the form χ_E, $E \in \Sigma$, and hence correspond naturally to classical events. The operation $'$ corresponds to complementation ^{c}, $\leq$ corresponds to inclusion $\subseteq$, $\vee$ corresponds to union $\cup$, and $\wedge$ corresponds to intersection $\cap$.

In this section we began with the states and observables and derived the set of generalized events $\mathcal{E}$ from these. In the next section we shall begin with the more fundamental set of generalized events and derive the states and observables. The states turn out to be analogous to probability measures on $\mathcal{E}$. An observable will be analogous to a map $E \mapsto \chi_E(x)$ from $\mathcal{B}(\mathbb{R})$ to $\mathcal{E}$ which associates a Borel set E with the generalized event $\chi_E(x)$ that x has a value in E. The framework of the present section can be extended by additional axioms [173] to obtain a framework which is equivalent to that of the next section. However, we shall proceed differently.

3.2. Event Structures

In this approach, the events of a physical system are taken as primitive axiomatic elements. The events correspond to physical phenomena which occur or do not occur. For example, suppose the physical system consists of a single particle and let Δ be a region of space. If a_Δ is a counter which is activated if and only if the particle enters the region Δ, then a_Δ corresponds to an event. Events can also be thought of as propositions corresponding to true–false experiments. This is the reason for the quantum logic terminology. Since we are stressing the stochastic properties of quantum mechanics we shall use the term event instead of proposition.

Let $\mathcal{E}$ be a set of elements called *events*. If $a \in \mathcal{E}$, then a occurs or does not occur depending upon the state of the system. Since in quantum mechanics, one can only predict the probability that an event will occur, a state s can be thought of as a function from $\mathcal{E}$ to the unit interval $[0, 1]$, and $s(a)$ is interpreted as the probability that a occurs when the system is in the state s. If $s(a) = 1$, then a occurs with certainty in the state s. Now suppose $a, b \in \mathcal{E}$ and $s(a) + s(b) \leq 1$ for every state s. Since $s(a) = 1$ implies that $s(b)=0$, whenever a occurs with certainty, b does not occur with certainty. In this case a and b are interpreted as corresponding to noninterfering experiments and their occurrence or nonoccurrence can be verified simultaneously. We can then consider the event c which occurs with certainty precisely when a and b do not occur with certainty. We should then have $s(c)+s(a)+s(b)=1$ for every state s. For example, in our counterexperiment suppose Δ and Γ are disjoint regions of space. Then for any state s, the probability that the particle is in Δ plus the probability that the particle is in Γ does not exceed 1; $s(a_\Delta)+s(a_\Gamma)\leq 1$. Now if $\Lambda=(\Delta\cup\Gamma)^c$ we should have $s(a_\Lambda)+s(a_\Delta)+s(a_\Gamma)=1$. Such considerations also carry over to sequences of events. These ideas lead to the following axioms.

Let $\mathcal{E}$ be a nonempty set and let S be a set of functions from $\mathcal{E}$ into $[0,1]$. We call $(\mathcal{E},S)$ an *event structure* if the following two axioms hold.

Axiom A. If $s(a)=s(b)$ for every $s\in S$, then $a=b$.

Axiom B. Let $a_1,a_2,\ldots\in\mathcal{E}$ satisfy $s(a_i)+s(a_j)\leq 1$, $i\neq j$, for every $s\in S$. Then there exists a $b\in\mathcal{E}$ such that $s(b)+s(a_1)+s(a_2)+\cdots=1$ for every $s\in S$.

If $(\mathcal{E},S)$ is an event structure, we call the elements of $\mathcal{E}$ *events* and the elements of S *states*. If $a,b\in\mathcal{E}$, define $a\leq b$ if $s(a)\leq s(b)$ for every $s\in S$. As in Section 3.1, it is easy to show that $\leq$ is a partial order relation, so $(\mathcal{E},\leq)$ is a poset. If $a\in\mathcal{E}$, since $s(a)\leq 1$ for every $s\in S$, by Axiom B there exists a $b\in\mathcal{E}$ such that $s(b)=1-s(a)$ for every $s\in S$. We then write $b=a'$ and call b the *orthocomplement* of a. We can interpret a' as the event which occurs if and only if a does not occur. If $a\leq b'$ we say that a and b are *orthogonal* and write $a\perp b$.

Before we give an important structure theorem, we need some definitions. An *orthocomplemented* poset is a poset with an orthocomplementation $a\mapsto a'$ as defined in Lemma 3.3. An orthocomplemented poset $(\mathcal{P},\leq,')$ is *σ-orthocomplete* if $a_i\in\mathcal{P}$, $a_i\perp a_j$, $i\neq j$, implies that $\bigvee a_i$ exists, $i=1,2,\ldots$. An orthocomplemented poset $(\mathcal{P},\leq,')$ is *orthomodular* if $a\leq b$ implies that $b=a\vee(b\wedge a')$. A *quantum logic* is a σ-orthocomplete orthomodular poset. A *probability measure* on a σ-orthocomplete poset $(\mathcal{P},\leq,')$ is a map $m:\mathcal{P}\to[0,1]$ which satisfies: $m(1)=1$; if $a_i\in\mathcal{P}$, $a_i\perp a_j$, $i\neq j$, then $m(\bigvee a_i)=\Sigma m(a_i)$, $i=1,2,\ldots$. A set of probability measures $\mathfrak{M}$ on $(\mathcal{P},\leq,')$ is *order determining* if $m(a)\leq m(b)$ for every $m\in\mathfrak{M}$ implies that $a\leq b$.

Theorem 3.4. *Let $\mathcal{E}$ be a nonempty set and let S be a set of functions from $\mathcal{E}$ into $[0,1]$. Then $(\mathcal{E},S)$ is an event structure if and only if $(\mathcal{E},\leq,')$ is a* *quantum logic and S is an order-determining set of probability measures on $\mathcal{E}$.*

PROOF. First suppose that $(\mathcal{E}, S)$ is an event structure. We have already noted that $(\mathcal{E}, \leq)$ is a poset. By definition, a' is the unique event satisfying $s(a') = 1 - s(a)$ for every $s \in S$. We now show that $'$ is an orthocomplementation. Since $s(a'') = s(a)$ for all $s \in S$, $a'' = a$, by Axiom A. If $a \leq b$, then

$$s(b') = 1 - s(b) \leq 1 - s(a) = s(a')$$

for every $s \in S$, so $b' \leq a'$. For $a \in \mathcal{E}$, we have $s(a) + s(a') = 1$ for every $s \in S$, so by Axiom B there is an element $0 \in \mathcal{E}$ such that $s(0) + s(a) + s(a') = 1$ for every $s \in S$. It follows that $s(0) = 0$ for every $s \in S$. Define $1 = 0'$. Notice that $0 \leq a \leq 1$ for every $a \in \mathcal{E}$ so 0 and 1 are the first and last elements of $\mathcal{E}$, respectively. If $b \geq a, a'$, then $s(a), s(a') \leq s(b)$. Hence, $s(a) + s(b') \leq 1$ and $s(a') + s(b') \leq 1$ for every $s \in S$. Then a, a', b' satisfy the hypothesis of Axiom B so there exists a $c \in \mathcal{E}$ such that $s(c) + s(a) + s(a') + s(b') = 1$ for every $s \in S$. But then $s(c) = s(b') = 0$ for every $s \in S$. Hence $s(b) = 1$ for every $s \in S$ and $b = 1$. It follows that $a \vee a' = 1$ and $'$ is an orthocomplementation. We next show that $(\mathcal{E}, \leq, ')$ is σ-orthocomplete. Now $a \perp b$ if and only if $s(a) + s(b) \leq 1$ for every $s \in S$. Thus if $a_i \perp a_j$, $i \neq j$, $i = 1, 2, \ldots$, then by Axiom B there exists a $b \in \mathcal{E}$ such that $s(b) + s(a_1) + s(a_2) + \cdots = 1$. Hence $s(b') = \Sigma s(a_i)$ for every $s \in S$. It follows that $b' \geq a_1, a_2, \ldots$. Now suppose that $c \geq a_1, a_2, \ldots$. Then $s(a_i) + s(c') \leq 1$ for every $s \in S$. Hence, the sequence $c', a_1, a_2, \ldots$ satisfies the hypothesis of Axiom B, so there exists a $d \in \mathcal{E}$ such that $s(d) + s(c') + \Sigma s(a_i) = 1$. It follows that $s(c) \geq \Sigma s(a_i) = s(b')$ for every $s \in S$ and $c \geq b'$. Hence $b' = \bigvee a_i$ and $(\mathcal{E}, \leq, ')$ is σ-orthocomplete. Moreover, since $s(\bigvee a_i) = s(b') = \Sigma s(a_i)$ and $s(1) = 1$, it follows that every $s \in S$ is a probability measure on $\mathcal{E}$. It is obvious that S is order determining. To show that $(\mathcal{E}, \leq, ')$ is orthomodular, suppose that $a \leq b$. Then $a \perp b'$ and since $a \leq a \vee b' = (a' \wedge b)'$, $a \perp a' \wedge b$. Hence, for every $s \in S$

$$s[a \vee (a' \wedge b)] = s(a) + s(a' \wedge b) = s(a) + 1 - s(a \vee b')$$

$$= s(a) + 1 - s(a) - s(b') = s(b).$$

Therefore $b = a \vee (a' \wedge b)$. Conversely, suppose that $(\mathcal{E}, \leq, ')$ is a quantum logic and S is an order-determining set of probability measures on $\mathcal{E}$. Then $s(a) = s(b)$ for every $s \in S$ implies that $a = b$, so Axiom A holds. Now let a_i be a sequence in $\mathcal{E}$ satisfying $s(a_i) + s(a_j) \leq 1$, $i \neq j$, for every $s \in S$. Then $s(a_i) \leq 1 - s(a_j) = s(a'_j)$ for every $s \in S$, so $a_i \perp a_j$, $i \neq j$. Hence, $b = \bigvee a_i$ exists and $s(b') + \Sigma s(a_i) = 1$. Thus, Axiom B holds and $(\mathcal{E}, S)$ is an event structure. $\square$

We thus see that there is no difference between an event structure and a quantum logic with an order-determining set of probability measures. Notice that an event structure need not be a lattice (that is, $a \vee b$ and $a \wedge b$ need not exist). For example, let $\Omega = \{1,2,3,4,5,6\}$ and let $\mathcal{E}$ be the collection of subsets of Ω with an even number of elements. Order $\mathcal{E}$ by inclusion and let $'$ be the usual set complementation. For $i = 1, \ldots, 6$ define for $a \in \mathcal{E}$, $s_i(a) = 1$ if $i \in a$ and $s_i(a) = 0$ if $i \notin a$. Then if we let $S = \{s_i : i = 1, \ldots, 6\}$ it is easily verified that $(\mathcal{E}, S)$ is an event structure. However, $\mathcal{E}$ is not a lattice since, for example, $\{1,2,3,4\} \wedge \{2,3,4,5\}$ does not exist.

We say that two events a, b are *compatible* (written $a \leftrightarrow b$) if there exist mutually orthogonal events a_1, b_1, and c such that $a = a_1 \vee c$, $b = b_1 \vee c$. We shall see that compatible events are ones that can be verified simultaneously; that is, events whose experiments do not interfere. Notice that if $a \perp b$, then $a \leftrightarrow b$, and that $0 \leftrightarrow a$, $1 \leftrightarrow a$ for every $a \in \mathcal{E}$. Physically, our interpretation of $a \leq b$ demands that $a \leftrightarrow b$ if $a \leq b$. This is indeed the case since if $a \leq b$, then by orthomodularity $b = a \vee (b \wedge a')$ and $a = a \vee 0$ where $a \perp (b \wedge a')$.

We now show how observables can be defined. If x is an arbitrary observable and $E \in \mathcal{B}(\mathbb{R})$, then the pair (x, E) corresponds to the event: "The observable x has a value in the set E." Thus if $(\mathcal{E}, S)$ is an event structure, an *observable* can be thought of as a map $x: \mathcal{B}(\mathbb{R}) \to \mathcal{E}$. Furthermore, an observable should satisfy:

1. $x(\mathbb{R}) = 1$;
2. if $E \cap F = \varnothing$, then $x(E) \perp x(F)$;
3. if $E_i \in \mathcal{B}(\mathbb{R})$ is a sequence of mutually disjoint sets, then $x(\cup E_i) = \bigvee x(E_i)$.

The reader can easily justify these three conditions.

Two observables x, y are *compatible* (written $x \leftrightarrow y$) if $x(E) \leftrightarrow y(F)$ for all $E, F \in \mathcal{B}(\mathbb{R})$. We shall show later that observables which are compatible may be thought of physically as being simultaneously measurable.

The reader should note that we have again constructed a generalized probability theory. Instead of being a Boolean σ-algebra of subsets of a set, our events are more general, belonging to a less restrictive structure. The usual probability measures are replaced by states and the random variables by observables. Notice that if x is an observable and s a state, then the probability that x has a value in $E \in \mathcal{B}(\mathbb{R})$ when the system is in state s is $s[x(E)]$. Before proceeding further, let us consider two examples of event structures.

Example 1. Let (Ω, Σ) be a measurable space, and let S be the set of probability measures on Σ. It is easily checked that (Σ, S) is an event structure. If x is an observable, then it follows from a theorem of Sikorski–Varadarajan [249, 268] that there exists a random variable f such that $x(E) = f^{-1}(E)$ for every $E \in \mathcal{B}(\mathbb{R})$. Since the converse also holds there is a natural correspondence between observables and random variables. We thus see that event structures generalize classical probability theory. It is easily checked that all events and all observables are compatible in this example.

Example 2. Let $\mathcal{H}$ be a complex Hilbert space and let $\mathcal{P}$ be the collection of all orthogonal projections on $\mathcal{H}$. Let S be the set of states of the form $P \mapsto \langle P\phi, \phi \rangle$, $P \in \mathcal{P}$, $\phi \in \mathcal{H}$, $\|\phi\| = 1$ (we could also include mixed states, but that will not be necessary here). Then $(\mathcal{P}, S)$ is an event structure. An observable becomes a projection-valued measure. Since, by the spectral theorem [67], there is a one-to-one correspondence between projection-valued measures and self-adjoint operators, we may identify observables with self-adjoint operators. Moreover, the probability that an observable A has a value in $E \in \mathcal{B}(\mathbb{R})$ when the system is in the state corresponding to the unit vector ψ is $\langle P^A(E)\psi, \psi \rangle$. Thus, the event structure, in this case, reduces to the von Neumann formulation of quantum mechanics. It is straightforward to show that $P_1, P_2 \in \mathcal{P}$ are compatible if and only if they commute. It follows that two observables are compatible if and only if they commute. Thus, our definition of compatibility reduces to the usual concept of noninterference in quantum mechanics.

It follows from the above two examples that event structures generalize both classical and quantum probability theory. Moreover, if $(\mathcal{E}, S)$ is an event structure and $\mathcal{O}$ is the set of all observables, then $(\mathcal{O}, S)$ satisfies Axioms 3.1 and 3.2, so the results of Section 3.1 are applicable here.

3.3. Compatibility

In this section we study compatibility in more detail. Throughout this section $(\mathcal{E}, S)$ will denote an event structure. We have seen in Theorem 3.4 that $\mathcal{E}$ is a quantum logic and S is an order-determining set of probability measures on $\mathcal{E}$. If x is an observable, we call the set $\{x(E) : E \in \mathcal{B}(\mathbb{R})\} \subseteq \mathcal{E}$ the *range* of x.

Lemma 3.5. *Two events a, b are compatible if and only if they are in the range of a single observable.*

45

PROOF. If a and b are compatible, there are mutually orthogonal events a_1, b_1, c such that $a = a_1 \vee c$, $b = b_1 \vee c$. Define the observable x by $x(\{0\}) = a_1$, $x(\{1\}) = b_1$, $x(\{2\}) = c$, $x(\{3\}) = (a_1 \vee b_1 \vee c)'$, and extend x to $\mathcal{B}(\mathbb{R})$ in the natural way. Then $a = x(\{0,2\})$, $b = x(\{1,2\})$, and thus a and b are in the range of x. Conversely, suppose there is an observable x and Borel sets E, F such that $a = x(E)$, $b = x(F)$. Then $x(E \cap F)$, $x(E \cap F^c)$, and $x(F \cap E^c)$ are mutually orthogonal events and $a = x(E) = x(E \cap F^c) \vee x(E \cap F)$, $b = x(F) = x(F \cap E^c) \vee x(E \cap F)$. Hence, $a \leftrightarrow b$. □

This last lemma justifies the fact that compatible events are simultaneously verifiable, since to verify two compatible events one need measure only a single observable. Let $a_1, a_2, \ldots$ be mutually orthogonal events. The method used in the proof of Lemma 3.5 can be used to define an observable with minimal range that includes $\{a_i : i = 1, 2, \ldots\}$. Indeed, let $\lambda_0, \lambda_1, \ldots$ be distinct real numbers and define $a_0 = (\vee a_i)'$ and

$$x(E) = \bigvee \{a_j : \lambda_j \in E\} \qquad E \in \mathcal{B}(\mathbb{R}). \tag{3.1}$$

It is not hard to show that x is an observable and $x(\{\lambda_i\}) = a_i$, $i = 1, 2, \ldots$. If x is an arbitrary observable and $u: \mathbb{R} \to \mathbb{R}$ is a Borel function, we have seen in Section 3.1 that the natural definition of $u(x)$ is $u(x)(E) = x[u^{-1}(E)]$ for all $E \in \mathcal{B}(\mathbb{R})$. It is straightforward to check that $u(x)$ is indeed an observable. Since the range of $u(x)$ is contained in the range of x, it follows from Lemma 3.5 that $u(x) \leftrightarrow x$.

If $a \leq b$, the orthomodularity of $\mathcal{E}$ implies that $b = a \vee (b \wedge a')$. We denote the event $b \wedge a'$ by $b - a$. The next two lemmas give useful technical results.

Lemma 3.6. (i) *If* $a \leftrightarrow b$, *then* $a \leftrightarrow b'$ *and* $a \vee b$, $a \wedge b$ *exist. If* a_1, b_1, c *are mutually orthogonal with* $a = a_1 \vee c$, $b = b_1 \vee c$, *then* a_1, b_1, c *are unique. In fact* $c = a \wedge b$, *and*

$$a_1 = a - a \wedge b = a \vee b - b, \qquad b_1 = b - a \wedge b = a \vee b - a.$$

(ii) $a \leftrightarrow b$ *if and only if* $a \wedge b$ *exists and* $(a - a \wedge b) \perp b$.

PROOF. (i) Let $d = a_1 \vee b_1 \vee c$. Now $a \leq d$ and $b \leq d$. If $a \leq e$ and $b \leq e$, then $d = a_1 \vee b_1 \vee c \leq e$. Thus $a \vee b$ exists and $a \vee b = d$. Now $d = a \vee b_1 = b \vee a_1$ and so $a_1 = d - b = a \vee b - b$ and $b_1 = d - a = a \vee b - a$. Moreover, $a = d - b_1$ and $b = d - a_1$ so $a' = b_1 \vee d'$ and $b' = a_1 \vee d'$. Thus $a \leftrightarrow b'$ and by the above argument $a' \vee b' = a_1 \vee b_1 \vee d' = [d - (a_1 \vee b_1)]' = c'$. Hence, $c = a \wedge b$.

(ii) If $a \leftrightarrow b$, then $a \wedge b$ exists and $(b - a \wedge b) \leq (a - a \wedge b)'$ by (i). Since $a \wedge b \leq (a - a \wedge b)'$ we have

$$b = (b - a \wedge b) \vee (a \wedge b) \leq (a - a \wedge b)'.$$

Hence, $(a-a\wedge b)\perp b$. Conversely, if $a\wedge b$ exists and $(a-a\wedge b)\perp b$, then $(b-a\wedge b)\perp(a-a\wedge b)$. Hence, $a=(a-a\wedge b)\vee(a\wedge b)$ and $b=(b-a\wedge b)\vee(a\wedge b)$ and $a\leftrightarrow b$. $\square$

Lemma 3.7. *Let* $\{a,a_1,a_2,\ldots\}\subseteq\mathcal{E}$, *suppose* $a\leftrightarrow a_i$, $i=1,2,\ldots$, *and suppose* $\bigvee a_i$ *and* $\bigvee(a\wedge a_i)$ *exist.* (i) $a\leftrightarrow\bigvee a_i$; (ii) $a\wedge(\bigvee a_i)=\bigvee(a\wedge a_i)$.

PROOF. Let $c=\bigvee(a\wedge a_i)$. Clearly, $c\leq a$ and $c\leq\bigvee a_i$. Since $a\leftrightarrow a_i$, it follows from Lemma 3.6(i) that $a-(a\wedge a_i)\leq a_i'$. Since $a\wedge a_i\leq c$ we have $a-c\leq a-(a\wedge a_i)\leq a_i'$. Hence, $(a-c)\perp a_i$ and $(a-c)\perp(\bigvee a_i)$. Therefore $(a-c)\perp(\bigvee a_i-c)$ and we have $a=c\vee(a-c)$ and $\bigvee a_i=c\vee(\bigvee a_i-c)$. It follows that $a\leftrightarrow\bigvee a_i$ and from Lemma 3.6(i) we have $c=\bigvee(a\wedge a_i)=a\wedge(\bigvee a_i)$. $\square$

We call $\mathcal{E}$ a *lattice* if $a\vee b$ and $a\wedge b$ exist for all $a,b\in\mathcal{E}$, and $\mathcal{E}$ is *distributive* if it is a lattice and the distributive law $a\wedge(b\vee c)=(a\wedge b)\vee(a\wedge c)$ holds for all $a,b,c\in\mathcal{E}$. If $\mathcal{E}$ is distributive, then $\mathcal{E}$ is a Boolean σ-algebra.

Theorem 3.8. $\mathcal{E}$ *is a Boolean σ-algebra if and only if its elements are mutually compatible.*

PROOF. If $\mathcal{E}$ is a Boolean σ-algebra, then

$$a-a\wedge b=a\wedge(a\wedge b)'=a\wedge(a'\vee b')=a\wedge b'\leq b'.$$

Since $(a-a\wedge b)\perp b$, it follows from Lemma 3.6(ii) that $a\leftrightarrow b$. Conversely, if all the elements of $\mathcal{E}$ are mutually compatible, then by Lemma 3.6(i) $\mathcal{E}$ is a lattice and by Lemma 3.7 $\mathcal{E}$ is distributive. $\square$

We thus see that Boolean σ-algebras give event structures in which no two measurements interfere. Since interfering measurements are characteristic of quantum mechanical effects, one would expect that Boolean σ-algebras describe purely classical phenomena, and this is true. Applying a representation theorem of Loomis [167], if $\mathcal{E}$ is a Boolean σ-algebra, there is a measurable space (Ω,Σ) and a map $h:\Sigma\to\mathcal{E}$ satisfying

(i) $h(\Omega)=1$;
(ii) if $\Lambda\cap\Gamma=\varnothing$, then $h(\Lambda)\perp h(\Gamma)$;
(iii) $h(\bigcup\Lambda_i)=\bigvee h(\Lambda_i)$, if $\Lambda_i\cap\Lambda_j=\varnothing$, $i\neq j$.

Now the measurable space (Ω,Σ) may be thought of as a phase space for some classical mechanic system and thus we are back to classical physics. If s is a state on $\mathcal{E}$, then s induces a probability measure μ on Σ defined by

$\mu(\Lambda) = s[h(\Lambda)]$, and thus we have a probability space (Ω, Σ, μ). Moreover, if x is an observable with range in $\mathcal{E}$, then it follows from the Sikorski–Varadarajan theorem (see Example 1) that there exists a random variable f on (Ω, Σ) such that $x(E) = h[f^{-1}(E)]$ for all $E \in \mathcal{B}(\mathbb{R})$. Thus observables reduce to random variables, and our event structure reduces to classical probability theory on a classical mechanical phase space.

It follows from Lemma 3.5 and Theorem 3.8 that the range of an observable is a Boolean sub-σ-algebra of $\mathcal{E}$. Hence, if only one observable is being considered, it can be treated like a random variable. If two or more noncompatible observables are being considered, the situation is entirely different.

If $S \subseteq \mathcal{E}$, the *compatant* of S is

$$S^0 = \{ b \in \mathcal{E} : b \leftrightarrow a \text{ for all } a \in S \}.$$

Clearly $S_1 \subseteq S_2$ implies $S_2^0 \subseteq S_1^0$. Denoting $(S^0)^0$ by S^{00} it is clear that $S \subseteq S^{00}$. A set $S \subseteq \mathcal{E}$ is *compatible* if the elements of S are mutually compatible. Of course, $S \subseteq S^0$ if and only if S is compatible. If $S_1, S_2 \subseteq \mathcal{E}$ we write $S_1 \leftrightarrow S_2$ if $a_1 \leftrightarrow a_2$ for all $a_1 \in S_1$, $a_2 \in S_2$. It is important to know when a collection of Boolean sub-σ-algebras of $\mathcal{E}$ is contained in a single Boolean sub-σ-algebra. For two Boolean sub-σ-algebras we have the following result.

Theorem 3.9. *If B_1 and B_2 are Boolean sub-σ-algebras of $\mathcal{E}$, there is a Boolean sub-σ-algebra containing both B_1 and B_2 if and only if $B_1 \leftrightarrow B_2$.*

PROOF. We first assume that B_1 and B_2 have a finite number of elements and $B_1 \leftrightarrow B_2$. Let $\{ a_i : i = 1, \ldots, p \}$ and $\{ b_i : i = 1, \ldots, q \}$ be the distinct non-zero minimal elements of B_1 and B_2, respectively. The a_i's are mutually disjoint, since otherwise $a_i \wedge a_j \neq 0$ for some i, j and $a_i \wedge a_j = a_i = a_j$, a contradiction. Also $\bigvee a_i = 1$, since otherwise we would be missing a minimal element. We also have the same properties for the b_i's. Let $c_{ij} = a_i \wedge b_j$, $i = 1, \ldots, p$, $j = 1, \ldots, q$. If $c_{ij} \neq c_{i'j'}$, then either $i \neq i'$ or $j \neq j'$ and hence either $c_{ij} \leq a_i$ and $c_{i'j'} \leq a_{i'}$, $i \neq i'$, or $c_{ij} \leq b_j$ and $c_{i'j'} \leq b_{j'}$, $j \neq j'$. Hence the c_{ij}'s are mutually orthogonal. Also

$$\bigvee_{j=1}^{q} c_{ij} = \bigvee \{ a_i \wedge b_j : j = 1, \ldots, q \} = a_i \wedge \left(\bigvee b_j \right) = a_i.$$

Similarly,

$$\bigvee_{i=1}^{p} c_{ij} = b_j \quad \text{and} \quad \bigvee_{i,j=1}^{p,q} c_{ij} = 1.$$

If B is 0 together with all suprema of the c_{ij}'s, then $B_1 \subseteq B$ and $B_2 \subseteq B$ and it is clear that B is a Boolean sub-σ-algebra with a finite number of elements. Denote B by $[B_1, B_2]$.

Now assume that B_1 and B_2 are arbitrary Boolean sub-σ-algebras and $B_1 \leftrightarrow B_2$. Let $B = \bigcup [A_1, A_2]$, where A_1 and A_2 run over all finite Boolean sub-σ-algebras such that $A_1 \subseteq B_1$ and $A_2 \subseteq B_2$. Now $0, 1 \in B$ and if $a \in B$, then $a' \in B$. If $a, b \in B$, then there are finite Boolean sub-σ-algebras $A_1, A_2, A_1\hat{}, A_2\hat{}$ such that $a \in [A_1, A_2]$, $b \in [A_1\hat{}, A_2\hat{}]$. Letting $A_1\hat{}\hat{} = [A_1, A_1\hat{}]$ and $A_2\hat{}\hat{} = [A_2, A_2\hat{}]$ we see that a and b are in $[A_1\hat{}\hat{}, A_2\hat{}\hat{}]$. Thus $a \vee b \in B$ and $a \leftrightarrow b$. It follows that B is a Boolean subalgebra (closed under *finite* suprema and infima). Now by Zorn's lemma, B is contained in a maximal Boolean subalgebra B_0. Let a_i be a sequence of mutually orthogonal elements of B_0. Let $a = \bigvee a_i$ and let $b \in B_0$. Then $b \leftrightarrow a_i$, $i = 1, 2, \ldots$, and hence, by Lemma 3.7, $b \leftrightarrow a$. Letting $C = \{0, 1, a, a'\}$, since B_0 is maximal, we have $B_0 = [B_0, C]$ and thus $a \in B_0$. Thus B_0 is a Boolean sub-σ-algebra containing B_1 and B_2. The converse follows from Theorem 3.8. $\square$

The following example, due to Ramsey [221], shows the surprising fact that Theorem 3.9 does not hold for more than two Boolean sub-σ-algebras. Let

$$\mathcal{E} = \{a : a \subseteq \{1, 2, \ldots, 8\}, \ a \text{ has an even number of elements}\}$$

with the usual set-theoretic operations and let S be the set of probability measures on $\mathcal{E}$. Then $(\mathcal{E}, S)$ is an event structure. Let $a = \{1, 2, 3, 4\}$, $b = \{1, 3, 5, 6\}$, and $c = \{1, 3, 6, 8\}$. Then a, b, and c are mutually compatible but $(a \vee b) \wedge c$ does not exist since $\{1, 3\}$ and $\{1, 6\}$ are lower bounds of $\{a \vee b, c\}$ and there is no lower bound larger than both of them. Thus $a \vee b \not\leftrightarrow c$ and $\{a, b, c\}$ is not contained in a Boolean sub-σ-algebra.

However, there is a simple, necessary and sufficient condition for Theorem 3.9 to hold for any collection of Boolean sub-σ-algebras. We say that $\mathcal{E}$ satisfies *condition* C (for compatibility) if for any three mutually compatible elements a, b, c we have $a \leftrightarrow b \vee c$. It follows from Lemma 3.7 that if $\mathcal{E}$ is a lattice it satisfies condition C.

Theorem 3.10. *Any compatible set $S \subseteq \mathcal{E}$ is contained in a Boolean sub-σ-algebra if and only if $\mathcal{E}$ satisfies condition C.*

PROOF. Necessity is trivial so we turn to sufficiency, and assume that $\mathcal{E}$ satisfies condition C. Since $S \subseteq S^{\infty}$, it suffices to show that S^{∞} is a Boolean sub-σ-algebra. Obviously $0, 1 \in S^{\infty}$ and if $a \in S^{\infty}$, then $a' \in S^{\infty}$.

Now suppose a_i is a sequence of mutually orthogonal elements of S^{00}, and suppose $b \in S^0$. Then $a_i \leftrightarrow b$, $i = 1, 2, \ldots$, so by Lemma 3.7 $b \leftrightarrow \bigvee a_i$ and thus $\bigvee a_i \in S^{00}$. If $b_1, b_2 \in S^{00}$, then by condition C, $b_1 \vee b_2 \in S^{00}$ so S^{00} is a lattice. Now from $S \subseteq S^0$ we get $S^{00} \subseteq S^0$ and hence $S^{00} \subseteq (S^{00})^0$. Thus S^{00} is compatible and again by Lemma 3.7 S^0 is distributive so S^0 is a Boolean sub-σ-algebra. $\qquad \square$

Corollary 3.11. *Any collection of mutually compatible Boolean sub-σ-algebras is contained in a single Boolean sub-σ-algebra if and only if $\mathcal{E}$ satisfies condition C.*

A Boolean sub-σ-algebra $B \subseteq \mathcal{E}$ is *separable* if there is a countable subset $S \subseteq B$ such that the smallest Boolean sub-σ-algebra containing S is B itself. We say that $\mathcal{E}$ is *separable* if every Boolean sub-σ-algebra in $\mathcal{E}$ is separable. Since $\mathfrak{B}(\mathbb{R})$ is separable it easily follows that the range of an observable is a separable Boolean sub-σ-algebra. The converse of this statement is proved in [268].

Lemma 3.12. *B is a separable Boolean sub-σ-algebra if and only if B is the range of an observable.*

Now let (Ω, Σ) be a measurable space, f a Σ-measurable function, and Σ_f the smallest σ-algebra with respect to which f is measurable. It is a standard measure-theoretic fact that if g is measurable with respect to Σ_f, then g is a Borel function of f. (For a proof, see the proof of Lemma 6.12.) Applying the Sikorski–Varadarajan and Loomis theorems, both of which were described earlier, we have:

Lemma 3.13. *If the range of an observable y is contained in the range of an observable x, then there is a Borel function u such that $y = u(x)$.*

We now prove the main result of this section.

Theorem 3.14. (i) *Two observables x_1, x_2 are compatible if and only if there is an observable x and Borel functions u_1, u_2 such that $x_1 = u_1(x)$ and $x_2 = u_2(x)$.* (ii) *If $\mathcal{E}$ satisfies condition C then a sequence of observables x_i is compatible if and only if there is an observable x and Borel functions u_i such that $x_i = u_i(x)$.* (iii) *If $\mathcal{E}$ is separable and satisfies condition C, then a collection of observables $\{x_\alpha\}$ is compatible if and only if there is an observable x and Borel functions u_α such that $x_\alpha = u_\alpha(x)$.*

PROOF.   (i) Sufficiency is trivial. For necessity suppose $x_1 \leftrightarrow x_2$. Denoting the range of an observable x by $R(x)$, by Theorem 3.9 there is a Boolean σ-algebra containing $R(x_1) \cup R(x_2)$. Intersecting all the Boolean sub-σ-algebras containing $R(x_1) \cup R(x_2)$, we see that there is a smallest Boolean sub-σ-algebra B containing $R(x_1) \cup R(x_2)$. Since $R(x_1)$ and $R(x_2)$ are separable, it follows that B is separable and applying Lemma 3.12 there is an observable x such that $R(x) = B$. Applying Lemma 3.13 there are Borel functions u_1 and u_2 such that $x_1 = u_1(x)$ and $x_2 = u_2(x)$. The proofs of (ii) and (iii) are similar.   □

This theorem justifies our contention that compatible observables correspond to noninterfering experiments since to measure compatible observables one need only measure a single observable. It also follows that compatible observables can be treated like random variables.

3.4. A Variance Theorem

Roughly speaking, the Heisenberg uncertainty principle states that if the variances of two observables can be made simultaneously small, then they are compatible. It is not known whether such a result holds in a general event structure. However, a partial result in this direction has been proved by Louton [168]. Before we give this result, we need some preliminaries.

The *spectrum* $\sigma(x)$ of an observable x is the smallest closed subset Λ of $\mathbb{R}$ such that $x(\Lambda) = 1$. The spectrum of x corresponds physically to the allowable values of x; that is, the values that x may attain (or approach arbitrarily closely). The *expectation* or *mean* of x in the state s, if it exists, is $s(x) = \int \lambda s[x(d\lambda)]$. It follows that if u is a Borel function, then $s[u(x)] = \int u(\lambda) s[x(d\lambda)]$. The *variance* of x in the state s, if it exists, is $V_s(x) = s[(x - s(x))^2]$. We call the probability measure $s_x = s[x(\cdot)]$ on $\mathcal{B}(\mathbb{R})$ the *distribution* of x in the state s. An important inequality of probability theory, called *Chebyshev's inequality*, says that if $s(x)$ and $V_s(x)$ exist, then for any $\varepsilon > 0$ we have

$$s_x\{\lambda : |\lambda - s(x)| \geq \varepsilon\} \leq V_s(x)/\varepsilon^2.$$

To prove this inequality, let $\Lambda = \{\lambda : [\lambda - s(x)]^2 \geq \varepsilon^2\}$. Then

$$V_s(x) = \int [\lambda - s(x)]^2 s_x(d\lambda) \geq \int_\Lambda [\lambda - s(x)]^2 s_x(d\lambda)$$
$$\geq \varepsilon^2 s_x(\Lambda) = \varepsilon^2 s_x[\{\lambda : |\lambda - s(x)| \geq \varepsilon\}].$$

For an event structure $(\mathcal{E}, S)$, we say that S is *convex* if for any $s_1, \ldots, s_n \in S$, $\lambda_1, \ldots, \lambda_n \geq 0$, with $\Sigma \lambda_i = 1$, we have $\Sigma \lambda_i s_i \in S$. We say that S is *strongly order determining* if $\{s \in S : s(a) = 1\} \subseteq \{s \in S : s(b) = 1\}$ implies that $a \leq b$. The *variance formula* states that if $s = \Sigma \lambda_i s_i$ with $\lambda_i \geq 0$, $\Sigma \lambda_i = 1$, then

$$V_s(x) = \sum \lambda_i V_{s_i}(x) + \sum_{i < j} \lambda_i \lambda_j [s_i(x) - s_j(x)]^2.$$

The proof of the variance formula is straightforward and is left to the reader.

Let x and y be observables. We write $x \rightarrow y$ if for any $\varepsilon > 0$ and $\lambda \in \sigma(x)$ there exists a $\delta(\varepsilon, \lambda) > 0$ such that whenever a state s satisfies $|s(x) - \lambda| < \delta$ and $V_s(x) < \delta$, then $V_s(y) < \varepsilon$. Roughly speaking, $x \rightarrow y$ means that $V_s(y)$ can be made arbitrarily small by making $V_s(x)$ sufficiently small. Louton's result states that if $\sigma(x)$ is countable and $x \rightarrow y$, then y is a function of x (and hence $x \leftrightarrow y$). Of course, it would be desirable to eliminate the countability condition (Louton has made some progress toward this [168]). Moreover, although the result is interesting as it stands, it would be desirable to have a symmetrical form more in accordance with the uncertainty principle. Such a result would state that if $V_s(x)$ and $V_s(y)$ can be made simultaneously small, then $x \leftrightarrow y$.

Theorem 3.15 (Louton [168]). *Let* $(\mathcal{E}, S)$ *be an event structure with* S *convex and strongly order determining. If* $\sigma(x)$ *is countable and* $x \rightarrow y$, *then there exists a Borel function* f *such that* $y = f(x)$.

PROOF. Suppose $\sigma(x)$ is countable and $x \rightarrow y$. We break the proof down into a number of steps.

Step 1. If s_i is a sequence of states satisfying $\lim s_i(x) = \lambda \in \sigma(x)$ and $\lim V_{s_i}(x) = 0$, then $s_i(y)$ is Cauchy.

PROOF. Let $\varepsilon_1 > 0$, choose ε such that $0 < \varepsilon < (\varepsilon_1/2)^2$, and let $0 < \delta(\varepsilon, \lambda) < 1$ be chosen as the definition of $x \rightarrow y$ allows. Moreover, let N be a positive integer such that $i > N$ implies that $|s_i(x) - \lambda| < \delta/2$ and $V_s(x) < \delta/2$. Let $s = \frac{1}{2} s_i + \frac{1}{2} s_j \in S$ for some $i, j > N$. By the variance formula, we have

$$V_s(x) = \frac{1}{2} V_{s_i}(x) + \frac{1}{2} V_{s_j}(x) + \frac{1}{4} |s_i(x) - s_j(x)|^2$$

$$\leq \delta/2 + \frac{1}{4} \big[|s_i(x) - \lambda| + |s_j(x) - \lambda| \big]^2$$

$$\leq \delta/2 + \delta^2/4 \leq \delta.$$

We also have

$$|s(x) - \lambda| \leq \frac{1}{2} |s_i(x) - \lambda| + \frac{1}{2} |s_j(x) - \lambda| < \delta.$$

We conclude that $V_s(y) < \varepsilon$. By the variance formula $\frac{1}{4}[s_i(y) - s_j(x)]^2 < \varepsilon$ and hence $|s_i(y) - s_j(y)| < \varepsilon_1$ for $i,j > N$, so $s_i(y)$ is Cauchy.

Step 2. Let s_i, t_i be sequences of states such that $\lim s_i(x) = \lim t_i(x) = \lambda$ $\in \sigma(x)$ and $\lim V_{s_i}(x) = \lim V_{t_i}(x) = 0$. Then $\lim s_i(y) = \lim t_i(y)$.

PROOF. Given $\varepsilon_1 > 0$, choose ε and δ as in the proof of Step 1. Choose N such that $i > N$ implies that $|s_i(x) - \lambda|$, $|t_i(x) - \lambda|$, $V_{s_i}(x)$, $V_{t_i}(x) < \delta/2$. Let $s = \frac{1}{2}s_i + \frac{1}{2}t_j$ for some $i,j > N$. As in the proof of Step 1, $|s(x) - \lambda| < \delta$ and $V_s(x) < \delta$. Hence, $V_s(y) < \varepsilon$ and as in the proof of Step 1, $|s_i(y) - t_j(y)| < \varepsilon_1$ whenever $i, j > N$. Thus, $\lim s_i(y) = \lim t_i(y)$.

Step 3. If s_i is a sequence of states as in Step 1, then $\gamma = \lim s_i(y) \in \sigma(y)$.

PROOF. For $\varepsilon > 0$, there exists an N_1 such that $i > N_1$ implies that $V_{s_i}(y) < \varepsilon^2/8$. For $i > N_1$, by Chebyshev's inequality we have

$$s_{iy}\{\alpha : |\alpha - s_i(y)| \geq \varepsilon/2\} \leq 4V_{s_i}(y)/\varepsilon^2 < \tfrac{1}{2}.$$

Hence,

$$s_{iy}(s_i(y) - \varepsilon/2, s_i(y) + \varepsilon/2) \geq \tfrac{1}{2}.$$

Now choose N_2 such that $i > N_2$ implies that $|s_i(y) - \gamma| < \varepsilon/2$. Then for $i > N_2$ we have

$$(s_i(y) - \varepsilon/2, s_i(y) + \varepsilon/2) \subseteq (\gamma - \varepsilon, \gamma + \varepsilon).$$

Hence for $i > \max\{N_1, N_2\}$, we have $s_{iy}(\gamma - \varepsilon, \gamma + \varepsilon) \geq \tfrac{1}{2}$. Therefore, $y(\gamma - \varepsilon, \gamma + \varepsilon) \neq 0$ and $\gamma \in \sigma(y)$.

Step 4. If $\lambda \in \sigma(x)$, since $x(\lambda - 1/i, \lambda + 1/i) \neq 0$ there exists states s_i such that

$$s_i[x(\lambda - 1/i, \lambda + 1/i)] = 1, \qquad i = 1, 2, \ldots.$$

Then $\lim s_i(x) = \lambda$ and $\lim V_{s_i}(x) = 0$. We define $f(\lambda) = \lim s_i(y)$. Notice that f is well defined.

Step 5. $f: \sigma(x) \to \sigma(y)$ is continuous in the topology on $\sigma(x)$ induced by the usual topology on $\mathbb{R}$.

PROOF. Let $\lambda \in \sigma(x)$ and let $\varepsilon > 0$ be given. Choose $0 < \delta < 1$ corresponding to λ and $\varepsilon^2/36$ as in the definition of $x \to y$. Let $\lambda_1 \in \sigma(x)$ satisfy $|\lambda_1 - \lambda| < \delta/2$. Choose two sequences of states s_i and t_i such that $\lim s_i(x) = \lambda$, $\lim t_i(x) = \lambda_1$, and $\lim V_{s_i}(x) = \lim V_{t_i}(x) = 0$. Let N be a positive integer large enough so that $i > N$ implies $|s_i(x) - \lambda|$, $|t_i(x) - \lambda_1| < \delta/3$ and $V_{s_i}(x)$, $V_{t_i}(x) <$

3. The Quantum Logic Approach

$\delta/2$. For $i > N$ let $u_i = \frac{1}{2}s_i + \frac{1}{2}t_i$. Then

$$|u_i(x) - \lambda| = |\left[\tfrac{1}{2}s_i(x) - \tfrac{1}{2}\lambda\right] + \left[\tfrac{1}{2}t_i(x) - \tfrac{1}{2}\lambda_1\right] + \tfrac{1}{2}(\lambda_1 - \lambda)| \leq \delta.$$

Moreover, by the variance formula

$$V_{u_i}(x) = \tfrac{1}{2}V_{s_i}(x) + \tfrac{1}{2}V_{t_i}(x) + \tfrac{1}{4}|s_i(x) - t_i(x)|^2 < \delta.$$

It follows that $V_{u_i}(y) < \varepsilon^2/36$. Again by the variance formula we have $|s_i(y) - t_i(y)| < \varepsilon/3$. Hence for i sufficiently large we obtain

$$|f(\lambda) - f(\lambda_1)| \leq |f(\lambda) - s_i(y)| + |s_i(y) - t_i(y)| + |f(\lambda_1) - t_i(y)| < \varepsilon.$$

Step 6. If $\lambda \in \sigma(x)$, then $x(\{\lambda\}) \leq y(\{f(\lambda)\})$.

PROOF. If $x(\{\lambda\}) = 0$, we are finished, so suppose that $x(\{\lambda\}) \neq 0$. Let s be a state satisfying $s[x(\{\lambda\})] = 1$. Clearly $s(x) = \lambda$ and $V_s(x) = 0$. It follows that $V_s(y) = 0$ and, by Step 4, $s(y) = f(y)$. Hence $s[y(\{f(\lambda)\})] = s[y(\{s(y)\})] = 1$ and $x(\{\lambda\}) \leq y(\{f(\lambda)\})$.

Step 7. $y = f(x)$.

PROOF. If $\gamma \in \sigma(y)$, since $\sigma(x)$ is countable we have

$$f(x)(\{\gamma\}) = x(f^{-1}\{\gamma\}) = \bigvee\{x(\{\lambda\}) : \lambda \in f^{-1}\{\gamma\}\}$$
$$\leq y(\{\gamma\}).$$

Suppose that $\sigma(x) = \{\lambda_1, \lambda_2, \ldots\}$ and $f(\lambda_i) = \gamma_i$, $i = 1, 2, \ldots$. Then $x(f^{-1}\{\gamma_i\}) \leq y(\{\gamma_i\})$ and

$$1 = \bigvee\{x(f^{-1}\{\gamma_i\}) : i = 1, 2, \ldots\} \leq \bigvee\{y(\{\gamma_i\}) : i = 1, 2, \ldots\}.$$

It follows that $y(E) = \bigvee\{y(\{\gamma_i\}) : \gamma_i \in E\}$ for all $E \in \mathscr{B}(\mathbb{R})$. Moreover, since

$$\bigvee\{x(f^{-1}\{\gamma\}) : \gamma \in \sigma(y) - \{\gamma'\}\} \leq \bigvee\{y(\{\gamma\}) : \gamma \in \sigma(y) - \{\gamma'\}\}$$

we have $x(f^{-1}\{\gamma'\}) \geq y(\{\gamma'\})$. Hence $x(f^{-1}\{\gamma\}) = y(\{\gamma\})$ for every $\gamma \in \sigma(y)$. Therefore, for any $E \in \mathscr{B}(\mathbb{R})$ we have

$$f(x)(E) = x\left[f^{-1}(E)\right] = \bigvee\{x(\{\lambda_i\}) : \lambda_i \in f^{-1}(E)\}$$
$$= \bigvee\{y(\{\gamma_i\}) : \gamma_i \in E\} = y(E). \qquad \square$$

3.5. Uniqueness and Existence

In this section we consider two important questions. The first question is whether the means of an observable determine the observable uniquely. The answer to this question is unknown in general, although we shall answer it affirmatively for some special cases. The second question asks whether the sum of any two observables exists in a natural sense. We shall see that the answer is no.

In this section we assume that $(\mathcal{E}, S)$ is an event structure with S strongly order determining and strongly convex (closed under countable convex combinations). An observable x is *bounded* if $\sigma(x)$ is a bounded set. We use the notation $S_x = \{s \in S : |s(x)| < \infty\}$.

Lemma 3.16. *An observable x is bounded if and only if $S_x = S$.*

PROOF. Necessity is trivial. To prove sufficiency, suppose that x is unbounded. Then there are distinct numbers $\lambda_n \in \sigma(x)$ such that $|\lambda_n| > 2^{n+1}$, $n = 1, 2, \ldots$. Let E_n be a sequence of disjoint open intervals of length less than one centered at λ_n and let $a_n = x(E_n) \neq 0$. Since S is strongly order determining, there exist $s_i \in S$ such that $s_i(a_i) = 1$. Since S is closed under countable convex combinations, $s = \Sigma 2^{-i} s_i \in S$. Since $a_i \perp a_j$, $s_i(a_j) = 0$ for $i \neq j$ and $s(a_j) = 2^{-j}$. Now suppose $|s(x)| < \infty$ and hence $\int |\lambda| s[x(d\lambda)] < \infty$. But

$$\int |\lambda| s[x(d\lambda)] \geq \sum \int_{E_i} |\lambda| s[x(d\lambda)] \geq \sum (2^{i+1} - 1) 2^{-i}$$
$$= +\infty.$$

This contradiction shows that $s(x)$ does not exist. $\square$

The means of an observable in different states are, physically, values obtained for the observable using a macroscopic experiment. These values are, in a sense, averages over the microscopic or "actual" values of the observable, which are the numbers in the spectrum. Thus one would expect the means to "smooth out" or "fill in" the spectral values. This is shown in the next lemma whose proof is left as an exercise.

Lemma 3.17. *The set $\{s(x) : s \in S_x\}$ is the smallest closed interval containing $\sigma(x)$.*

We say that $(\mathcal{E}, S)$ satisfies *condition* U if the means of two bounded observables are equal in every state implies that the observables are equal. Although we cannot define the sum of two bounded observables in a direct way, we can define it indirectly as follows. We say that an observable z is the *sum* of two bounded observables x and y if $s(z) = s(x) + s(y)$ for every state s. If the sum of any two bounded observables exists we say that $(\mathcal{E}, S)$ satisfies *condition* E. This existence property is so important that it is postulated in some models for quantum mechanics [72, 238]. We now ask whether an event structure satisfies conditions U and E.

Let us consider Examples 1 and 2 of Section 3.2 to illustrate our problem. Let (Ω, Σ) be a measurable space. As we have seen, the states of this event structure are the probability measures on (Ω, Σ), and the observables are essentially the random variables on (Ω, Σ). Suppose S is the set of all probability measures on (Ω, Σ). Condition U may be stated as follows: If f and g are bounded random variables, does $\int f d\mu = \int g d\mu$ for every $\mu \in S$ imply $f = g$? The answer, which is yes, may be seen as follows. Let $\omega \in \Omega$ and let μ be a probability measure concentrated at ω. Now it easily follows that $\int_\Lambda f d\mu = \int_\Lambda g d\mu$ for every $\Lambda \in \Sigma$. Hence, $f = g$ a.e. $[\mu]$ and in particular $f(\omega) = g(\omega)$. The existence condition holds even more trivially since $f + g$ is a bounded random variable and $\int (f + g) d\mu = \int f d\mu + \int g d\mu$ for every $\mu \in S$. Now consider Example 2. The uniqueness condition becomes: If A and B are bounded self-adjoint operators, does $\langle A\phi, \phi \rangle = \langle B\phi, \phi \rangle$ for all $\phi \in \mathcal{H}$, $\|\phi\| = 1$, imply $A = B$? It is well known that the answer is yes. Condition E is again trivially satisfied since $A + B$ is a bounded self-adjoint operator and $\langle (A + B)\phi, \phi \rangle = \langle A\phi, \phi \rangle + \langle B\phi, \phi \rangle$ for every $\phi \in \mathcal{H}$.

We now consider condition U more closely.

Lemma 3.18. *Let x and y be bounded observables and suppose $s(x) = s(y)$ for every $s \in S$. Then $\lambda_0 = \max\{\lambda : \lambda \in \sigma(x)\} = \max\{\lambda : \lambda \in \sigma(y)\}$ and $x(\{\lambda_0\}) = y(\{\lambda_0\})$.*

PROOF. That $\max\{\lambda : \lambda \in \sigma(x)\} = \max\{\lambda : \lambda \in \sigma(y)\}$ follows from Lemma 3.17. Now suppose that $s[x(\{\lambda_0\})] = 1$ and $s[y(\{\lambda_0\})] \neq 1$. Then there is a number $\lambda_1 < \lambda_0$ such that $s[y(-\infty, \lambda_1]] > 0$. We now have

$$\lambda_0 = s(x) = s(y) = \int_{(-\infty, \lambda_0]} \lambda s[y(d\lambda)]$$

$$= \left(\int_{(-\infty, \lambda_1)} + \int_{[\lambda_1, \lambda_0]} \right) \lambda s[y(d\lambda)]$$

$$\leq \lambda_1 s[y(-\infty, \lambda_1)] + \lambda_0 s(y[\lambda_1, \lambda_0]) < \lambda_0,$$

which is a contradiction. Thus, $s[y(\{\lambda_0\})] = 1$ whenever $s[x(\{\lambda_0\})] = 1$ and hence $x(\{\lambda_0\}) \leq y(\{\lambda_0\})$. By symmetry $x(\{\lambda_0\}) = y(\{\lambda_0\})$. □

Theorem 3.19. *Let x and y be bounded observables and suppose that $\sigma(x)$ has at most one limit point. If $s(x) = s(y)$ for every $s \in S$, then $x = y$.*

PROOF. The most general such x has a point $\lambda_0 \in \sigma(x)$ which is a limit point from both above and below of elements of $\sigma(x)$. The other cases will follow in a similar manner. We can also assume without loss of generality that $\lambda_0 = 0$. Let the points of $\sigma(x)$ be ordered as follows: $\mu_1 < \mu_2 < \cdots < \lambda_0 < \cdots \lambda_2 < \lambda_1$. Now by Lemma 3.18, $\max\{\lambda : \lambda \in \sigma(y)\} = \lambda_1$ and $y(\{\lambda_1\}) = x(\{\lambda_1\})$. Let f be the identity function $f(\lambda) = \lambda$ and let

$$x_1 = (f - \lambda_1\chi_{\{\lambda_1\}})(x), \qquad y_1 = (f - \lambda_1\chi_{\{\lambda_1\}})(y).$$

Then for all $E \in \mathcal{B}(\mathbb{R})$ we have

$$x_1(E) = x\left[(f - \lambda_1\chi_{\{\lambda_1\}})^{-1}(E)\right]$$

$$= \begin{cases} x(E) \wedge [x(\{\lambda_1\})]' & \text{if } 0 \notin E \\ x(E) \vee [x(\{\lambda_1\})] & \text{if } 0 \in E \end{cases} \tag{3.2}$$

It is easy to see that $\sigma(x_1) = \sigma(x) - \{\lambda_1\}$, $x_1(\{\lambda_i\}) = x(\{\lambda_i\})$, $i = 2, 3, \ldots$, and $x_1(\{\mu_i\}) = x(\{\mu_i\})$, $i = 1, 2, \ldots$. Now

$$s(x_1) = s(x) - \lambda_1 s[x(\{\lambda_1\})] = s(y) - \lambda_1 s[y(\{\lambda_1\})] = s(y_1).$$

Applying Lemma 3.18 we have $\lambda_2 = \max\{\lambda : \lambda \in \sigma(y_1)\}$ and $y_1(\{\lambda_2\}) = x_1(\{\lambda_2\}) = x(\{\lambda_2\})$. It now follows by applying Eq. (3.2) to y_1 and y that λ_2 is the second largest number in $\sigma(y)$ and $y(\{\lambda_2\}) = y_1(\{\lambda_2\}) = x(\{\lambda_2\})$. Continuing this process with the λ_i's and also the μ_i's we have

$$\{\lambda_i, \mu_i : i = 1, 2, \ldots\} \subseteq \sigma(y) \qquad \text{and} \qquad y(\{\lambda_i\}) = x(\{\lambda_i\}),$$

$y(\{\mu_i\}) = x(\{\mu_i\})$, $i = 1, 2, \ldots$. Since λ_0 is a limit point of the λ_i's, it follows that $\lambda_0 \in \sigma(y)$, and

$$y(\{\lambda_0\}) = y(\{\lambda_i, \mu_i : i = 1, 2, \ldots\}^c)$$

$$= \left[\bigvee y(\{\lambda_i\}) \vee y(\{\mu_i\})\right]' = \left[\bigvee x(\{\lambda_i\}) \vee x(\{\mu_i\})\right]'$$

$$= x(\{\lambda_0\}).$$

Hence $y = x$. $\qquad\qquad\qquad\qquad\qquad\qquad\qquad\qquad\qquad\qquad\qquad\qquad \square$

We next consider condition E. First of all it is not hard to show that if $x \leftrightarrow y$, then $x + y$ exists. We now construct a class of examples in which the sum of no two noncompatible observables exists. An *antilattice* is an orthocomplemented lattice in which the supremum of any two nonzero elements is 1. It is easily seen that an antilattice $\mathcal{E}$ together with its set of probability measures S forms an event structure $(\mathcal{E}, S)$ in which S is strongly order determining.

Theorem 3.20. *Let $\mathcal{E}$ be an antilattice, S the set of probability measures on $\mathcal{E}$, and form the event structure $(\mathcal{E}, S)$. (i) A Boolean subalgebra of $\mathcal{E}$ can have at most four elements. (ii) Every observable on $\mathcal{E}$ is of the form $x = (\lambda_1 - \lambda_2)\chi_{\{\lambda_1\}}(x) + \lambda_2 I$, where $\sigma(x) = \{\lambda_1, \lambda_2\}$ and I is the identity* *observable (i.e., the unique observable satisfying $\sigma(I) = \{1\}$). (iii) If x and y are noncompatible observables on $\mathcal{E}$, then $x + y$ does not exist.*

PROOF. (i) Let a, b be distinct elements of $\mathcal{E}$ not equal to 0 or 1 and $a \neq b'$. We now show that $a \leftrightarrow b$. Suppose, on the contrary, that $a = a_1 \vee c$ and $b = b_1 \vee c$, where $a_1 \perp b_1$. Now neither a_1 nor b_1 is 0, hence $c = 0$, since otherwise $a_1 \vee c = 1$. Therefore, $a \perp b$. But this is impossible since then $b' = a \vee b' = 1$. The result follows since the elements of a Boolean subalgebra must be compatible. (ii) It follows from (i) that if x is an observable, then x has one or two point spectrum so suppose $\sigma(x) = \{\lambda_1, \lambda_2\}$. (If x has one point in its spectrum we let $\lambda_1 = \lambda_2$.) Thus x has the form $x = \lambda_1 \chi_{\{\lambda_1\}}(x) + \lambda_2 \chi_{\{\lambda_2\}}(x)$. Since $I = \chi_{\{\lambda_1\}}(x) + \chi_{\{\lambda_2\}}(x)$, the result follows. (iii) Suppose $z = x + y$ exists. Applying (ii) we have

$$z = x + y = (\lambda_1 - \lambda_2)\chi_{\{\lambda_1\}}(x) + (\mu_1 - \mu_2)\chi_{\{\mu_1\}}(y) + (\lambda_2 + \mu_2)I,$$

where $\sigma(x) = \{\lambda_1, \lambda_2\}$, $\sigma(y) = \{\mu_1, \mu_2\}$. Since $(\lambda_2 + \mu_2)I$ is compatible with every observable, it follows that $z_0 = \chi_{\{\lambda_1\}}(x) + \beta\chi_{\{\mu_1\}}(y)$ must exist, where $\beta = (\mu_1 - \mu_2)/(\lambda_1 - \lambda_2)$ (notice that $\lambda_1 \neq \lambda_2$). Now if $\sigma(z_0) = \{\omega_1, \omega_2\}$ we have

$$(\omega_1 - \omega_2)\chi_{\{\omega_1\}}(z_0) + \omega_2 I = \chi_{\{\lambda_1\}}(x) + \beta\chi_{\{\mu_1\}}(y).$$

Hence,

$$(\omega_1 - \omega_2)s\big[z_0(\{\omega_1\})\big] + \omega_2 = s\big[x(\{\lambda_1\})\big] + \beta s\big[y(\{\mu_1\})\big]$$

for every $s \in S$. Now it is easily seen that on an antilattice $\mathcal{E}$, given $0 < \alpha < 1$ and $c \in \mathcal{E}$, there is a probability measure s such that $s(c) = \alpha$ and $s(d) = 0$, where d is any event not equal to c, c', or 1. Now suppose $z_0(\{\omega_1\}) = x(\{\lambda_1\})$. Letting s be a state which is 0 on $y(\{\mu_1\})$ and 1 on $x(\{\lambda_1\})$ we have $\omega_1 = 1$. Letting s be a state which is 0 on $x(\{\lambda_1\})$ and 1 on $y(\{\mu_1\})$ we have $\omega_2 = \beta$ and hence $s[x(\{\lambda_1\})] + s[y(\{\mu_1\})] = 1$ for all $s \in S$, which is impossible. Hence, $z_0(\{\omega_1\}) \neq x(\{\lambda_1\})$, and in a similar way $z_0(\{\omega_1\}) \neq x(\{\lambda_1\})', y(\{\mu_1\}), y(\{\mu_1\})', 0,$ or 1. Now let s satisfy $s[x(\{\lambda_1\})] = \alpha \neq \omega_2$ and $s[y(\{\mu_1\})] = s[z_0(\{\omega_1\})] = 0$. This gives a contradiction and hence z_0 does not exist. $\qquad\square$

In Section 3.7 we shall give examples of event structures in which the sum of some noncompatible observables exists while the sum of others does not. We now show that a necessary condition for $(\mathcal{E}, S)$ to satisfy condition E is that $\mathcal{E}$ be a lattice. If $a \in \mathcal{E}$ we denote by x_a the unique observable which satisfies $x_a(\{1\}) = a$, $x_a(\{0\}) = a'$. Notice that $\chi_E(y) = x_{y(E)}$ for any observable y and any $E \in \mathcal{B}(\mathbb{R})$.

Theorem 3.21. *Let* $(\mathcal{E}, S)$ *be an event structure in which* S *is strongly order determining and which satisfies condition* E. *Then* $\mathcal{E}$ *is a lattice, and in fact* $a \wedge b = (x_a + x_b)(\{2\})$. *If* $s(a) = s(b) = 1$, *then* $s(a \wedge b) = 1$.

PROOF. Let $z = x_a + x_b$. Notice that $0 \leq s(z) \leq 2$ for every $s \in S$ and hence by Lemma 3.17, $\sigma(z) \subseteq [0, 2]$. Let $c = z(\{2\})$ and suppose that $s(c) = 1$. Then $s(z) = 2$ and $s(a) = s(b) = 1$. Hence, $c \leq a$, $c \leq b$. Now suppose that $d \leq a$, $d \leq b$. Then $s(d) = 1$ implies $s(z) = 2$ which implies $s(c) = 1$. Thus $d \leq c$ and $c = a \wedge b$. For the last statement suppose that $s(a) = s(b) = 1$. Then $s(x_a + x_b) = s(a) + s(b) = 2$ and hence, $1 = s[(x_a + x_b)(\{2\})] = s(a \wedge b)$. $\qquad\square$

3.6. Hidden Variables

The question of hidden variables in quantum mechanics is, in a sense, a philosophical question which asks whether probability is really necessary in quantum theory. The situation, according to the hidden variable proponents, is similar to that occurring in statistical mechanics. If we have a large number of molecules in a container, then it is impractical, and in fact nearly impossible, to describe the motion of each particle individually. For this reason one gives a macroscopic description of the system using macroscopic quantities such as volume, temperature, and pressure. In terms of these macroscopic quantities one can make only statistical predictions about the motions of the individual particles. Thus probabilities enter only because of human deficiencies and not because of any intrinsic, inherent properties of the system. If one could describe the condition of the system more precisely, then theoretically the probabilistic nature of the system would vanish. According to some physicists and philosophers this could be the case in quantum mechanics. It is possible that there are certain "hidden variables," as yet unknown, which would determine the condition of a quantum mechanical system so precisely that statistical considerations would vanish and exact phenomenological predictions could be made. Once the values of these hidden variables are known one could verify with certainty whether any event occurs or not instead of merely giving probabilities for such occurrences. In the present axiomatic system, the existence of hidden variables would imply the existence of "dispersion-free states," that is, states which have only the values 0 or 1.

In this section we shall consider both sides of the hidden variable question. We shall prove a result which shows that hidden variables cannot exist and another result which shows that hidden variables can exist in quantum mechanics. The reason that these two theorems do not contradict each other is that they involve different interpretations as to what hidden variables mean. Such differences of interpretation lie at the heart of the hidden variables controversy that has raged since the early beginnings of quantum mechanics.

Von Neumann [207] has shown that the usual Hilbert space model for quantum mechanics does not admit hidden variables in a certain sense, and Jauch and Piron [140] have shown that more general models do not. We shall show that the present event structure model, which is more general than that of Jauch and Piron, does not admit hidden variables in this certain sense. One would then conclude that we are forced to contend with statistical considerations in quantum mechanics.

A state s on an event structure $(\mathcal{E}, S)$ is *dispersion-free* if s has only the values 0 and 1 and satisfies:

D1. if $s(a) = s(b) = 1$ and if $a \wedge b$ exists, then $s(a \wedge b) = 1$;

D2. if $\{a_\alpha : \alpha \in A\}$ are mutually compatible, $\bigwedge a_\alpha$ exists, and $s(a_\alpha) = 1$, $\alpha \in A$, then $s(\bigwedge a_\alpha) = 1$.

An event structure $(\mathcal{E}, S)$ is *compatibly complete* if for any compatible set $\{a_\alpha : \alpha \in A\} \subseteq \mathcal{E}$, $\bigwedge a_\alpha$ exists; and $(\mathcal{E}, S)$ is *complete* if $\bigwedge b_\alpha$ exists for any collection of events b_α. We say that $(\mathcal{E}, S)$ *admits hidden variables in the von Neumann sense*, if it is compatibly complete and if for any nonzero $a \in \mathcal{E}$, there is a dispersion-free state s such that $s(a) = 1$. The *center* Z of $\mathcal{E}$ is the set of events which are compatible with every event. Notice that $\{0, 1\} \subseteq Z$ and that $\mathcal{E}$ is compatible if and only if $\mathcal{E} = Z$. We say that $\mathcal{E}$ is *coherent* or *trivial* if $Z = \{0, 1\}$ or $\mathcal{E} = \{0, 1\}$, respectively. An event a is an *atom* if $a \neq 0$ and $b \leq a$ implies that $b = a$ or $b = 0$. We say that $\mathcal{E}$ is *atomic* if every nonzero event contains an atom.

Theorem 3.22. *A compatibly complete event structure $(\mathcal{E}, S)$ has a dispersion-free state if and only if $\mathcal{E}$ has an atom in its center.*

PROOF. To prove necessity, let s be a dispersion-free state and let $\mathcal{E}_s = \{a \in \mathcal{E} : s(a) = 1\}$. Note that $\mathcal{E}_s \neq \varnothing$ and let T be a totally ordered subset of $\mathcal{E}_s$. Then since the elements of T are compatible, $a_0 = \bigwedge \{a : a \in T\}$ exists and by D2, $s(a_0) = 1$. Thus, $a_0 \in \mathcal{E}_s$ and $a_0 \leq a$ for all $a \in T$. By Zorn's lemma $\mathcal{E}_s$ has a minimal element a_1. We now show that a_1 is a lower bound for $\mathcal{E}_s$ (i.e., $a_1 \leq a$ for all $a \in \mathcal{E}_s$). Let $a_2 \in \mathcal{E}_s$. Now there is a nonzero $a_3 \in \mathcal{E}$ such that $a_3 \leq a_1, a_2$ since otherwise $a_1 \wedge a_2 = 0$ and $s(a_1 \wedge a_2) = 0$ which contradicts D1. Now suppose that $s(a_3) = 0$. Then $s(a_1 - a_3) = 1$ and $a_1 - a_3 \in \mathcal{E}_s$. Since $a_1 - a_3 \leq a_1$, $a_1 - a_3 = a_1$, and hence $a_3 = a_3 \wedge a_1 = 0$, which is a contradiction. Thus, $s(a_3) = 1$, $a_3 \in \mathcal{E}_s$, and hence, $a_3 = a_1$. Therefore, $a_1 \leq a_2$ and a_1 is a lower bound for $\mathcal{E}_s$. We next show that

$a_1 \in Z$. If $b \in \mathcal{E}_s$, then $a_1 \leq b$ and $a_1 \leftrightarrow b$. If $b \notin \mathcal{E}_s$, then $b' \in \mathcal{E}_s$ and $a_1 \leq b'$. Thus $a_1 \leftrightarrow b'$ and it follows from Lemma 3.6(i) that $a_1 \leftrightarrow b$. To show that a_1 is an atom, notice that $a_1 \neq 0$ and suppose $b \leq a_1$. If $b \in \mathcal{E}_s$, then $b = a_1$. If $b \notin \mathcal{E}_s$, then $b' \in \mathcal{E}_s$ and $a_1 \leq b'$. Thus $b \leq b'$ and hence, $b = b \wedge b \leq b \wedge b' = 0$. To prove sufficiency, let $a_1 \in Z$ be an atom and let $b \in \mathcal{E}$. We now show that either $a_1 \leq b$ or $a_1 \leq b'$. Since $b \leftrightarrow a_1$, $a_1 \wedge b$ exists and $a_1 \wedge b \leq a_1$. Therefore, either $a_1 \wedge b = a_1$ and $a_1 \leq b$ or $a_1 \wedge b = 0$, which implies that $a_1 \leq b'$. Now let s be a state satisfying $s(a_1) = 1$. Then either $a_1 \leq b$ and $s(b) = 1$ or $b \leq a_1'$ and $s(b) = 0$. Thus, s is dispersion free. $\qquad\square$

Corollary 3.23. *A nontrivial, coherent, compatibly complete event structure* $(\mathcal{E}, S)$ *has no dispersion-free states.*

PROOF. Suppose that $(\mathcal{E}, S)$ has a dispersion-free state. Then by Theorem 3.22, there is an atom in the center $Z = \{0, 1\}$. Since 0 is not an atom 1 must be an atom. But then $\mathcal{E}$ is trivial which is a contradiction. $\qquad\square$

Theorem 3.24. *An event structure* $(\mathcal{E}, S)$ *admits hidden variables in the von Neumann sense if and only if* $\mathcal{E}$ *is complete, atomic and compatible.*

PROOF. Suppose $(\mathcal{E}, S)$ admits hidden variables in the von Neumann sense. If $0 \neq a \in \mathcal{E}$, then there is a dispersion-free state s such that $s(a) = 1$. By the proof of Theorem 3.22, there is an atom $a_1 \in Z$ such that $a_1 \leq a$. Let Z_1 be the collection of atoms in Z, and let a, b be two events. Let

$$a_1 = \bigvee\{z \in Z_1 : z \leq a, z \leq b'\},$$
$$b_1 = \bigvee\{z \in Z_1 : z \leq b, z \leq a'\},$$
$$c = \bigvee\{z \in Z_1 : z \leq a, z \leq b\}.$$

Now a_1, b_1, c are mutually disjoint and $a = a_1 \vee c$, $b = b_1 \vee c$. Hence, $\mathcal{E}$ is atomic, compatible, and complete. To prove sufficiency, the states concentrated on atoms are dispersion free. $\qquad\square$

Since we know from experiment that the event structure describing a quantum mechanical system is not compatible, we see that the event structure cannot admit hidden variables in the von Neumann sense. We have shown that the admission of hidden variables in the von Neumann sense is equivalent to $\mathcal{E}$ being a complete, atomic, Boolean algebra. But these are the essential features of a phase space description of a classical mechanical system. We thus see that $(\mathcal{E}, S)$ admits hidden variables in the von Neumann sense if and only if $(\mathcal{E}, S)$ describes a classical system.

3. The Quantum Logic Approach

We have seen that quantum mechanical systems do not admit hidden variables in a certain sense. However, if we use a different interpretation of what hidden variables mean, we can show that a hidden variables theory is always possible and is unique in a certain sense. This latter interpretation seems to encompass the ones advocated by hidden variables proponents [25, 26, 71, 103].

We first give an English-language version of what we feel the hidden variables advocates mean by a hidden variables theory.

"The state s of a quantum-mechanical system is not complete in the sense that another variable ξ can be adjoined to s so that the pair (s, ξ) completely determines the system. That is, a knowledge of (s, ξ) enables one to predict precisely the outcome of any single measurement. Furthermore, an average over the values of ξ gives the usual quantum state s."

We now attempt to translate the above version of a hidden variables theory to a mathematical-language version. Let $(\mathcal{E}, S)$ be an event structure. As we have seen, single measurements correspond to Boolean sub-σ-algebras of $\mathcal{E}$. A 0–1 *state s* on $\mathcal{E}$ is a map $s: \mathcal{E} \to \{0, 1\}$ such that $s(1) = 1$ and $s(\bigvee a_i) = \Sigma s(a_i)$ where $a_i \perp a_j$, $i = j$, $i = 1, \ldots, n$. To say that the results of a measurement (corresponding to a Boolean sub-σ-algebra $B \subseteq \mathcal{E}$) are completely determined means that one has a 0–1 state s on B. We denote the set of 0–1 states on B by S_B. We are now ready to formulate our definition of a hidden variables theory.

Let $(\mathcal{E}, S)$ be an event structure where S is strongly order determining. We say that $(\mathcal{E}, S)$ *admits a hidden variables theory* if there is a probability space (Ω, Σ, μ) with the property that, for any maximal Boolean sub-σ-algebra $B \subseteq \mathcal{E}$, there is a map H_B from $S \times \Omega$ onto S_B such that:

(i) $\omega \mapsto H_B(s, \omega)(a)$ is measurable for every $s \in S$, $a \in B$;
(ii) $\int H_B(s, \omega)(a) \, d\mu(\omega) = s(a)$ for every $s \in S$, $a \in B$.

Denote the set of maximal Boolean sub-σ-algebras on $\mathcal{E}$ by $\mathfrak{B}$. We call $((\Omega, \Sigma, \mu), \{H_B : B \in \mathfrak{B}\})$ a *hidden variables theory* for $(\mathcal{E}, S)$. A hidden variables theory $((\Omega, \Sigma, \mu), \{H_B : B \in \mathfrak{B}\})$ is *minimal* if $H_B(s, \omega_1) = H_B(s, \omega_2)$ for every $s \in S$ and $B \in \mathfrak{B}$ implies that $\omega_1 = \omega_2$.

We consider only maximal Boolean sub-σ-algebras so that the theory does not get too cumbersome. This is only a technicality since any Boolean sub-σ-algebra is contained in a maximal one. The probability space (Ω, Σ, μ) may be thought of as the space of hidden variables. If a hidden variables theory is minimal there are just enough hidden variables to give all the 0–1 states.

Let B_2 denote the Boolean algebra with two elements 0, 1. A *two-valued homomorphism* f on $\mathcal{E}$ is a map $f: \mathcal{E} \to B_2$ such that $f(a') = f(a)'$; and if

$a \leftrightarrow b$, then $f(a \vee b) = f(a) \vee f(b)$ and $f(a \wedge b) = f(a) \wedge f(b)$. We identify B_2 with the pair of real numbers $\{0, 1\} \subseteq \mathbb{R}$ in the natural way. In the literature [158, 286] two-valued homomorphisms are often considered instead of 0–1 states. However, we now show that these concepts are equivalent.

Lemma 3.25. f *is a* 0–1 *state on* $\mathcal{E}$ *if and only if* f *is a two-valued homomorphism on* $\mathcal{E}$.

PROOF. Suppose f is a two-valued homomorphism on $\mathcal{E}$. Notice first that there is an $a \in \mathcal{E}$ such that $f(a) = 1$ since if $f(a) = 0$, then $f(a') = 1$. Now if $a \in \mathcal{E}$,

$$f(1) = f(1 \vee a) = f(1) \vee f(a) \geq f(a).$$

Hence $f(1) = 1$. Next, we notice that f preserves order since if $a \leq b$, then

$$f(b) = f(a \vee b) = f(a) \vee f(b) \geq f(a).$$

It follows that if a and b are orthogonal, so are $f(a)$ and $f(b)$. Let a_i, $i = 1, \ldots, n$, be mutually orthogonal. If $f(\vee a_i) = 1$, then $\vee f(a_i) = 1$. It follows that there is an a_j such that $f(a_j) = 1$ and $f(a_i) = 0$, $i \neq j$. Hence $\Sigma f(a_i) = 1 = f(\vee a_i)$. If $f(\vee a_i) = 0$, then $\vee f(a_i) = 0$ and $f(a_i) = 0$, $i = 1, \ldots, n$. Hence, $\Sigma f(a_i) = 0 = f(\vee a_i)$ and f is a 0–1 state. Conversely, suppose f is a 0–1 state on $\mathcal{E}$. Then

$$1 = f(a \vee a') = f(a) + f(a')$$

and hence,

$$f(a') = 1 - f(a) = f(a)'.$$

If $a \leftrightarrow b$ then we can write

$$a \vee b = [a \wedge b] \vee [b \wedge (a \vee b)'] \vee [a \wedge (a \vee b)']$$

where the terms in brackets are mutually orthogonal. It then follows that

$$f(a \vee b) = f(a) + f(b) - f(a \wedge b).$$

If $f(a \vee b) = 1$, then either $f(a) = 1$ or $f(b) = 1$, so

$$f(a) \vee f(b) = 1 = f(a \vee b).$$

If $f(a \vee b) = 0$, then $f(a) = 0$ and $f(b) = 0$, so

$$f(a) \vee f(b) = 0 = f(a \vee b).$$

It then follows that $f(a \wedge b) = f(a) \wedge f(b)$ and thus, f is a two-valued homomorphism. $\square$

We are now ready to prove our main theorem.

3. The Quantum Logic Approach

Theorem 3.26. *Any event structure* $(\mathcal{E}, S)$, *where* S *is strongly order determining, admits a hidden variables theory* $((\Omega, \Sigma, \mu), \{H_B : B \in \mathcal{B}\})$. *Moreover,* $((\Omega, \Sigma, \mu), \{H_B : B \in \mathcal{B}\})$ *is the unique minimal hidden variables theory in the sense that if* $((\Omega', \Sigma', \mu'), \{H'_B : B \in \mathcal{B}\})$ *is another hidden variables theory, there is a measurable map* τ *from* Ω' *onto* Ω *such that*

$$H_B(s, \tau\omega')(a) = H'_B(s, \omega')(a)$$

for every $B \in \mathcal{B}$, $\omega' \in \Omega'$, $s \in S$, *and* $a \in B$,

$$\mu'[\tau^{-1}(\Lambda)] = \mu(\Lambda)$$

for every $\Lambda \in \Sigma$; *and if* $((\Omega', \Sigma', \mu'), \{H'_B : B \in \mathcal{B}\})$ *is minimal, then* τ *is injective.*

PROOF. If B is a maximal Boolean sub-σ-algebra, it is in particular a Boolean algebra and hence, by Stone's representation theorem [18], there is an isomorphism h_B from B onto a Boolean algebra G_B of open and closed subsets of a compact space Ω'_B. If $s \in S$, define $\tilde{s}_0$ on G_B by $\tilde{s}_0(\Lambda) = s[h_B^{-1}(\Lambda)]$. We now show that $\tilde{s}_0$ is countably additive on G_B. Let $\Lambda_i \in G_B$, $i = 1, 2, \ldots$, be mutually disjoint and suppose that $\bigcup \Lambda_i \in G_B$. Since $\bigcup \Lambda_i$ is closed and hence, compact, all but finitely many of the Λ_i's are empty. We thus have

$$\tilde{s}_0(\bigcup \Lambda_i) = s[h_B^{-1}(\bigcup \Lambda_i)] = s[\bigvee h_B^{-1}(\Lambda_i)]$$
$$= \Sigma s[h_B^{-1}(\Lambda_i)] = \Sigma \tilde{s}_0(\Lambda_i).$$

Hence, $\tilde{s}_0$ is countably additive on G_B and by the Hahn extension theorem [123], $\tilde{s}_0$ has a unique extension to a measure $\tilde{s}$ on the σ-algebra K_B generated by G_B. Notice that K_B is contained in the Borel field of Ω'_B. Thus, $(\Omega'_B, K_B, \tilde{s})$ is a probability space. We now form the product probability space $(\Omega_B, \Sigma_B, \mu_B)$ where $\Omega_B = \Pi\{\Omega_s : s \in S\}$, $\Sigma_B = \Pi\{K_s : s \in S\}$, $\mu_B = \Pi\{\tilde{s} : s \in S\}$, and where $\Omega_s = \Omega'_B$ and $K_s = K_B$ for all $s \in S$. That is, Ω_B is the Cartesian product of Ω'_B with itself to the cardinality of S, Σ_B is the σ-algebra generated by the cylinder sets according to Kolmogorov's construction [160], and μ_B is the product measure defined in terms of these cylinder sets. The space $(\Omega_B, \Sigma_B, \mu_B)$ exists by Kolmogorov's theorem [160]. By Tychonov's theorem, note that Ω_B is compact and Σ_B is contained in the Borel field. We can then use Kolmogorov's theorem again to form the product probability space (Ω, Σ, μ), where $\Omega = \Pi\{\Omega_B : B \in \mathcal{B}\}$, $\Sigma = \Pi\{\Sigma_B : B \in \mathcal{B}\}$, and $\mu = \Pi\{\mu_B : B \in \mathcal{B}\}$. Now for $B \in \mathcal{B}$, define the map $H_B : S \times \Omega \to S_B$ by $H_B(s, \omega)(a) = 1$, if $\omega(B)(s) \in h_B(a)$ and $H_B(s, \omega)(a) = 0$ otherwise. We now show that $H_B(s, \omega)(\cdot)$ is a 0–1 state. Certainly, $H_B(s, \omega)(1) = 1$. Suppose that the a_i are mutually orthogonal, $i = 1, \ldots, n$. If $H_B(s, \omega)(\bigvee a_i) =$

0, then
$$\omega(\dot{B})(s) \notin h_B(\vee a_i) = \cup h_B(a_i).$$
Thus, $\omega(B)(s) \notin h_B(a_i)$ and $H_B(s, \omega)(a_i) = 0$, $i = 1, \ldots, n$, and
$$0 = H_B(s, \omega)(\vee a_i) = \cup H_B(a_i).$$
If $H_B(s, \omega)(\vee a_i) = 1$, then
$$\omega(B)(s) \in h_B(\vee a_i) = \cup h_B(a_i).$$
Since the $h_B(a_i)$ are are mutually orthogonal, $i = 1, \ldots, n$, there is a $1 \leq j \leq n$ such that $\omega(B)(s) \in a_j$, $\omega(B)(s) \notin a_i$, $i \neq j$. Hence, $H_B(s, \omega)(a_i) = \delta_{ij}$ and
$$1 = H_B(s, \omega)(\vee a_i) = \Sigma H_B(s, \omega)(a_i).$$
Therefore, $H_B(s, \omega)(\cdot)$ is a 0–1 state. Now $\omega \mapsto H_B(s, \omega)(a)$ is measurable since it has the value one on the cylinder set
$$\Lambda = \{ \omega \in \Omega : \omega(B) \in h(a) \times \Pi \{ \Omega_{s'} : \Omega_{s'} = \Omega'_B, s' \neq s \} \}$$
$$= \{ h(a) \times \Pi \{ \Omega_{s'} : \Omega_{s'} = \Omega'_B, s' \neq s \} \times \Pi \{ \Omega_{B'} : B' \neq B \} \}$$
and the value zero otherwise. Also,
$$\int_\Omega H_B(s, \omega)(a) \, d\mu(\omega) = \int_\Lambda d\mu(\omega) = \mu(\Lambda)$$
$$= \mu [h_B(a) \times \Pi \{ \Omega_{s'} : \Omega_{s'} = \Omega'_B, s' \neq s \}]$$
$$= \tilde{s} [h_B(a)] = s(a)$$
for all $s \in S$, $a \in B$. Now the map $\omega \mapsto \omega(B)(s)$ is onto Ω'_B by definition. It can be shown [18] that the Stone space Ω'_B is isomorphic to the set of two-valued homomorphisms on B. By Lemma 3.25, this latter set is the same as S_B and hence, H_B is surjective. Now suppose that $((\Omega', \Sigma', \mu'),$ $\{ H'_B : B \in \mathfrak{B} \})$ is another hidden variables theory. Define $\tau : \Omega \to \Omega'$ by $\tau(\omega')(B)(s) = H'_B(s, \omega')$. Then τ is surjective since H'_B is surjective. The measurability of τ follows from the measurability of $\omega' \mapsto H'_B(s, \omega')$, and $H_B(s, \tau\omega')(a) = H'_B(s, \omega')(a)$ by definition. The rest of the proof is left to the reader. □

Although, as the above theorem shows, hidden variable theories are possible in quantum mechanics, they do not seem to evoke any simplifications or advantages and are rarely used.

3.7. Generalized Probability Spaces

In this section we give an example of an event structure which is more general than classical probability theory yet less general than the poset of an abstract event structure. This type of event structure is general enough to exhibit the noncompatibility of quantum interference effects which are not present in classical probability theory. It is based on a slight weakening of the σ-algebra Σ of events on a set Ω. We begin with some physical motivation.

Suppose we are making length measurements with a micrometer and that this micrometer is accurate to within 10^{-4} cm and can measure lengths up to 1 cm. Since we must round off measurements to the nearest 10^{-4} cm, the micrometer in effect is able to give lengths only of the form $n10^{-4}$ cm, where n is an integer between 1 and 10^4. Let Ω be the interval $[0, 1]$ and let C be the class of subsets of Ω to which we can attribute definite lengths. Now any interval of the form $[m10^{-4}, n10^{-4}]$, $m \leq n$, $m, n = 1, 2, \ldots, 10^4$, has the definite length $(n - m)10^{-4}$, so such intervals are in C. In the same way any interval of the form $[\alpha + m10^{-4}, \alpha + n10^{-4}]$, $0 \leq \alpha \leq 1$, $\alpha + n10^{-4} \leq 1$, $m \leq n = 1, 2, \ldots, 10^4$, is in C and has length $(n - m)10^{-4}$. We can also admit noninterval sets in C. For example, the set $[0, 10^{-4}] \cup [2 \times 10^{-4}, 3 \times 10^{-4}]$ is in C and has definite length 2×10^{-4}. Although the length of the set $A = [0, \frac{1}{2} \times 10^{-4}] \cup [\frac{3}{2} \times 10^{-4}, 2 \times 10^{-4}]$ cannot be accurately measured directly, we can measure its complement $[\frac{1}{2} \times 10^{-4}, \frac{3}{2} \times 10^{-4}] \cup [2 \times 10^{-4}, 1]$ (we assume the length of a point is zero) and conclude the length of A is one. Thus $A \in C$. We thus conclude that C contains all intervals of the form $[\alpha + m10^{-4}, \alpha + n10^{-4}]$, as above, together with all sets that can be obtained from these using the operations of finite disjoint unions (ignoring points) and complementations. Notice that C is not closed under formations of unions or intersections of two arbitrary sets of C. The length of any member of C must have the form $n10^{-4}$, $n = 1, \ldots, 10^4$.

Notice that if the length is to have decent properties we are forced to assume that C has the above structure. For example, suppose we were to assume that any subinterval of $[0, 1]$ is in C and that the length of a subinterval is that measured by the micrometer rounded off to the nearest 10^{-4} cm. We would then conclude that the length of the interval $[0, \frac{1}{3} \times 10^{-4}]$ is zero, for example. Similarly, the lengths of $[\frac{1}{3} \times 10^{-4}, \frac{2}{3} \times 10^{-4}]$ and $[\frac{2}{3} \times 10^{-4}, 10^{-4}]$ are zero. If the length is to have the eminently reasonable property of additivity we would then be faced with the contradictory conclusion that the length of $[0, 10^{-4}]$ is zero. As another example of what can go wrong in this case, suppose we assume the length of $[0, \frac{1}{2} \times 10^{-4}]$ is one. Using additivity we would conclude that the length of $(\frac{1}{2} \times 10^{-4}, 1]$ is zero and similarly the length of $[0, \frac{1}{2} \times 10^{-4})$ is zero so that the length of the point $\frac{1}{2} \times 10^{-4}$ is one!

One might argue that the situation we have considered is unrealistic in that we have unduly restricted ourselves. We do not have to use micrometers with an accuracy of only 10^{-4} cm. We can, in principle, construct length-measuring apparata with increasingly fine accuracy. We can then attribute to any interval an arbitrarily precise length, go through a similar construction as before (using countable unions instead of finite disjoint unions) and obtain the Lebesgue measurable sets and Lebesgue measure on $[0, 1]$. Our answer to this argument is twofold. First, if we are to describe a particular measuring apparatus, we must be content with its inherent accuracy. Second, it is possible that nature has forced upon us an intrinsic limit to accurate length measurement. There is experimental as well as theoretical evidence [11, 98, 128] pointing toward the existence of an elementary length. This would be a length λ (about 10^{-15} cm) such that no smaller length measurement is attainable and all length measurements must be an integer multiple of λ. If an "ultimate" apparatus could be constructed which can measure this length then upon replacing 10^{-4} cm by λ we would be forced to a construction similar to the above.

There are, in fact, instances in nature in which an ultimate accuracy is known to obtain. It is accepted that the charge on the electron is the smallest charge obtainable. All charge measurements must give an integer multiple of e. If we think, for example, of a charge of $10^4 e$ as uniformly distributed over the interval $[0, 10^4 e]$ then a description of a charge measurement would begin with a construction similar to the above.

There are important situations in which the apparatus under consideration has inherent accuracy limitations. For example, all high-speed computers have such limitations. The computer will accept only a certain quantity of significant digits and all numbers must be rounded off to this quantity. Numbers are rounded off with each internal operation performed by the computer, which sometimes results in considerable round-off error. This type of phenomenon also occurs in pattern recognition studies [272–274].

Another example of the above phenomenon is motivated by quantum mechanics. Suppose we are considering a particle which is constrained to move along the x axis. According to classical mechanics, if we form a two-dimensional phase space Ω with coordinate axes x, p, where p is the x momentum of the particle, then the mechanics of the particle is completely described by a point in Ω. If we want to describe quantum mechanical effects, however, we would have to contend with the Heisenberg uncertainty principle. In fact, if Δx, Δp are the errors made in an x and p measurement respectively, then $\Delta x \Delta p \geq \hbar/2$, where $\hbar$ is Planck's constant h divided by 2π. For this reason a point in Ω is physically meaningless since one can never determine whether the phase space coordinates (x, p) of a particle are at a particular point. The best one can do is determine whether (x,p) is contained in a rectangle with sides of length $\Delta x, \Delta p$, where $\Delta x \Delta p = \hbar/2$. Let us suppose our measuring instruments have ultimate precision so we can determine whether (x,p) is in a rectangle of area $\hbar/2$. Physically, these elementary rectangles in Ω become the basic elements of the theory instead of the points in Ω. Generalizing this slightly, we say that a set E is *admissible* if the area of E (strictly speaking, we mean the Lebesgue measure of E) is an integer multiple of $\hbar/2$. An admissible set E is one for which it is physically meaningful to say that $(x,p) \in E$. For simplicity, let us assume that Ω is a large rectangle of area $n\hbar/2$ where n is a large integer. Let us now consider the mathematical structure of the class C of admissible sets. First, it is clear that $\Omega \in C$ and also $\varnothing \in C$. It is clear that if $E \in C$ then the complement $E^c \in C$. Further, if $E, F \in C$ are disjoint, then $E \cup F \in C$. Note, however, that in general if $E, F \in C$, then $E \cup F$ and $E \cap F$ need not be in C.

We now give the general definitions. Let Ω be a nonempty set. A σ-*class* C of subsets of Ω is a collection of subsets which satisfy:

(i) $\Omega \in C$;
(ii) if $a \in C$, then $a^c \in C$;
(iii) if $a_i \in C$ are mutually disjoint, then $\bigcup a_i \in C$, $i = 1, 2, \ldots$.

A *probability measure* μ on a σ-class C is a nonnegative set function μ on C such that $\mu(\Omega) = 1$, and $\mu(\bigcup a_i) = \Sigma \mu(a_i)$ if a_i are mutually disjoint elements of C, $i = 1, 2, \ldots$. A *generalized probability space* is a triple (Ω, C, μ), where C is a σ-class of subsets of Ω and μ is a probability measure on C. The proof of the following lemma is straightforward.

Lemma 3.27. *Let C be a σ-class and let S be the set of probability measures on C. (i) (C, S) is an event structure. (ii) If $a, b \in C$, then $a \leftrightarrow b$ if and only if $a \cap b \in C$.*

The simplest σ-class that is not a σ-algebra is given as follows. Let $\Omega = \{1, 2, 3, 4\}$ and let C be the class of subsets of Ω with an even number of elements. Then C is a σ-class but $\{1,2\} \cap \{2,3\} = \{2\} \notin C$ so C is not a σ-algebra. A generalization of this example is the collection of Lebesgue measurable subsets of the real line with even measure. More generally, let $\lambda > 0$ and let $\Omega = [0, n\lambda]$ where n is a positive integer. If C is the collection of Lebesgue measurable subsets of Ω whose Lebesgue measures are an integer multiple of λ, then C is a σ-class but not a σ-algebra.

Let C be a σ-class and let (C, S) be the event structure as defined in Lemma 3.27. It is not hard to show that the event structure order $a \leq b$ is equivalent to the order $a \subseteq b$. Now C may not be a lattice but if $a \cap b \in C$ or $a \cup b \in C$, it is easily seen that $a \wedge b = a \cap b$ and $a \vee b = a \cup b$. If $a \cap b \notin$

C (or $a \cup b \notin C$) we still may have $a \wedge b$ (or $a \vee b$) existing. For instance, in the first example of the previous paragraph, $\{1,2\} \cap \{2,3\} = \{2\} \notin C$ but $\{1,2\} \wedge \{2,3\} = \varnothing$. Also, if $a \perp b$, then $a \cap b = a \wedge b = \varnothing$, but $a \wedge b = \varnothing$ does not imply $a \perp b$. Indeed, in the above example, $\{1,2\} \wedge \{2,3\} = \varnothing$, yet $\{1,2\} \not\perp \{2,3\}$. Since (C,S) is an event structure, we have seen in Theorem 3.4 that C is a quantum logic (i.e., a σ-orthocomplete orthomodular poset). The next theorem characterizes those quantum logics that are isomorphic to a σ-class.

Theorem 3.28. *A quantum logic $\mathcal{P}$ is isomorphic to a σ-class if and only if the set S_0 of 0–1 states on $\mathcal{P}$ is order determining. In this case $\mathcal{P}$ is isomorphic to a σ-class of subsets of S_0.*

PROOF. Suppose $\mathcal{P}$ is isomorphic to a σ-class C of subsets of Ω under the isomorphism $\phi \colon \mathcal{P} \to C$. Suppose that $a,b \in \mathcal{P}$ and $s(a) \leq s(b)$ for all $s \in S_0$. In particular if $\omega \in \phi(a)$ and μ_ω is the probability measure concentrated at ω, then $c \mapsto \mu_\omega[\phi(c)]$ is a 0–1 state so $\mu_\omega[\phi(a)] \leq \mu_\omega[\phi(b)]$. Hence, $\omega \in \phi(b)$ and $\phi(a) \leq \phi(b)$. It follows that $a \leq b$ so S_0 is order determining. Conversely, suppose that S_0 is order determining. Let h be the map from $\mathcal{P}$ to the collection of subsets of S_0 defined by $h(a) = \{s \in S_0 : s(a) = 1\}$, and let C be the range of h. We first show that C is a σ-class. Now $S_0 = h(1)$ and $\varnothing = h(0)$ so $S, \varnothing \in C$. For $h(a) \in C$ we have $h(a)^c = h(a') \in C$. Also, if the $h(a_i)$ are mutually disjoint we have $\cup h(a_i) = h(\vee a_i) \in C$. If $a \leq b$ then clearly $h(a) \leq h(b)$. To show that h is injective, suppose that $a \neq b$. Since S_0 is order determining, there is an $s \in S_0$ such that $s(a) \neq s(b)$. Thus, $s(a) = 1$, $s(b) = 0$, or vice versa, so $h(a) \neq h(b)$. Finally, if $h(a) \leq h(b)$, then $s(a) \leq s(b)$ for all $s \in S_0$ and hence, $a \leq b$. $\qquad\square$

The above theorem shows that the distinguishing feature of σ-classes among general quantum logics is that σ-classes have an order-determining set of 0–1 states. This might be interpreted as corresponding to the existence of hidden variables (see Section 3.6). This seems to be close to what hidden variable advocates are proposing since the existence of many 0–1 states means that there is determinism in the sense that exact predictions can be made. On the other hand, interference effects and noncompatibility are still possible.

A σ-class event structure (C,S) is not isomorphic to a Hilbert space event structure $(\mathcal{P}, S_1)$, where $\mathcal{P}$ is the lattice of orthogonal projections on a Hilbert space $\mathcal{H}$ with dim $\mathcal{H} > 2$. This is because $\mathcal{P}$ does not admit any 0–1 states. In fact, suppose s is a 0–1 state on $\mathcal{P}$. By Gleason's theorem, which we shall consider in detail later, there is a density operator T such that $s(P)=\mathrm{tr}(PT)$ for all $P\in\mathcal{P}$. If λ_i and ϕ_i are the eigenvalues and mutually orthogonal eigenvectors of T, it follows that $s(P)=\Sigma\lambda_i\langle P\phi_i,\phi_i\rangle$ for all $P\in\mathcal{P}$. If P_{ϕ_j} is the one-dimensional projection onto ϕ_j, we have

$$\lambda_j=\lambda_j|\langle\phi_j,\phi_j\rangle|^2=\Sigma\lambda_i\langle\phi_i,P_{\phi_j}\phi_i\rangle=s(P_{\phi_j}).$$

Hence, $\lambda_j=0$ or 1. Since $1=s(T)=\Sigma\lambda_i$ we see that all the λ_i's are 0 except one, so T is itself a one-dimensional projection P_ϕ. Now if ψ is a unit vector we have

$$|\langle\phi,\psi\rangle|^2=\mathrm{tr}(P_\phi P_\psi)=s(P_\psi)=0 \quad \text{or} \quad 1.$$

This is clearly impossible. Figure 4 summarizes the relationships between the different structures.

Let μ be a probability measure on a σ-class C. Let us briefly compare some of the properties of μ to those of a probability measure on a σ-algebra. Now a σ-algebra probability measure ν always satisfies $\nu(A\cup B)+\nu(A\cap B)=\nu(A)+\nu(B)$. The probability measure μ always satisfies this condition when the left-hand side is defined. Indeed, we would then have $A\cap B\in C$, so by Lemma 3.27, $A\leftrightarrow B$. It follows that A and B are contained in a sub-σ-algebra Σ of C and the restriction of μ to Σ is an ordinary probability measure. Now a probability measure ν on a σ-algebra Σ is always subadditive; that is, $\nu(\cup A_i)\le\Sigma\nu(A_i)$ for any sequence $A_i\in\Sigma$. We now show that μ does not necessarily have this property when the left-hand side is defined. Consider the following subsets of the

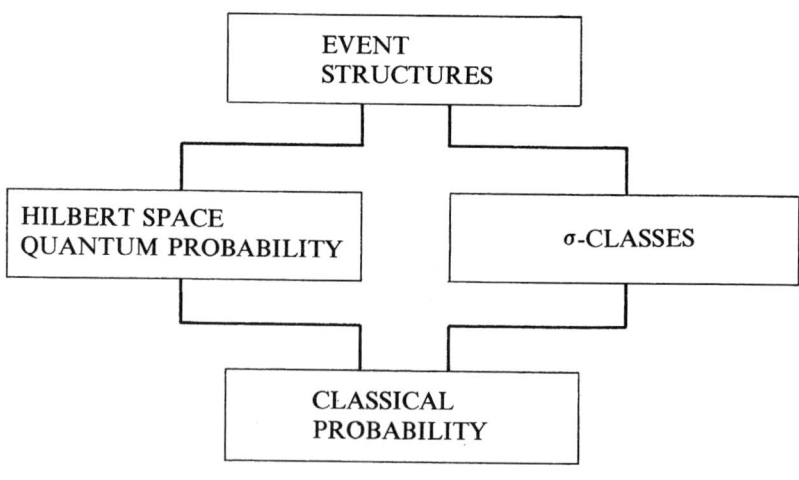

Figure 4

real line: $A = [0, 4]$, $B = [2, 5]$, $D = [0, 1] \cup [2, 3] \cup [5, 6]$. Let $C = \{\varnothing, [0, 6], A, B, D, A^c, B^c, D^c\}$. Then C is a σ-class of subsets of $[0, 6]$. Define the probability measure μ on C as follows:

$$\mu(\varnothing) = 0, \qquad \mu(A) = \mu(B) = \mu(D) = \tfrac{1}{4},$$

$$\mu(A^c) = \mu(B^c) = \mu(D^c) = \tfrac{3}{4}, \qquad \mu([0, 1]) = 1 \qquad (3.3)$$

Now $A \cup B \cup D = [0, 6] \in C$ and yet

$$\mu([0, 6]) = 1 > \tfrac{3}{4} = \mu(A) + \mu(B) + \mu(D).$$

A probability measure μ on a σ-class C is always subadditive on two sets; that is, $\mu(A \cup B) \leq \mu(A) + \mu(B)$ for all $A, B \in C$ for which $A \cup B \in C$. Indeed, if $A \cup B \in C$, then $A^c \cap B^c = (A \cup B)^c \in C$ so $A^c \leftrightarrow B^c$. It follows from Lemma 3.6(i) that $A \leftrightarrow B$ and hence, A and B are contained in a sub-σ-algebra of C. It follows from the nonsubadditivity that a probability measure μ on a σ-class C cannot in general be extended to the σ-algebra generated by C.

We now briefly consider integration in this framework. Let (Ω, C, μ) be a generalized probability space. A function $f: \Omega \to \mathbb{R}$ is *measurable* if $f^{-1}(E) \in C$ for every open set $E \subseteq \mathbb{R}$. It is not hard to show that the σ-class generated by the open sets of $\mathbb{R}$ is $\mathscr{B}(\mathbb{R})$. We can then conclude that if f is measurable, the $f^{-1}(E) \in C$ for every $E \in \mathscr{B}(\mathbb{R})$. Thus, the measurable functions correspond to the observables of a general event structure. It is well known that in probability spaces the sum of any two measurable functions is measurable. It follows that if f and g are compatible measurable functions [i.e., $f^{-1}(E) \leftrightarrow g^{-1}(F)$ for all $E, F \in \mathscr{B}(\mathbb{R})$] on (Ω, C, μ), then $f + g$ is measurable. However, the sum of two noncompatible measurable functions need not be measurable. This follows from the general fact that C is a σ-algebra if and only if the sum of any two measurable functions is measurable. This gives an example of an event structure which does not satisfy the existence condition E of Section 3.5. One can show that the sum of two noncompatible measurable characteristic functions is never measurable. However, there are noncompatible measurable functions whose sum is measurable. For example, let $\Omega = \{1, 2, \dots, 8\}$ and let C be the σ-class of subsets of Ω with an even number of elements. Define the measurable functions f and g by

$$f(1) = f(2) = f(7) = f(8) = 0, \qquad f(3) = f(4) = 1, \qquad f(5) = f(6) = 2,$$

$$g(1) = g(6) = 1, \qquad g(2) = g(4) = 2, \qquad g(3) = g(5) = g(7) = g(8) = 0. \quad (3.4)$$

We leave it to the reader to show that $f + g$ is measurable, and that $f \not\leftrightarrow g$.

Let f be a measurable function on a generalized probability space (Ω, C, μ), and let $A_f = \{f^{-1}(E) : E \in \mathscr{B}(\mathbb{R})\}$. It is easily seen that A_f is a sub-σ-algebra of C. Hence, μ restricted to A_f is an ordinary probability measure and (Ω, A_f, μ) becomes a probability space. We define the integral $\int f d\mu$ in the usual way. Let us discuss the basic properties of this integral. It is clear that $\int f d\mu \geq 0$ if $f \geq 0$ and $\int \alpha f d\mu = \alpha \int f d\mu$ for any $\alpha \in \mathbb{R}$. Also if $f \leftrightarrow g$, then $\int (f + g) d\mu = \int f d\mu + \int g d\mu$ (unless $\int f d\mu = \infty$ and $\int g d\mu = -\infty$, or vice versa). Of course, in ordinary integration theory on a σ-algebra, the integral is always additive. We now inquire whether this property holds for our generalized integral. We say that two measurable functions f and g are *summable* if $f + g$ is measurable. We say that the integral is *additive* if for summable functions f and g we have $\int (f + g) d\mu = \int f d\mu + \int g d\mu$ (unless $\int f d\mu = \infty$, $\int g d\mu = -\infty$, or vice versa). It is not known whether the integral is additive in general. Although some progress has been made [99, 106] this important problem is far from solved.

3.8. Notes and References

The approach taken in Section 3.1 is mainly due to Mackey [173]. For investigations into generalized transition probabilities we refer the reader to [35, 187–189, 261]. The methods used in Section 3.2 are mainly due to Maczynski [177–180]. The work on compatibility in Section 3.3 was done originally by Varadarajan [267]. This section contains some modifications and simplifications due to Ramsey and the author [221, 101]. The variance theorem was proved by Louton [168], although the presentation given here contains some simplifications of the author. For more details on uniqueness and existence see [93].

The hidden variables controversy has a long and varied history. The argument has at times been quite heated and has included work by giants such as von Neumann, Einstein, Bohr, Heisenberg, and Wiener. Hundreds of papers have been written on this subject. As a sample, we recommend [12, 15, 25, 26, 27, 71, 103, 140, 157, 159, 286]. For more details on the results presented here, the reader is referred to [96, 103, 201].

The germ of the ideas presented in Section 3.7 on generalized probability spaces may be traced to Suppes [259]. Further developments are carried out in [99, 106].

For a general lattice-theoretic background to this chapter, the reader can consult [18, 19, 134, 181].

3.9. Exercises

1. Show that $x \in \mathcal{O}$ is a generalized event if and only if $x^2 = x$.

2. Show that $x \in \mathcal{O}$ is a generalized event if and only if $x = \chi_E(y)$ for some $y \in \mathcal{O}$, $E \in \mathcal{B}(\mathbb{R})$.

3. Using Axioms 3.1 and 3.2, show that if $a, b \in \mathcal{E}$ and $s(a) = s(b)$ for every $s \in S$, then $a = b$.

4. Using Axioms 3.1 and 3.2, show that if $s_1, s_2 \in S$ and $s_1(a) = s_2(a)$ for every $a \in \mathcal{E}$, then $s_1 = s_2$.

5. Using Axioms 3.1 and 3.2, show that $\leq$ is a partial order relation on $\mathcal{E}$.

6. Prove Lemma 3.3.

7. Verify all the statements in the second to last paragraph of Section 3.1.

8. Using Axioms A and B show that $\leq$ is a partial order relation on $\mathcal{E}$.

9. Verify all the statements for the example in the paragraph after Theorem 3.4.

10. Show that (Σ, S) in Example 1 is an event structure.

11. If f is a random variable, show that f^{-1} is an observable.

12. Prove that all events and all observables in Example 1 are compatible.

13. Show that $(\mathcal{P}, S)$ is an event structure in Example 2.

14. In Example 2, show that $P_1, P_2 \in \mathcal{P}$ are compatible if and only if they commute.

15. Show that Eq. (3.1) defines an observable.

16. If x is an observable and u is a real Borel function, show that $u(x)$ is an observable.

17. Prove that the range of an observable is a Boolean sub-σ-algebra of $\mathcal{E}$.

18. In the example after Theorem 3.9, show that $(\mathcal{E}, S)$ is an event structure.

19. Prove that $\mathcal{B}(\mathbb{R})$ is separable. Prove that the range of an observable is separable.

20. Supply the details for the proof of Lemma 3.13.

21. Show that $\lambda \in \sigma(x)^c$ if and only if there is an open set U containing λ such that $x(U) = 0$.

22. Show that if $\mathcal{E}$ has an infinite number of orthogonal nonzero events, then any nonempty closed subset of $\mathbb{R}$ is the spectrum of some observable.

23. The *point spectrum* $\sigma_p(x)$ of an observable x is the set $\sigma_p(x) = \{\lambda \in \mathbb{R} : x(\{\lambda\}) \neq 0\}$. Show that $\sigma_p(x) \subseteq \sigma(x)$ but that $\sigma_p(x)$ need not be closed. Show that if λ is an isolated point of $\sigma(x)$, then $\lambda \in \sigma_p(x)$.

24. Prove the variance formula.

25. Prove Lemma 3.17.

26. If A and B are bounded self-adjoint operators on $\mathcal{H}$ and $\langle A\phi, \phi \rangle = \langle B\phi, \phi \rangle$ for every $\phi \in \mathcal{H}$, $\|\phi\| = 1$, prove that $A = B$.

3. The Quantum Logic Approach

27. Let $(\mathcal{E}, S)$ be an event structure with S strongly order determining and strongly convex. For bounded observables x and y, prove the following statements. (i) If $s[u(x)] = s[u(y)]$ for every Borel function u and every $s \in S$, then $x = y$. (ii) If $x \leftrightarrow y$ and $s(x) = x(y)$ for every $s \in S$, then $x = y$. (iii) If x or y has one or two point spectrum and $s(x) = x(y)$ for every $s \in S$, then $x = y$.

28. Show that if $x \leftrightarrow y$, then $x + y$ exists.

29. If $\mathcal{E}$ is an antilattice and S is the set of all probability measures on $\mathcal{E}$, show that $(\mathcal{E}, S)$ is an event structure and S is strongly order determining.

30. Show that on an antilattice $\mathcal{E}$, given $0 < \alpha < 1$ and $c \in \mathcal{E}$, there is a probability measure s such that $s(c) = \alpha$ and $s(d) = 0$, where d is any event not equal to c, c', or 1.

31. Show that the set of states in Examples 1 and 2 are strongly order determining.

32. Complete the proof of Theorem 3.26.

33. Prove Lemma 3.27.

34. Let C be a σ-class and let (C, S) be the event structure as defined in Lemma 3.27. Show that $a \leq b$ if and only if $a \subseteq b$. Show that if $a \cap b \in C$, then $a \cap b = a \wedge b$. Show that $a \perp b$ if and only if $a \cap b = \varnothing$.

35. Show that μ defined by (3.3) is a probability measure and that C is a σ-class.

36. Show that the σ-class generated by the open sets of $\mathbb{R}$ is $\mathcal{B}(\mathbb{R})$.

37. Prove that if f is a measurable function on a σ-class C, then $f^{-1}(E) \in C$ for every $E \in \mathcal{B}(\mathbb{R})$.

38. Prove that a σ-class C is a σ-algebra if and only if the sum of any two measurable functions on C is measurable.

39. Prove that the sum of two noncompatible measurable characteristic functions is never measurable.

40. Show that f and g defined in Eq. (3.4) are measurable and noncompatible, and that $f + g$ is measurable.

4

The Operational Approach

In the previous chapter we considered the quantum logic approach to axiomatic quantum mechanics. In that approach the main framework is an event structure in which the primitive axiomatic elements are the quantum mechanical events. In the operational approach, the primitive axiomatic elements are the quantum mechanical states. The resulting framework is called a state space or convex structure. This approach is actually more general than that of Chapter 3 and the emphasis is quite different. An important role is played by the convexity properties of states.

4.1. Convex Structures

The important property of states, as far as the operational approach is concerned, is that they are closed under the formation of mixtures or convex combinations. That is, if s_1 and s_2 are states then so is $\lambda s_1 + (1-\lambda)s_2$, where $0 < \lambda < 1$. It is easy to define a convex combination of elements in a linear space, but unfortunately a linear space is artificial and devoid of physical meaning for states. One cannot add states or multiply them by scalars to get other states. Only the operation of forming mixtures of states has meaning. For this reason, an abstract definition of mixtures is defined that is independent of the concept of linearity. We call the resulting framework a convex structure. This approach to convexity originated with von Neumann, Morgenstern, and Stone [253, 271] and later developed into the operational approach of Mielnik [187–189], Ludwig [170, 171], and Davies and Lewis [54, 55].

A *convex structure* is a set S with the following two properties:

C1. for any finite set of positive numbers $\lambda_1,\ldots,\lambda_n$ satisfying $\Sigma\lambda_i = 1$ and any $s_1,\ldots,s_n \in S$ there exists a unique element $\langle \lambda_1,\ldots,\lambda_n; s_1,\ldots,s_n \rangle$ $\in S$;

C2. $\langle \lambda_1,\ldots,\lambda_n; s,s,\ldots,s \rangle = s$.

We assume that the set of (normalized) states for a physical system form a convex structure S and interpret $\langle \lambda_1,\ldots,\lambda_n; s_1,\ldots,s_n \rangle$ as a mixture of the states $s_1,\ldots,s_n$ in proportion $\lambda_1,\ldots,\lambda_n$, respectively. Property C1 ensures that S is closed under mixtures and C2 is the obvious property that mixtures containing only the state s equal s. For mixtures of two states we use the simpler notation $\langle \lambda, 1-\lambda; s, t \rangle = \langle \lambda; s, t \rangle$. A state s is *pure* (or *extreme*) if s cannot be written in the form $s = \langle \lambda; s_1, t_1 \rangle$ where $s_1 \neq t_1$. An example of a convex structure is a convex subset of a real linear space where

$$\langle \lambda_1,\ldots,\lambda_n; s_1,\ldots,s_n \rangle = \sum \lambda_i s_i. \tag{4.1}$$

A convex subset of a real linear space will always be assumed to have its natural convex structure (4.1).

A state of a physical system $\mathbb{S}$ is supposed to completely describe the statistical properties of $\mathbb{S}$. If a measurement A is made on $\mathbb{S}$ and Δ is a set of values, then the state s gives the probability that the result of a measurement of A has a value in Δ. More precisely, if one prepares an ensemble of systems identical to $\mathbb{S}$ and subjects each member of the ensemble to the measurement A, then s determines the proportion of the members in which A has a value in Δ. Another way to describe this situation is to think of $\mathbb{S}$ as a particle. Then prepare a beam of noninteracting particles identical to $\mathbb{S}$. Suppose that one particle is emitted at each unit of time so that the beam strength (or intensity) is 1. A filter placed in the beam path transmits only those particles for which A has a value in Δ. Then s determines the beam strength of the transmitted beam. If $\lambda_1,\ldots,\lambda_n$ > 0, $\Sigma\lambda_i = 1$, and $s_1,\ldots,s_n$ are states, then $\langle \lambda_1,\ldots,\lambda_n; s_1,\ldots,s_n \rangle$ represents the state of an ensemble composed of systems in the states $s_1,\ldots,s_n$ in proportion $\lambda_1, \ldots, \lambda_n$, respectively. Or in optical language, $\langle \lambda_1,\ldots,\lambda_n; s_1,\ldots,s_n \rangle$ describes the state of a beam of unit strength consisting of n types of noninteracting particles p_i in the states s_i and of beam strengths λ_i, $i = 1,\ldots,n$.

In order to form infinite countable mixtures and discuss other analytic concepts, we introduce a natural distance function on S. The closeness of $s, t \in S$ can be measured by comparing mixtures $\langle \lambda; s_1, s \rangle$ and $\langle \lambda; t_1, t \rangle$ of s and t with other states. If s and t are very close one would expect to find a mixture containing mostly s equal to a mixture containing mostly t; that is,

$$\langle \lambda; s_1, s \rangle = \langle \lambda; t_1, t \rangle, \tag{4.2}$$

with λ very small. Thus, the parameter λ such that (4.2) holds for some $s_1, t_1 \in S$ gives a measure of the closeness of s to t. We hence define a distance function $\sigma(s, t)$ on $S \times S$ as follows. If there exist $s_1, t_1 \in S$ such that (4.2) holds, then

$$\sigma(s, t) = \inf \{ 0 < \lambda \le 1 : \langle \lambda; s_1, s \rangle = \langle \lambda; t_1, t \rangle, s_1, t_1 \in S \};$$

otherwise, $\sigma(s, t) = \frac{1}{2}$. In general, σ need not be a metric; however, we shall later see that in important special cases σ is a metric.

A *σ-convex structure* is a convex structure S on which the following two properties hold:

C3. if $s_n \in S$ satisfies $\lim_{m, n \to \infty} \sigma(s_n, s_m) = 0$ then there exists a unique $s \in S$ such that $\lim_{n \to \infty} \sigma(s_n, s) = 0$.

C4. If $\lambda_i > 0$, $\Sigma \lambda_i = 1$, $t_1, t_2, \ldots \in S$, and

$$s_n = \langle \lambda_1, \ldots, \lambda_n, 1 - \sum_{i=n+1}^{\infty} \lambda_i; t_1, \ldots, t_{n+1} \rangle,$$

then $\lim_{m, n \to \infty} \sigma(s_m, s_n) = 0$.

Property C3 is a "completeness" axiom and C4 enables us to treat infinite countable mixtures. We denote the limit of s_n in C4 by $\langle \lambda_1, \lambda_2, \ldots; t_1, t_2, \ldots \rangle$.

If S and T are convex structures, a map $F: S \to T$ is *affine* if

$$F(\langle \lambda_1, \ldots, \lambda_n; s_1, \ldots, s_n \rangle) = \langle \lambda_1, \ldots, \lambda_n; F(s_1), \ldots, F(s_n) \rangle$$

for all $\lambda_1, \ldots, \lambda_n > 0$ such that $\Sigma \lambda_i = 1$ and $s_1, \ldots, s_n \in S$. We denote the set of affine maps from S to T by $Af(S, T)$ and we use the notation $Af(S) = Af(S, S)$. If there is an affine bijection $J: S \to T$ we say that S and T are *isomorphic* and J is an *isomorphism*. The maps $f \in Af(S, \mathbb{R})$ are called *affine functionals* and we use the notation $S^* = Af(S, \mathbb{R})$. We denote by 0 and τ the affine functionals satisfying $0(s) = 0$ and $\tau(s) = 1$ for every $s \in S$, respectively. Notice that S^* is a real linear space under the usual pointwise operations. If $f, g \in S^*$ we define $f \le g$ if $f(s) \le g(s)$ for every $s \in S$. An *effect* is a map $f \in S^*$ satisfying $0 \le f \le \tau$ and the set of effects is denoted by $\mathcal{E}(S)$. If $F \in Af(S)$ we define the linear map $F^*: S^* \to S^*$ by $(F^*f)(s) = f[F(s)]$ for every $f \in S^*$, $s \in S$. It is easy to show that $(FG)^* = G^*F^*$. The following lemma shows that effects and elements of $Af(S)$ are "continuous."

4. The Operational Approach

Lemma 4.1. *Let S be a σ-convex structure and let $f \in \mathcal{E}(S)$, $F \in Af(S)$.*

(a) *If $\lim \sigma(s_n, s) = 0$, then $\lim f(s_n) = f(s)$.*

(b) *$f(\langle \lambda_1, \lambda_2, \ldots; t_1, t_2, \ldots \rangle) = \Sigma \lambda_i f(t_i)$.*

(c) *$\sigma(F(s), F(t)) \leq \sigma(s, t)$ and if F is bijective, then $\sigma(F(s), F(t)) = \sigma(s, t)$ for all $s, t \in S$.*

PROOF. (a) Since $\lim \sigma(s_n, s) = 0$, for any $\varepsilon > 0$ there exists an integer $N > 0$ such that $n \geq N$ implies that there exist a $0 < \lambda \leq \varepsilon/(\varepsilon + 2)$, and $t_1, t_2 \in S$ such that $\langle \lambda; t_1, s_n \rangle = \langle \lambda; t_2, s \rangle$. Hence

$$\lambda f(t_1) + (1 - \lambda) f(s_n) = \lambda f(t_2) + (1 - \lambda) f(s)$$

and

$$|f(s) - f(s_n)| = \frac{\lambda}{1 - \lambda} |f(t_2) - f(t_1)| \leq \frac{2\lambda}{1 - \lambda} < \varepsilon.$$

(b) This easily follows from (a).

(c) $\sigma(F(s), F(t)) = \inf\{0 < \lambda \leq 1 : \langle \lambda; s_1, F(s) \rangle = \langle \lambda; t_1, F(t) \rangle, s_1, t_1 \in S\}$

$\leq \inf\{0 < \lambda \leq 1 : F(\langle \lambda; s_1, s \rangle) = F(\langle \lambda; t_1, t \rangle), s_1, t_1 \in S\}$

$\leq \inf\{0 < \lambda \leq 1 : \langle \lambda; s_1, s \rangle = \langle \lambda; t_1, t \rangle, s_1, t_1 \in S\}$

$= \sigma(s, t)$.

The second part of (c) now follows easily. $\qquad\qquad\square$

An effect f corresponds to the result of a measurement. If the system is in the state s, then $f(s)$ is the probability that the effect f is observed. The functional τ corresponds to an effect which is always observed and 0 to an effect which is never observed. The set of effects $\mathcal{E}(S)$ form a convex subset of the linear space S^* and the extreme effects are called *propositions*. This corresponds to the fact that a proposition describes the simplest kind of measurement, namely, one that cannot be decomposed into other measurements. The set of propositions $\mathcal{P}(S) \subseteq S^*$ inherits the order of S^* and so is a poset with least element 0 and greatest element τ. If $f \in \mathcal{P}(S)$, let $f' = \tau - f$. It is clear that $f' \in \mathcal{E}(S)$ and to show that $f' \in \mathcal{P}(S)$ suppose f' is not extreme. Then there exist $g_1 \neq g_2 \in \mathcal{E}(S)$ and $\lambda > 0$ such that $f' = \lambda g_1 + (1 - \lambda) g_2$. Hence

$$f = \tau - \lambda g_1 - (1 - \lambda) g_2 = \lambda(\tau - g_1) + (1 - \lambda)(\tau - g_2).$$

Thus f is not extreme, which is contradiction. It follows that $'$ maps $\mathcal{P}(S)$ into $\mathcal{P}(S)$ and it is clear that $0, \tau \in \mathcal{P}(S)$, $0' = \tau$, $f \leq g$ implies $g' \leq f'$, and $f'' = f$. Thus $\mathcal{P}(S)$ is almost an orthocomplemented poset. In general, $\mathcal{P}(S)$ is not an orthomodular poset.

We shall later consider an important special case in which $\mathscr{P}(S)$ is a complete orthomodular lattice, and shall show that $\mathscr{P}(S)$ reduces to the usual event structure in the Hilbert space case. Thus, we have a connection between the present approach and the quantum logic approach.

We now endow S^* with the weak*-topology; that is, if $f_0 \in S^*$, a neighborhood of f_0 is a set of the form

$$N(f_0; s_1, \ldots, s_n; \varepsilon) = \{ f \in S^* : |f(s_i) - f_0(s_i)| < \varepsilon, \, i = 1, \ldots, n \},$$

where $s_1, \ldots, s_n \in S$, $\varepsilon > 0$. Physically, this is the natural topology for S^* since in this topology a sequence of effects f_n converges to an effect f if and only if $f_n(s) \to f(s)$ for every state s. This topology will be useful for defining observables.

Let x be a physical observable for a quantum system whose states are represented by the convex structure S. Corresponding to a triple (s, x, E), $s \in S$, $E \in \mathscr{B}(\mathbb{R})$ we should be able to associate a number $p(s, x, E) \in [0, 1]$ which gives the probability that x has a value in E when the system is in state s. It then follows that $E \mapsto p(s, x, E)$ is a probability measure on $\mathscr{B}(\mathbb{R})$. It is also evident that

$$p(\langle \lambda_1, \ldots, \lambda_n; s_1, \ldots, s_n \rangle, x, E) = \sum \lambda_i p(s_i, x, E)$$

so $s \mapsto p(s, x, E)$ is an effect. We thus define an *observable* to be a map $x: \mathscr{B}(\mathbb{R}) \to \mathscr{E}(S)$ satisfying:

(i) $x(\mathbb{R}) = \tau$;
(ii) for any mutually disjoint sequence $E_i \in \mathscr{B}(\mathbb{R})$, $x(\cup E_i) = \sum x(E_i)$, where the sum converges in the weak*-topology of S^*.

In short, an observable x is an effect-valued measure on $\mathscr{B}(\mathbb{R})$. For $E \in \mathscr{B}(\mathbb{R})$, $x(E)$ is interpreted as the effect observed when x has a value in E. If $s \in S$, then $E \mapsto x(E)s$ is the probability distribution of x in the state s. One can define observables based on σ-algebras other than $\mathscr{B}(\mathbb{R})$ in the obvious way. To be precise the observables defined above should be called $\mathscr{B}(\mathbb{R})$-based observables, but we shall not use this longer terminology.

4.2. Operations and Instruments

If S is a convex structure, we define $S_+ = \{(\alpha, s) : \alpha \geq 0, s \in S\}$. We define $(\alpha, s) = (\beta, t)$ if $\alpha = \beta \neq 0$ and $s = t$ and $(0, s) = (0, t) = 0$ for all $s, t \in S$. If S is a set of states we call S_+ the set of *generalized states*. If $s \in S$, then s describes the statistical properties of a beam or ensemble with unit intensity, while (α, s) describes the same beam amplified or attenuated to beam strength $\alpha \geq 0$. We make S_+ into a convex structure by defining

$$\langle \lambda_1, \ldots, \lambda_n; (\alpha_1, s_1), \ldots, (\alpha_n, s_n) \rangle$$

$$= \left(\sum \lambda_i \alpha_i, \langle \lambda_1 \alpha_1 / \sum \lambda_i \alpha_i, \ldots, \lambda_n \alpha_n / \sum \lambda_i \alpha_i; s_1, \ldots, s_n \rangle \right) \quad (4.3)$$

if $\sum \lambda_i \alpha_i \neq 0$ and 0 otherwise [if $\alpha_j = 0$, that term is excluded from the expression on the right-hand side of (4.3)]. It is easy to check that S_+ satisfies C1 and C2. We identify an element of the form $(1, s) \in S_+$ with the element $s \in S$ and in this way S can be thought of as a subconvex structure of S_+. Define $S_+^* = \{ f \in Af(S_+, \mathbb{R}) : f(0) = 0 \}$. Actually, this is an abuse of notation but it should not cause confusion.

Lemma 4.2. (a) *If* $f \in S^*$, *then there is a unique extension* $\hat{f} \in S_+^*$. (b) *If* $f \in S_+^*$, *then* $f((\alpha, s)) = \alpha f(s)$ *for all* $(\alpha, s) \in S_+$.

PROOF. (a) For $f \in S^*$, define $\hat{f}((\alpha, s)) = \alpha f(s)$. Then $\hat{f}((1, s)) = f(s)$, so f is an extension of f and clearly $f(0) = 0$. To show that $\hat{f} \in Af(S_+, \mathbb{R})$, if $\sum \lambda_i \alpha_i \neq 0$, then

$$\hat{f}(\langle \lambda_1, \ldots, \lambda_n; (\alpha_1, s_1), \ldots, (\alpha_n, s_n) \rangle)$$

$$= \hat{f} \left(\left(\sum \lambda_i \alpha_i, \langle \lambda_1 \alpha_1 / \sum \lambda_i \alpha_i, \ldots, \lambda_n \alpha_n / \sum \lambda_i \alpha_i; s_1, \ldots, s_n \rangle \right) \right) \quad (4.4)$$

$$= \sum \lambda_i \alpha_i f(s_i) = \sum \lambda_i \hat{f}((\alpha_i, s_i)).$$

If $\sum \lambda_i \alpha_i = 0$, then the two extreme sides of Eq. (4.4) equal 0. To prove uniqueness, suppose $g \in S_+^*$ is an extension of f. If $\alpha = 0$ or 1, then $g((\alpha, s)) = \alpha f(s)$. If $0 < \alpha < 1$, then

$$g((\alpha, s)) = g(\langle \alpha; (1, s), (0, s) \rangle) = \alpha g((1, s)) = \alpha f(s).$$

If $\alpha > 1$, then

$$f(s) = g((1, s)) = g(\langle \alpha^{-1}; (\alpha, s), (0, s) \rangle) = \alpha^{-1} g((\alpha, s)).$$

Hence $g((\alpha, s)) = \hat{f}((\alpha, s))$ for every $(\alpha, s) \in S_+$.

(b) If $f \in S_+^*$, then $f | S \in S^*$. Hence by part (a) $f = (f | S)^\wedge$ and the result follows from the proof of (a). $\qquad \square$

As a particular case of Lemma 4.2, τ has the unique extension $\hat{\tau} \in S_+^*$, where $\hat{\tau}((\alpha, s)) = \alpha$. In the sequel we shall drop the $\wedge$ on $\hat{\tau}$. The functional $\tau \in S_+^*$ gives the beam strength of a beam (or ensemble) in a generalized state (α, s) and is sometimes called the quantum counter [187, 188].

We now consider one of the fundamental concepts of this approach, namely an operation. An operation describes the change of state associated with a measurement which passes only a portion of the beam (or ensemble) tested. If a beam is in state s, then an operation F maps s to a generalized state (α, s'), where s' is the state and $0 \leq \alpha \leq 1$ is the beam strength of the subbeam which passes after the measurement corresponding to F is made. Moreover, it is reasonable that F preserves mixtures. We thus define an *operation* as a map $F \in Af(S_+)$ satisfying $\tau[F(w)] \leq \tau(w)$ for all $w \in S_+$. We denote the set of operations by $\mathcal{O}(S)$.

If $F \in \mathcal{O}(S)$ then for $(\alpha, t) \in S_+$, $F((\alpha, t)) = (\alpha', t')$, so there are two parts to an operation $\alpha \mapsto \alpha'$, $t \mapsto t'$. The part $t \mapsto t'$ represents the "distortion" or "collapse" of the beam or ensemble and $\alpha \mapsto \alpha'$ the reduction of beam strength due to a measurement. For $s \in S$, we interpret $\tau[F(s)]$ as the probability of transmission of a beam in state s conditioned by the operation F.

Associated with any operation F is its effect defined by $f = F^*(\tau) | S$. Since $\tau[F(s)] = (F^*\tau)(s) = f(s)$ for every $s \in S$, the effect f determines the probability of transmission but not the form of the transmitted state. Thus an operation contains more information than its effect. Although an operation determines a unique effect, every effect is so determined by many operations. Indeed, if $f \in \mathcal{E}(S)$ and $s \in S$, define $F: S_+ \to S_+$ by $F(w) = (\hat{f}(w), s)$. To show that $F \in Af(S_+)$ we have

$$F(\langle \lambda_1, \ldots, \lambda_n ; (\alpha_1, s_1), \ldots, (\alpha_n, s_n) \rangle)$$
$$= \left(\hat{f}(\langle \lambda_1, \ldots, \lambda_n ; (\alpha_1, s_1), \ldots, (\alpha_n, s_n) \rangle), s \right)$$
$$= (\Sigma \lambda_i \alpha_i f(s_i), s) = \langle \lambda_1, \ldots, \lambda_n ; (\alpha_1 f(s_1), s), \ldots, (\alpha_n f(s_n), s) \rangle$$
$$= \langle \lambda_1, \ldots, \lambda_n ; (\hat{f}(\alpha_1, s_1), s), \ldots, (\hat{f}(\alpha_n, s_n), s) \rangle$$
$$= \langle \lambda_1, \ldots, \lambda_n ; F((\alpha_1, s_1)), \ldots F((\alpha_n, s_n)) \rangle.$$

Moreover, $F \in \mathcal{O}(S)$ since if $w = (\alpha, t)$ we have

$$\tau[F(w)] = \tau[(\hat{f}(w), s)] = \hat{f}(w) = \alpha f(t) \leq \alpha = \tau(w).$$

Finally, F determines f since for any $t \in S$ we have

$$(F^*\tau)(t) = \tau[F(t)] = \tau[(f(t), s)] = f(t).$$

Suppose a measurement is made on a system in the generalized state w. If the measurement results in a value in $E \in \mathcal{B}(\mathbb{R})$, denote the output generalized state conditioned by this result by $\mathcal{I}(E)w$. By our previous interpretation of operations we conclude that $\mathcal{I}(E) \in \mathcal{O}(S)$. Motivated by this, we define an *instrument* on a σ-convex structure S to be a map $\mathcal{I}: \mathcal{B}(\mathbb{R}) \to Af(S_+)$ satisfying:

(i) $\tau[\mathcal{I}(\mathbb{R})w] = \tau(w)$ for all $w \in S_+$;

(ii) for mutually disjoint $E_i \in \mathcal{B}(\mathbb{R})$ and $s \in S$, $\beta = \Sigma \tau[\mathcal{I}(E_i)s] < \infty$, and if $\beta \neq 0$, then

$$\mathcal{I}(\cup E_i)(\alpha, s) = (\alpha\beta, \langle \alpha_1/\beta, \alpha_2/\beta, \ldots; s_1, s_2, \ldots \rangle),$$

where $\mathcal{I}(E_i)s = (\alpha_i, s_i)$ and if $\beta = 0$, then $\mathcal{I}(\cup E_i) = 0$.

Condition (ii) has the following interpretation. If the input state is s and an instrument $\mathcal{I}$ determines a value in the set $\cup E_i$, then the output generalized state is a mixture of the generalized states $\mathcal{I}(E_1)s, \mathcal{I}(E_2)s, \ldots$, in proportion

$$\tau[\mathcal{I}(E_1)s]/\Sigma \tau[\mathcal{I}(E_i)s], \qquad \tau[\mathcal{I}(E_2)s]/\Sigma \tau[\mathcal{I}(E_i)s], \ldots,$$

respectively.

Lemma 4.3. Let $\mathcal{I}: \mathcal{B}(\mathbb{R}) \to Af(S_+)$ be an instrument. (a) $E \mapsto \tau[\mathcal{I}(E)w]$ is a measure for every $w \in S_+$. (b) $\mathcal{I}(E) \in \mathcal{O}(S)$ for every $E \in \mathcal{B}(\mathbb{R})$.

PROOF. (a) Since $w \mapsto \tau[\mathcal{I}(E)w]$ is affine, it follows from Lemma 4.2 (b) that $\tau[\mathcal{I}(E)(\alpha, s)] = \alpha\tau[\mathcal{I}(E)s]$ for every $\alpha \geq 0$, $s \in S$. Using the notation of (ii) we have for $w = (\alpha, s)$

$$\tau[\mathcal{I}(\cup E_i)w] = \alpha\beta = \Sigma \alpha\tau[\mathcal{I}(E_i)s]$$

$$= \Sigma \tau[\mathcal{I}(E_i)(\alpha, s)] = \Sigma \tau[\mathcal{I}(E_i)w].$$

(b) For any $E \in \mathcal{B}(\mathbb{R})$, $w \in S_+$ we have

$$\tau[\mathcal{I}(E)w] \leq \tau[\mathcal{I}(E)w] + \tau[\mathcal{I}(E')w]$$

$$= \tau[\mathcal{I}(\mathbb{R})w] = \tau(w),$$

where E' is the complement of E. $\qquad \square$

Hence, an instrument is an operation-valued measure. As with observables we define instruments based on σ-algebras other than $\mathcal{B}(\mathbb{R})$ in the obvious way.

Corresponding to the relation between operations and effects there is a relation between instruments and observables. If $\mathcal{I}$ is an instrument, then there exists a unique observable x such that $\tau[\mathcal{I}(E)s] = x(E)s$ for all $s \in S$, $E \in \mathcal{B}(\mathbb{R})$. Indeed, we now show that $x(E)s = \tau[\mathcal{I}(E)s]$ defines an observable. Since $\mathcal{I}(E)$ is an operation, we have shown previously that $s \mapsto \tau[\mathcal{I}(E)s]$ is an effect. Also $x(\mathbb{R})s = \tau[\mathcal{I}(\mathbb{R})s] = \tau(s)$. To show that $x(\cup E_i) = \Sigma x(E_i)$ for mutually disjoint $E_i \in \mathcal{B}(\mathbb{R})$, suppose that $\mathcal{I}(E_i)s = (\alpha_i, t_i)$. Then

$$\Sigma x(E_i)s = \Sigma \tau[\mathcal{I}(E_i)s] = \Sigma \alpha_i$$

$$= \tau(\Sigma \alpha_i, \langle \alpha_1/\Sigma \alpha_i, \alpha_2/\Sigma \alpha_i, \ldots; t_1, t_2, \ldots \rangle)$$

$$= \tau[\mathcal{I}(\cup E_i)s] = x(\cup E_i)s.$$

Thus the observable x is defined by $x = \mathcal{I}(\cdot)^*\tau \,|\, S$. Conversely, an observable x is determined in the above manner by at least one instrument. Indeed, let $s \in S$ be fixed and define $\mathcal{I}(E): S_+ \to S_+$ by $\mathcal{I}(E)(\alpha, t) = (\alpha x(E)t, s)$. We leave it to the reader to verify that $\mathcal{I}(E) \in Af(S_+)$. Condition (i) holds since

$$\tau[\mathcal{I}(\mathbb{R})(\alpha, t)] = \tau[(\alpha x(\mathbb{R})t, s)] = \tau((\alpha, s)) = \alpha = \tau((\alpha, t)).$$

Condition (ii) holds since for mutually disjoint $E_i \in \mathcal{B}(\mathbb{R})$ we have

$$\mathcal{I}(\cup E_i)(\alpha, t) = (\alpha x(\cup E_i)t, s) = (\alpha \Sigma x(E_i)t, s)$$
$$= (\alpha \Sigma x(E_i)t, \langle x(E_1)t / \Sigma x(E_i)t, \dots, s, s, \dots \rangle).$$

Finally, $\tau[\mathcal{I}(E)t] = \tau[(x(E)t, s)] = x(E)t$.

More complicated instruments can be defined which determine an observable x. For example, if $s_i \in S$ and $E_i \in \mathcal{B}(\mathbb{R})$ are mutually disjoint, $i = 1, \dots, n$, define

$$\mathcal{I}(E)(\alpha, t) = \left(\alpha \Sigma x(E \cap E_i)t, \left\langle \frac{x(E \cap E_1)t}{\Sigma x(E \cap E_i)t}, \dots, \frac{x(E \cap E_n)t}{\Sigma x(E \cap E_i)t}; s_1, \dots, s_n \right\rangle \right). \tag{4.5}$$

If $F, G \in Af(S_+)$, then their composition $FG \in Af(S_+)$. Thus $Af(S_+)$ forms a semigroup under composition. Similarly, $\mathcal{O}(S)$ is a semigroup under composition. If $\mathcal{I}$ and $\mathcal{J}$ are instruments on $(\mathbb{R}, \mathcal{B}(\mathbb{R}))$ and there exists an instrument $\mathcal{K}$ on $(\mathbb{R}^2, \mathcal{B}(\mathbb{R}^2))$ such that $\mathcal{K}(E \times F) = \mathcal{I}(E)\mathcal{J}(F)$ for all $E, F \in \mathcal{B}(\mathbb{R})$, we call $\mathcal{K}$ the *composition of $\mathcal{I}$ following $\mathcal{J}$* and use the notation $\mathcal{K} = \mathcal{I} \circ \mathcal{J}$. For example, let Z be the integers and let 2^Z be the power set for Z. If $\mathcal{I}$ and $\mathcal{J}$ are instruments based on $(Z, 2^Z)$, then the composition of $\mathcal{I}$ following $\mathcal{J}$ exists and equals

$$\mathcal{K}(M)(\alpha, s) = \left(\alpha \Sigma \alpha_{mn}, \left\langle \frac{\alpha_{00}}{\Sigma \alpha_{mn}}, \dots; s_{00}, \dots \right\rangle \right), \tag{4.6}$$

where $\mathcal{I}(m)\mathcal{J}(n)s = (\alpha_{mn}, s_{mn})$. In general, it is not known whether the composition $\mathcal{K}$ of two arbitrary instruments exists although it is easily seen that $\mathcal{K}$ is unique if it exists.

We shall later show that the composition always exists in what we call a total σ-convex structure. Let $\mathcal{I}$ and $\mathcal{J}$ be instruments which determine observables x and y, respectively. The observable $E \mapsto \mathcal{J}(\mathbb{R})^*[x(E)]$ is called the observable x *conditioned by the measurement of y with instrument $\mathcal{J}$*. If $\mathcal{I} \circ \mathcal{J}$ exists, then the observable z based on $\mathcal{B}(\mathbb{R}^2)$ and defined as $z(M) = [\mathcal{I} \circ \mathcal{J}(M)]^*(\tau)$ is called the *joint distribution of $\mathcal{I}$ following $\mathcal{J}$*. Thus, in a total σ-convex structure the joint distribution of $\mathcal{I}$ following $\mathcal{J}$ always exists. We now show that the joint distribution gives the desired marginal distributions.

Lemma 4.4. $z(\mathbb{R} \times F) = y(F), z(E \times \mathbb{R}) = \mathcal{J}(\mathbb{R})^*[x(E)]$.

PROOF. Since $[\mathcal{J}(E)\mathcal{J}(F)]^* = \mathcal{J}(F)^*\mathcal{J}(E)^*$ we have

$$z(E \times F) = [\mathcal{J} \circ \mathcal{J}(E \times F)]^*(\tau) = [\mathcal{J}(E)\mathcal{J}(F)]^*(\tau)$$
$$= \mathcal{J}(F)^*\mathcal{J}(E)^*\tau.$$

Letting $E = \mathbb{R}$, we have

$$z(\mathbb{R} \times F) = \mathcal{J}(F)^*\mathcal{J}(\mathbb{R})^*\tau = \mathcal{J}(F)^*\tau = y(F).$$

Letting $F = \mathbb{R}$, we have

$$z(E \times \mathbb{R}) = \mathcal{J}(\mathbb{R})^*\mathcal{J}(E)^*\tau = \mathcal{J}(\mathbb{R})^*[x(E)]. \qquad \square$$

4.3. Examples

In this section we consider the abstract concepts of Sections 4.1 and 4.2 in two concrete cases and show that they have a familiar form. The concrete cases will be the standard ones of classical probability theory and Hilbert space quantum mechanics.

Classical Probability Theory

Let (Ω, Σ) be a measurable space and let S be the set of probability measures on (Ω, Σ). With the usual definition of convex combinations, S is a convex structure. It will follow from a later result (see Section 4.5) that σ is a metric on S and that S is complete under σ. Thus, S becomes a σ-convex structure. Let S_1' be the set of random variables f on (Ω, Σ) satisfying $0 \le f \le 1$. For $f \in S_1'$, $\mu \in S$, define $f(\mu) = \int f d\mu$. It is clear that $f: S \to \mathbb{R}$ is an affine functional and hence $S_1' \subseteq \mathcal{E}(S)$ [in general $S_1' \ne \mathcal{E}(S)$]. The functional τ is represented by the function 1 which is identically 1. If $f \in S_1'$, then it is easy to show that f is a proposition if and only if f has only values 0 and 1; that is, f is a characteristic function.

Let g be a random variable on (Ω, Σ) and for $E \in \mathcal{B}(\mathbb{R})$ define $x_g(E) = \chi_{g^{-1}(E)}$. Then $x_g(E) \in S_1'$ and it is easy to show that x_g is an observable. Thus, the random variables of classical probability theory are represented in a natural way by observables. However, even in this example the concept of observable is more general than that of a random variable since some observables do not come from random variables. In fact, the observables that do come from random variables are precisely the proposition-valued observables. A general observable is what one might call a "fuzzy" random variable [5, 273].

We identify elements $(\alpha, \mu) \in S_+$ with measures $\alpha\mu$. In this way S_+ can be thought of as the set of bounded nonnegative measures on (Ω, Σ). We then have that $\tau(w)=w(\Omega)$ for all $w\in S_+$. If we let S' be the set of random variables on (Ω,Σ), then $S'\subseteq\mathscr{E}(S_+)$ [in general, $S'\neq\mathscr{E}(S_+)$]. We now consider an important class of operations. If $A\in\Sigma$, define $F_A:S_+\to S_+$ by $F_A(w)=w(A\cap(\cdot))$. It is clear that $F_A\in Af(S_+)$ and since

$$\tau[F_A(w)]=w(A)\leq w(\mathbb{R})=\tau(w)$$

we have $F_A\in\mathcal{O}(S)$. In particular, if $\mu\in S$ and $\mu(A)\neq 0$, then $F_A(\mu)/\mu(A)$ $=\mu(A\cap(\cdot))/\mu(A)$ is the conditional probability of μ given A. This connection between operations and conditional probabilities will be exploited in the next section.

Let g be a random variable and let x_g be the corresponding observable. We now construct an instrument corresponding to x_g. For each $E\in\mathscr{B}(\mathbb{R})$ define $\mathscr{I}_g(E)\in\mathcal{O}(S)$ by

$$\mathscr{I}_g(E)w=F_{g^{-1}(E)}(w)=w[g^{-1}(E)\cap(\cdot)].\qquad(4.7)$$

It is not hard to show that $\mathscr{I}_g$ is an instrument. Moreover, $\mathscr{I}_g$ corresponds to the observable x_g since

$$\tau[\mathscr{I}_g(E)w]=w[g^{-1}(E)]=\int_{g^{-1}(E)}dw$$

$$=\int\chi_{g^{-1}(E)}dw=x_g(E)w.$$

Now let f and g be random variables and let $\mathscr{I}_f$ and $\mathscr{I}_g$ be the instruments just defined. We shall show later that $\mathscr{I}_f\circ\mathscr{I}_g$ exists. Let $z_{f,g}(M)=$ $[\mathscr{I}_f\circ\mathscr{I}_g(M)]^*(\tau)$ be the joint distribution of $\mathscr{I}_f$ following $\mathscr{I}_g$. If $\mu\in S$, then $M\mapsto z_{f,g}(M)\mu$ is a probability measure on $\mathscr{B}(\mathbb{R}^2)$ and for $E,F\in\mathscr{B}(\mathbb{R})$ we have

$$z_{f,g}(E\times F)=\tau[\mathscr{I}_f\circ\mathscr{I}_g(E\times F)\mu]=\tau[\mathscr{I}_f(E)\mathscr{I}_g(F)\mu]$$

$$=\tau[\mu(f^{-1}(E)\cap g^{-1}(F)\cap(\cdot))]=\mu[f^{-1}(E)\cap g^{-1}(F)].$$

Thus $z_{f,g}(\cdot)\mu$ is the joint probability distribution of the random variables f,g on the probability space (Ω,Σ,μ). Of course, in this example $z_{f,g}=z_{g,f}$. In the next example this need not be the case.

Hilbert Space Quantum Mechanics

Let $\mathcal{H}$ be a complex separable Hilbert space and let S be the set of density operators on $\mathcal{H}$; that is, S is the set of positive trace class operators of trace 1 on $\mathcal{H}$. Again, it can be shown that S is a σ-convex structure. Let S_1' be the set of bounded self-adjoint operators A on $\mathcal{H}$ satisfying $0\leq A\leq I$. For $A\in S_1'$, $s\in S$, define $A(s)=\text{tr}(As)$. Then $A:S\to\mathbb{R}$ is an affine functional and since τ is represented by the identity operator I, we have

$S_1' \subseteq \mathcal{E}(S)$. It can be shown [195] that in this case $S_1' = \mathcal{E}(S)$. The following lemma shows that the operational approach generalizes the usual quantum mechanical framework.

Lemma 4.5 (Davies [54]). *An effect A is a proposition if and only if A is a projection operator.*

PROOF. Let P be a projection operator and suppose that $A_1, A_2 \in \mathcal{E}(S)$, $0 < \lambda < 1$, and $P = \lambda A_1 + (1-\lambda)A_2$. If $P\phi = 0$, then

$$\lambda \langle A_1\phi, \phi \rangle + (1-\lambda) \langle A_2\phi, \phi \rangle = 0.$$

Since $A_1, A_2 \geq 0$ we have $\langle A_1\phi, \phi \rangle = \langle A_2\phi, \phi \rangle = 0$ and hence, $A_1\phi = A_2\phi = 0$. Moreover, since

$$I - P = \lambda(I - A_1) + (1-\lambda)(I - A_2),$$

$P\psi = \psi$ implies that $(I - A_1)\psi = (I - A_2)\psi = 0$. It follows that $A_1 = A_2 = P$ and P is an extreme effect. Hence, P is a proposition. Conversely, suppose that $A \in \mathcal{E}(S)$ is not a projection. Then by the spectral theorem there exists a $c \in \sigma(A)$ with $0 < c < 1$. Let f be a continuous function on $[0,1]$ such that $0 \leq t \pm f(t) \leq 1$ for all $t \in [0,1]$ and $f(c) \neq 0$. By the spectral theorem if $A_1 = A + f(A)$ and $A_2 = A - f(A)$ we conclude that $A_1, A_2 \in \mathcal{E}(S)$ and $A_1, A_2 \neq A$. Since $A = \frac{1}{2}A_1 + \frac{1}{2}A_2$, A is not an extreme effect and hence, not a proposition. $\qquad\square$

Since an observable is an effect-valued measure, in this case observables are represented by POV (positive operator-valued) measures. Thus, observables in this approach are more general than the traditional observables which are represented by PV (projection-valued) measures. For example, let Q_i be mutually orthogonal projections on $\mathcal{H}$ such that $\Sigma Q_i = I$ and let a_i be distinct real numbers, $i = 1, 2, \dots$. Then the map $x(E) = \Sigma\{Q_i : a_i \in E\}$ gives a PV measure on $\mathcal{B}(\mathbb{R})$ and thus represents a traditional observable with spectrum $\{a_i : i = 1, 2, \dots\}$. Now suppose that P_j is another set of mutually orthogonal projections such that $\Sigma P_j = I$. Then the map

$$x_1(E) = \sum_{j=1}^{\infty} \sum_{a_i \in E} P_j Q_i P_j \tag{4.8}$$

is a POV measure (and hence an observable), but is not a PV measure unless the Q_i's and P_j's commute.

We identify elements $(\alpha, s) \in S_+$ with operators αs. Thus S_+ can be thought of as the set of positive trace-class operators. If S' denotes the set of bounded self-adjoint operators on $\mathcal{H}$ then it follows from our previous work that $S' = \mathcal{E}(S_+) - \mathcal{E}(S_+)$. Moreover, $\tau(w) = \text{tr}\, w$ for all $w \in S_+$. It is easy to show that if P is a projection, then the map $F_P(w) = PwP$ is an operation. Now let x be the observable $x(E) = \Sigma\{Q_i : a_i \in E\}$ defined in the previous paragraph. Then for $E \in \mathcal{B}(\mathbb{R})$, $w \in S_+$,

$$\mathcal{I}_x(E)w = \sum \{Q_i w Q_i : a_i \in E\} \tag{4.9}$$

defines an instrument. Moreover, $\mathcal{I}_x$ corresponds to x since

$$\tau[\mathcal{I}_x(E)w] = \text{tr}\Big(\sum \{Q_i w Q_i : a_i \in E\}\Big)$$
$$= \sum \{\text{tr}(Q_i w) : a_i \in E\} = \text{tr}\Big(\sum \{Q_i w : a_i \in E\}\Big) = x(E)w.$$

Now let $x(E) = \Sigma\{Q_i : a_i \in E\}$ and $y(E) = \Sigma\{P_i : b_i \in E\}$ be observables (in fact, PV measures) and let $\mathcal{I}_x$ and $\mathcal{I}_y$ be the corresponding instruments defined above. Let

$$z_{x,y}(M) = [\mathcal{I}_x \circ \mathcal{I}_y(M)]^*(\tau)$$

be the joint distribution of $\mathcal{I}_x$ following $\mathcal{I}_y$. If $s \in S$, then $M \mapsto z_{x,y}(M)s$ is a probability measure on $\mathcal{B}(\mathbb{R}^2)$ and for $E, F \in \mathcal{B}(\mathbb{R})$ we have

$$z_{x,y}(E \times F)s = \tau[\mathcal{I}_x \circ \mathcal{I}_y(E \times F)s] = \tau[\mathcal{I}_x(E)\mathcal{I}_y(F)s]$$
$$= \tau\Big[\mathcal{I}_x(E) \sum \{P_i s P_i : b_i \in F\}\Big]$$
$$= \tau\Big[\sum \{Q_j P_i s P_i Q_j : (a_i, b_j) \in E \times F\}\Big]$$
$$= \sum \{\text{tr}(Q_j P_i s P_i Q_j) : (a_i, b_j) \in E \times F\}.$$

It turns out that

$$z_{x,y}(M) = \sum \{P_i Q_j P_i : (a_i, b_i) \in M\}. \tag{4.10}$$

Indeed, by the cyclicity of the trace we have

$$\sum \{P_i Q_j P_i s : (a_i, b_j) \in E \times F\}$$
$$= \sum \{\text{tr}(P_i Q_j P_i s) : (a_i, b_j) \in E \times F\}$$
$$= \sum \{\text{tr}(Q_j^2 P_i s P_i) : (a_i, b_j) \in E \times F\}$$
$$= \sum \{\text{tr}(Q_j P_i s P_i Q_j) : (a_i, b_j) \in E \times F\} = z_{x,y}(E \times F)s.$$

The next lemma shows that results are independent of the order of measuring x and y if and only if x and y are compatible.

Lemma 4.6 (Davies [54]). $z_{x,y} = z_{y,x}$ if and only if P_i and Q_j commute for every i, j.

PROOF. If P_i and Q_j commute for all i,j, then $P_iQ_jP_i = Q_jP_iQ_j$ for all i,j. It follows from (4.10) that $z_{x,y} = z_{y,x}$. Conversely, if $z_{x,y} = z_{y,x}$ then by (4.10) $P_iQ_jP_i = Q_jP_iQ_j$ for all i,j. Then

$$P_i = \sum_k P_iQ_kP_i = \sum_k Q_kP_iQ_k.$$

Hence,

$$Q_jP_i = \sum_k Q_jQ_kP_iQ_k = Q_jP_iQ_j = \sum Q_kP_iQ_kQ_j = P_iQ_j. \qquad \square$$

Moreover, one can show that $z_{x,y}$ is not a PV measure unless x and y are compatible [54].

4.4. Conditional Expectations

Let S be a σ-convex structure representing the set of states of a physical system. We have seen that S^* is a real linear space, and if we equip S^* with the weak*-topology, then S^* becomes a topological linear space. We have seen that if $F \in Af(S)$, then $F^*: S^* \to S^*$ is a linear map and it is clear that F^* is continuous; that is, $F^* \in \mathcal{L}(S^*)$. Similarly, if $F \in Af(S_+)$, then $F^*: S_+^* \to S_+^*$ is a continuous linear map; that is, $F^* \in \mathcal{L}(S_+^*)$. We use the notation $\mathcal{E}(S)^{\hat{}} = \{\hat{f}: f \in \mathcal{E}(S)\}$. If $F \in \mathcal{O}(S)$, then $F^* \in \mathcal{L}(S_+^*)$ has the additional property $F^*: \mathcal{E}(S)^{\hat{}} \to \mathcal{E}(S)^{\hat{}}$; that is, F^* maps effects into effects. Indeed, if $f \in \mathcal{E}(S)$, then for every $s \in S$ we have $(F^*\hat{f})(s) = \hat{f}(F(s)) \in [0, 1]$. Now we have interpreted $F \in \mathcal{O}(S)$ as giving the change of state conditioned on a measurement corresponding to the operation F. We may then interpret F^* as giving the change of effects conditioned on a measurement corresponding to F. If $\mathcal{I}$ is an instrument, then $\mathcal{I}(E)^*$, $E \in \mathcal{B}(\mathbb{R})$, can be interpreted as giving the change of effects conditioned on the result of a measurement lying in E. Thus, in a certain sense, $\mathcal{I}(\cdot)^*$ corresponds to a conditional expectation of $\mathcal{I}$. In fact, $\mathcal{I}(\cdot)^*$ is a generalization of the usual conditional expectation for random variables in classical probability theory.

We now examine the form of $\mathcal{I}(\cdot)^*$ for the instruments in the examples of Section 4.3. Let $\mathcal{I}_g(E)w = w[g^{-1}(E) \cap (\cdot)]$ be the instrument defined for the classical probability example. If $f \in S'$, then for all $w \in S_+$ we have

$$\mathcal{I}_g(E)^*f(w) = f[\mathcal{I}_g(E)w] = \int f d[\mathcal{I}(E)w]$$

$$= \int f(\omega)\,dw[g^{-1}(E) \cap \{\omega\}] = \int_{g^{-1}(E)} f\,dw.$$

Let Σ_0 be the sub-σ-algebra of Σ generated by $\{g^{-1}(E): E \in \mathcal{B}(\mathbb{R})\}$ and let $E(f|\Sigma_0)$ be the usual conditional expectation of f given Σ_0. We now show that $\mathcal{I}_g(E)^*f = E(f|\Sigma_0)\chi_{g^{-1}(E)}$. Indeed, for any $w \in S_+$ we have

$$E(f|\Sigma_0)\chi_{g^{-1}(E)}(w) = \int E(f|\Sigma_0)\chi_{g^{-1}(E)}\,dw$$
$$= \int_{g^{-1}(E)} E(f|\Sigma_0)\,dw = \int_{g^{-1}(E)} f\,dw = \mathcal{I}_g(E)^*f(w).$$

Thus $\mathcal{I}_g(E)^*f$ is closely related to the usual conditional expectation $E(f|\Sigma_0)$ of f given g.

Now let $\mathcal{I}_x(E)w = \Sigma\{Q_i w Q_i : a_i \in E\}$ be the instrument defined in the Hilbert space example of Section 4.3. We now show $\mathcal{I}_x(E)^*A = \Sigma\{Q_i A Q_i : a_i \in E\}$ for all $A \in S'$.

$$\mathcal{I}_x(E)^*A(w) = A[\mathcal{I}_x(E)w] = \mathrm{tr}[A\mathcal{I}_x(E)w]$$
$$= \mathrm{tr}\Big(A\sum\{Q_i w Q_i : a_i \in E\}\Big) = \sum\{\mathrm{tr}(A Q_i w Q_i) : a_i \in E\}$$
$$= \mathrm{tr}\Big(\sum\{Q_i A Q_i w : a_i \in E\}\Big) = \Big(\sum\{Q_i A Q_i : a_i \in E\}\Big)(w).$$

It turns out that S_+^* is too large mathematically. Moreover, the only elements of S_+^* which are physically significant are the extensions of effects; that is, the members of $\mathcal{E}(S)^\frown$. Hence, we introduce the subspace $S'_+ \subseteq S_+^*$ defined by $S'_+ = \mathrm{span}\,\mathcal{E}(S)^\frown$. It is easy to show that any $f \in S'_+$ has the form $f = \alpha f_1 - \beta f_2$, where $\alpha, \beta \geq 0$, $f_1, f_2 \in \mathcal{E}(S)^\frown$.

If $\mathcal{I}$ is an instrument, we define the *dual instrument* $\mathcal{I}^*: \mathfrak{B}(\mathbb{R}) \to \mathcal{L}(S'_+)$ by $\mathcal{I}^*(E) = \mathcal{I}(E)^*|S'_+$. The following lemma summarizes the properties of dual instruments.

Lemma 4.7. *If $\mathcal{I}$ is an instrument, then $\mathcal{I}^*$ has the following properties*:

(a) $\mathcal{I}^*(E): \mathcal{E}(S)^\frown \to \mathcal{E}(S)^\frown$ *for every* $E \in \mathfrak{B}(\mathbb{R})$;

(b) $\mathcal{I}^*(\mathbb{R})\tau = \tau$;

(c) *for mutually disjoint* $E_i \in \mathfrak{B}(\mathbb{R})$ *and* $f \in S'_+$ $\mathcal{I}^*(\cup E_i)f = \Sigma\mathcal{I}^*(E_i)f$ *where the sum converges in the weak*-topology of* S'_+.

PROOF. (a) has already been demonstrated and (b) follows directly from the definition. (c) For any $s \in S$, $f \in \mathcal{E}(S)$, we have

$$\mathcal{I}^*(\cup E_i)\hat{f}(s) = \hat{f}(\mathcal{I}(\cup E_i)s)$$
$$= \hat{f}((\beta, \langle \alpha_1/\beta, \alpha_2/\beta, \ldots; s_1, s_2, \ldots\rangle)),$$

where $\beta = \Sigma\tau[\mathcal{I}(E_i)s]$ and $\mathcal{I}(E_i)s = (\alpha_i, s_i)$. It follows from Lemmas 2.1(b) and 3.1(b) that

$$\mathcal{I}^*(\cup E_i)\hat{f}(s) = \sum \alpha_i f(s_i) = \sum \hat{f}((\alpha_i, s_i))$$
$$= \sum \hat{f}(\mathcal{I}(E_i)s) = \sum \mathcal{I}^*(E_i)\hat{f}(s).$$

Replacing s by $w = (\alpha, s)$ gives the result. $\square$

We shall need slightly more general maps than those in $\mathfrak{L}(S'_+)$. We say that a linear map T on S'_+ is *monotonically weak*-continuous* if for any monotone increasing sequence $\{f_n\}\subseteq S'_+$ converging to an $f\in S'_+$ in the weak*-topology, we have Tf_n converging to Tf in the weak*-topology. We denote the set of monotonically weak*-continuous linear maps on S'_+ by $\mathfrak{L}_0(S'_+)$. Following Davies [54] we define an *expectation* to be a map $\varepsilon:\mathfrak{B}(\mathbb{R})\to\mathfrak{L}_0(S'_+)$ satisfying:

(a) $\varepsilon(E):\mathfrak{E}(S)^\wedge\to\mathfrak{E}(S)^\wedge$ for every $E\in\mathfrak{B}(\mathbb{R})$;
(b) $\varepsilon(\mathbb{R})\tau=\tau$;
(c) for any mutually disjoint sequence $E_i\in\mathfrak{B}(\mathbb{R})$ and $f\in S'_+$, $\varepsilon(\cup E_i)f=\Sigma\varepsilon(E_i)f$, where the sum converges in the weak*-topology of S'_+.

We see that expectations are slightly more general than dual instruments. An expectation need not have the two crucial properties of a classical conditional expectation [48]. Cycon and Hellwig [48] call these two properties the averaging property and the mean value property.

If $\mathcal{I}$ is an instrument, we denote by $S'_\mathcal{I}$ the weak*-closure of the span of the range of $\mathcal{I}^*(\cdot)\tau$ in S'_+; that is,

$$S'_\mathcal{I}=\overline{\text{span}}\{\mathcal{I}^*(E)\tau:E\in\mathfrak{B}(\mathbb{R})\}. \tag{4.11}$$

If $x_\mathcal{I}$ is the observable corresponding to $\mathcal{I}$ then we have

$$S'_\mathcal{I}=\overline{\text{span}}\{x_\mathcal{I}(E):E\in\mathfrak{B}(\mathbb{R})\}. \tag{4.12}$$

Notice from (4.12) that $S'_\mathcal{I}$ is analogous to the sub-σ-algebra generated by a random variable. An expectation ε has the *averaging property* with respect to $\mathcal{I}$ if $\varepsilon(E)f\in S'_\mathcal{I}$ for every $E\in\mathfrak{B}(\mathbb{R})$ and $f\in S'_+$. An expectation ε has the *mean value property* with respect to $\mathcal{I}$ and $s\in S$ if for every $f\in S'_+$, $E\in\mathfrak{B}(\mathbb{R})$ we have

$$\varepsilon(E)f(s)=f[\mathcal{I}(E)s]. \tag{4.13}$$

If $f\in\mathfrak{E}(S)^\wedge$, both sides of Eq. (4.13) are interpreted as the probability of observing the effect f in the state s conditioned on the result of a measurement lying in E. We can rewrite (4.13) in the form

$$\varepsilon(E)f(s)=\mathcal{I}^*(E)f(s) \tag{4.14}$$

for all $f\in S'_+$, $E\in\mathfrak{B}(\mathbb{R})$. Thus (4.13) says that $\varepsilon(E)$ and $\mathcal{I}^*(E)$ agree for every $E\in\mathfrak{B}(\mathbb{R})$ and $f\in S'_+$ at the state s.

It is clear that the dual instrument $\mathcal{I}^*$ is an expectation satisfying the mean value property with respect to $\mathcal{I}$ and s for every $s\in S$. However, $\mathcal{I}^*$ need not satisfy the averaging property with respect to $\mathcal{I}$. This will be seen in a later example (Section 4.6). If $\mathcal{I}$ is an instrument and $s\in S$, then an expectation ε is a *conditional expectation* with respect to $(\mathcal{G}, s)$ if ε satisfies both the averaging and mean value properties.

We say that $s \in S$ is *effective* with respect to $\mathcal{G}$ if $\mathcal{G}^*(E)\tau(s) = 0$ implies that $\mathcal{G}^*(E)\tau = 0$. This is equivalent to $\mathcal{G}(E)s = 0$ implies $\mathcal{G}(E)t = 0$ for all $t \in S$.

Theorem 4.8. *If s is effective with respect to $\mathcal{G}$, then a conditional expectation exists with respect to $(\mathcal{G}, s)$.*

PROOF. Let s be effective with respect to $\mathcal{G}$. For $f \in S'_+$, let μ_f be the finite measure on $\mathcal{B}(\mathbb{R})$ defined by $\mu_f(E) = f[\mathcal{G}(E)s]$ for all $E \in \mathcal{B}(\mathbb{R})$. If $\mu_\tau(E) = 0$, then $\tau[\mathcal{G}(E)s] = 0$ so $\mathcal{G}(E)s = 0$ and hence $\mu_f(E) = 0$. Thus μ_f is absolutely continuous relative to μ_τ for all $f \in S'_+$. By the Radon–Nikodym theorem there exists a real-valued function $h_f \in L^1(\mathbb{R}, \mathcal{B}(\mathbb{R}), \mu_\tau)$ such that $\mu_f(E) = \int_E h_f \, d\mu_\tau$ for every $E \in \mathcal{B}(\mathbb{R})$. For $w \in S_+$ define $[\varepsilon(E)f](w) = \int_E h_f(\lambda)\tau[\mathcal{G}(d\lambda)w]$. This map is well defined since the effectiveness of s implies that the measure $\tau[\mathcal{G}(\cdot)w]$ is absolutely continuous relative to μ_τ and hence $h_f = g$ a.e. $[\mu_\tau]$ implies that $h_f = g$ a.e. $[\tau(\mathcal{G}(\cdot)w)]$. We first show that $\varepsilon(E): S'_+ \to S'_+$ for every $E \in \mathcal{B}(\mathbb{R})$. If $f \in S'_+$, then

$$\varepsilon(E)f(\langle \lambda_1, \ldots, \lambda_n; w_1, \ldots, w_n \rangle)$$

$$= \int_E h_f(\lambda)\tau[\mathcal{G}(d\lambda)\langle \lambda_1, \ldots, \lambda_n; w_1, \ldots, w_n \rangle]$$

$$\int_E h_f(\lambda)\tau[\langle \lambda_1, \ldots, \lambda_n; \mathcal{G}(d\lambda)w_1, \ldots, \mathcal{G}(d\lambda)w_n \rangle]$$

$$= \int_E h_f(\lambda) \sum \lambda_i \tau[\mathcal{G}(d\lambda)w_i] = \sum \lambda_i \varepsilon(E)f(w_i).$$

It is now clear that $\varepsilon(E): \mathcal{E}(S)^\wedge \to \mathcal{E}(S)^\wedge$ and that $\varepsilon(E)$ is linear. Moreover, $\varepsilon(\mathbb{R})\tau = \tau$ since for $t \in S$ we have

$$\varepsilon(\mathbb{R})\tau(t) = \int h_\tau(\lambda)\tau[\mathcal{G}(d\lambda)t] = \tau[\mathcal{G}(\mathbb{R})t] = 1.$$

We now show that $\varepsilon(E)$ is monotonically weak*-continuous. Let $f_n \in S'_+$ be a monotone increasing sequence converging to $f \in S'_+$ in the weak*-topology. Since $f_{n+1} \geq f_n$, we have $\mu_{f_{n+1}}(E) \geq \mu_{f_n}(E)$ for every $E \in \mathcal{B}(\mathbb{R})$. Hence $\int_E h_{f_{n+1}} d\mu_\tau \geq \int_E h_{f_n} d\mu_\tau$ and we conclude that $h_{f_{n+1}} \geq h_{f_n}$ a.e. $[\mu_\tau]$. It follows from effectiveness that $h_{f_{n+1}} \geq h_{f_n}$ a.e. $[\tau(\mathcal{G}(\cdot)w)]$ for every $w \in S_+$. Moreover, since $f_n \uparrow f$, we have $f_n(\mathcal{G}(E)s) \uparrow f(\mathcal{G}(E)s)$ and hence $\mu_{f_n}(E) \uparrow \mu_f(E)$ for

4. The Operational Approach

every $E \in \mathfrak{B}(\mathbb{R})$. Applying the monotone convergence theorem, we have

$$\int_E \lim h_{f_n}\, d\mu_\tau = \lim \int_E h_{f_n}\, d\tau = \lim \mu_{f_n}(E)$$

$$= \mu_f(E) = \int_E h_f\, d\mu_\tau.$$

Hence $h_{f_n} \uparrow h_f$ and again using the monotone convergence theorem

$$\lim \varepsilon(E) f_n(w) = \lim \int_E h_{f_n}(\lambda)\tau[\,\mathcal{G}(d\lambda)w\,]$$

$$= \int_E h_f(\lambda)\tau[\,\mathcal{G}(d\lambda)w\,] = \varepsilon(E)f(w).$$

For countable additivity, if $E_i \in \mathfrak{B}(\mathbb{R})$ are mutually disjoint, $f \in S'_+$ and $w \in S_+$, then

$$\varepsilon(\cup E_i)f(w) = \int_{\cup E_i} h_f(\lambda)\tau[\,\mathcal{G}(d\lambda)w\,]$$

$$= \sum \int_{E_i} h_f(\lambda)\tau[\,\mathcal{G}(d\lambda)w\,] = \sum \varepsilon(E_i)f(w).$$

We have thus shown that ε is an expectation. To show that ε has the averaging property with respect to $\mathcal{G}$ suppose $f \in \mathcal{E}(S)^{\wedge}$. We shall show that $\varepsilon(E)f \in S'_{\mathcal{G}}$. Since $h_f \chi_E \geq 0$, there exists a monotone increasing sequence of nonnegative simple functions $h_n = \sum_i \lambda_i^n \chi_{A_i^n}$ such that $\lim h_n = h_f \chi_E$. Let $f_n = \sum_i \lambda_i^n \mathcal{G}*(A_i^n)\tau \in S'_{\mathcal{G}}$. Then $\lim f_n = \varepsilon(E)f$ in the weak*-topology since by the monotone convergence theorem

$$\lim f_n(w) = \lim \sum \lambda_i^n \tau[\,\mathcal{G}(A_i^n)w\,] = \lim \int h_n(\lambda)\tau[\,\mathcal{G}(d\lambda)w\,]$$

$$= \int_E h_f(\lambda)\tau[\,\mathcal{G}(d\lambda)w\,] = \varepsilon(E)f(w).$$

The averaging property follows by linearity. To prove the mean value property, if $f \in S'_+$, $E \in \mathfrak{B}(\mathbb{R})$, then

$$\varepsilon(E)f(s) = \int_E h_f(\lambda)\tau[\,\mathcal{G}(d\lambda)s\,] = \int_E h_f\, d\mu_\tau$$

$$= \mu_f(E) = f[\,\mathcal{G}(E)s\,]. \qquad \square$$

An instrument $\mathcal{G}$ is *nuclear* if for every $f \in \mathcal{E}(S)^{\wedge}$ there exists a real Borel function h_f such that

$$f[\,\mathcal{G}(E)w\,] = \int_E h_f(\lambda)\tau[\,\mathcal{G}(d\lambda)w\,] \qquad \text{for every} \quad E \in \mathfrak{B}(\mathbb{R}),$$

$w \in S_+$. This is a generalization of the definition due to Cycon and Hellwig [48]. Cycon and Hellwig have shown that many important instruments are nuclear.

Corollary 4.9. *If $\mathcal{I}$ is nuclear, then $\mathcal{I}^*$ is a conditional expectation with respect to $(\mathcal{I}, s)$ for every $s \in S$.*

PROOF. We know that $\mathcal{I}^*$ is an expectation satisfying the mean value property. Moreover, $\mathcal{I}^*$ satisfies the averaging property as in the proof of Theorem 4.8. $\qquad\square$

We now briefly discuss quantum stochastic processes. In classical probability theory, a *stochastic process* is a family of random variables $f_t, t \in T$, where T is usually a subset of $\mathbb{R}$; for example $T = [0, 1]$ or $T = [0, \infty)$. The *finite-dimensional distributions* of a stochastic process f_t on (Ω, Σ, μ) are the probability measures $\mu_{t_1, \ldots, t_n}$ defined on $\mathcal{B}(\mathbb{R}^n)$ by

$$\mu_{t_1, \ldots, t_n}(E_1 \times \cdots \times E_n) = \mu \left[f_{t_1}^{-1}(E_1) \cap \cdots \cap f_{t_n}^{-1}(E_n) \right],$$

where n is any positive integer and $t_1, \ldots, t_n \in T$, $E_1, \ldots, E_n \in \mathcal{B}(\mathbb{R})$. The finite-dimensional distributions satisfy the following consistency conditions:

(1) $\mu_{t_1, \ldots, t_n}(E_1 \times \cdots \times E_{n-1} \times \mathbb{R}) = \mu_{t_1, \ldots, t_{n-1}}(E_1 \times \cdots \times E_{n-1})$;

(2) $\mu_{t_1, \ldots, t_n}(E_1 \times \cdots \times E_n) = \mu_{t_1, \ldots, t_{n-1}}(E_1 \times \cdots \times E_{n-1} \cap E_n)$

  if $t_{n-1} = t_n$;

(3) $\mu_{\pi(t_1), \ldots, \pi(t_n)}(E_{\pi(1)} \times \cdots \times E_{\pi(n)}) = \mu_{t_1, \ldots, t_n}(E_1 \times \cdots \times E_n)$,

  where $\pi: \{1, \ldots, n\} \to \{1, \ldots, n\}$ is a permutation.

We define a *quantum stochastic process* to be a family of instruments $\mathcal{I}_t$ on a fixed σ-convex structure S. For concreteness we assume that $T = [0, \infty)$ and think of T as time. If $\mathcal{I}_t, t \in [0, \infty)$, is a quantum stochastic process, the *finite-dimensional distributions* of $\mathcal{I}_t$ are the probability measures $\mu_{t_1, \ldots, t_n}$ defined on $\mathcal{B}(\mathbb{R}^n)$ by

$$\mu_{t_1, \ldots, t_n}(E_1 \times \cdots \times E_n) = \tau \left[\mathcal{I}_{t_n}(E_n) \mathcal{I}_{t_{n-1}}(E_{n-1}) \cdots \mathcal{I}_{t_1}(E_1) s \right],$$

where n is any positive integer, $t_1 \leq t_2 \leq \cdots \leq t_n \in [0, \infty)$, $E_1, \ldots, E_n \in \mathcal{B}(\mathbb{R})$. The finite-dimensional distributions for a quantum stochastic process do not satisfy (2) and (3) in general. For (2) this is essentially because $\mathcal{I}_t(E \cap F) \neq \mathcal{I}_t(E) \mathcal{I}_t(F)$ in general. For (3) this is because the $\mathcal{I}_t$'s do not commute in general. However, it is clear that (1) does hold.

If $\mathcal{G}_t$ is a quantum stochastic process with finite-dimensional distributions $\mu_{t_1,\ldots,t_n}$, we define the conditional probabilities

$$W(E_n,t_n|E_{n-1},t_{n-1};\ldots;E_1,t_1) = \frac{\mu_{t_1,\ldots,t_n}(E_1\times\cdots\times E_n)}{\mu_{t_1,\ldots,t_{n-1}}(E_1\times\cdots\times E_{n-1})} \quad (4.15)$$

[assuming $\mu_{t_1,\ldots,t_{n-1}}(E_1\times\cdots\times E_{n-1})\neq 0$; otherwise define $W=0$]. Notice that $E\mapsto W(E,t_n|E_{n-1},t_{n-1};\ldots;E_1,t_1)$ is a probability measure on $\mathcal{B}(\mathbb{R})$ if $\mu_{t_1,\ldots,t_{n-1}}(E_1\times\cdots\times E_{n-1})\neq 0$. The *transition probabilities* are the probabilities $W(E_2,t_2|E_1,t_1)$, $t_1\leq t_2$. We say that $\mathcal{G}_t$ is *Markovian* in the state s if

$$W(E_n,t_n|E_{n-1},t_{n-1};\ldots;E_1,t_1) = W(E_n,t_n|E_{n-1},t_{n-1}) \quad (4.16)$$

for every $E_1,\ldots,E_n\in\mathcal{B}(\mathbb{R})$, $t_1\leq\cdots\leq t_n\in[0,\infty)$. Equation (4.16) is equivalent to

$$\mu_{t_1,\ldots,t_n}(E_1\times\cdots\times E_n)$$
$$= \mu_{t_1}(E_1)W(E_2,t_2|E_1,t_1)W(E_3,t_3|E_2,t_2)\cdots W(E_n,t_n|E_{n-1},t_{n-1}).$$
$$(4.17)$$

If $\mathcal{G}_t$ has a conditional expectation $\varepsilon_{\mathcal{G}_{t,s}}$ with respect to s, and $\varepsilon_{\mathcal{G}_{t,s}}$ is the dual instrument of a Markovian quantum stochastic process in the state s, we say that $\mathcal{G}_t$ is *dual Markovian* in the state s. We shall give examples of Markovian, non-Markovian, and dual Markovian quantum stochastic processes in Section 4.6.

4.5. Total Convex Structures

In this section we examine a class of convex structures which have a property we call total and show that such structures reduce to a linear space that is commonly studied in the operational approach.

Let $s,t\in S$ where S is a convex structuere. If $x(E)s=x(E)t$ for every observable x and $E\in\mathcal{B}(\mathbb{R})$, then s and t cannot be distinguished by observables. Notice that this is equivalent to $f(s)=f(t)$ for every $f\in\mathcal{E}(S)$. When this situation occurred in Chapter 3, we postulated that $s=t$. One can give simple examples of σ-convex structures in which this postulate does not hold. Moreover, since it is conceivable that there exist quantum systems in which this axiom does not hold [189], we have postponed postulating it until now. A *total convex structure* is a σ-convex structure on which $f(s)=f(t)$ for every $f\in\mathcal{E}(S)$ implies that $s=t$.

We have seen that if S is a convex structure, then S^* is a real linear space and hence a convex structure in the natural way. Similarly, S^{**} is a real linear space and there is a canonical imbedding $J:S\to S^{**}$, namely, $J(s)f=f(s)$ for all $s\in S$, $f\in S^*$.

We now need some definitions. Let S_0 be a convex subset of a real linear space V. The *hyperplane*, *cone* and *subspace* respectively, generated by S_0 are defined as

$$H(S_0) = \left\{ \sum_1^n \lambda_i x_i : \sum_1^n \lambda_i = 1, x_i \in S_0 \right\},$$

$$K(S_0) = \left\{ \sum_1^n \lambda_i x_i : \lambda_i \geq 0, x_i \in S_0 \right\},$$

$$V(S_0) = \left\{ \sum_1^n \lambda_i x_i : \lambda_i \in \mathbb{R}, x_i \in S_0 \right\},$$

where n varies over the positive integers. Two real linear spaces are *isomorphic* if there is a linear bijection from one to the other. Our first result shows that if S is a total convex structure, then (S, σ) is a complete metric space and J is the essentially unique isomorphism from S to a convex subset of a linear space.

Theorem 4.10. *Let S be a total convex structure. (a) The map $J: S \to J(S) \subseteq S^{**}$ is an affine isomorphism from S onto the convex subset $J(S)$ of the linear space S^{**}. (b) If T is an affine isomorphism from S onto a convex subset S_0 of a real linear space V such that $0 \notin H(S_0)$, then $V(S_0)$ and $V(J(S))$ are isomorphic and the isomorphism maps S_0 onto $J(S)$. (c) (S, σ) is a complete metric space.*

PROOF. (a) It is clear that $J: S \to J(S)$ is affine. To show that $J(S)$ is a convex set, suppose that $J(s), J(t) \in J(S)$, and $\lambda \in [0, 1]$. For $f \in S^*$ we have

$$[\lambda J(s) + (1-\lambda)J(t)] f = \lambda f(s) + (1-\lambda)f(t) = f(\langle \lambda; s, t \rangle)$$
$$= J(\langle \lambda; s, t \rangle)f.$$

Hence $\lambda J(s) + (1-\lambda)J(t) = J(\langle \lambda; s, t \rangle) \in J(S)$. To show that J is injective, suppose $s \neq t \in S$. Since S is total there exists an $f \in S^*$ such that $f(s) \neq f(t)$, so $J(s) \neq J(t)$. (b) Let T and S_0 satisfy the hypotheses of (b). The function $g = J \circ T^{-1}$ is an affine bijection from S_0 to $J(S)$. We now extend g to $K(S_0)$. First, if $0 \neq y \in K(S_0)$, then $y = \sum \lambda_i x_i, \lambda_i > 0, x_i \in S_0$. Hence $y = (\sum_j \lambda_j) \sum_i (\lambda_i / \sum_j \lambda_j) x_i = \lambda x$ where $\lambda > 0$, $x \in S_0$. We now show that the representation $y = \lambda x$ is unique. Suppose $y = \lambda x = \mu z$, where $\lambda, \mu > 0$, $x, z \in S_0$. Then if $\lambda \neq \mu$ we have

$$0 = \lambda x - \mu z = (\lambda - \mu)[\lambda(\lambda - \mu)^{-1}x - \mu(\lambda - \mu)^{-1}z]. \tag{4.18}$$

Since $0 \notin H(S_0)$ the second factor on the right-hand side of (4.18) is not 0. Hence $\lambda = \mu$, which is a contradiction. Thus $\lambda = \mu$ and hence $x = z$. Define

95

$g(y) = \lambda g(x)$. It is easy to see that the extended $g: K(S_0) \to K(J(S))$ is a bijection. The following shows that g is additive on $K(S_0)$:

$$g(\lambda x + \mu y) = g\{(\lambda + \mu)[\lambda(\lambda + \mu)^{-1}x + \mu(\lambda + \mu)^{-1}y]\}$$
$$= \lambda g(x) + \mu g(y) = g(\lambda x) + g(\mu y).$$

Also g is homogeneous on $K(S_0)$ since

$$g(\lambda(\mu x)) = g(\lambda \mu x) = \lambda \mu g(x) = \lambda g(\mu x) \quad \text{for} \quad \lambda, \mu > 0, x \in S_0.$$

We finally extend g to $V(S_0)$. Suppose $y \in V(S_0)$ and $y = \Sigma \lambda_i x_i$, $\lambda_i \in \mathbb{R}$, $x_i \in S_0$. Then the positive and negative coefficients can be grouped so that $y = \lambda x - \mu z$, where $\lambda, \mu \geq 0$, $x, z \in S_0$. Thus y has the form $y = u - v$, where $u, v \in K(S_0)$. Define g on $V(S_0)$ by $g(y) = g(u) - g(v)$. This extended g is well defined since if $u - v = u_1 - v_1$, $u_1, v_1 \in K(S_0)$, then $u + v_1 = u_1 + v$, so by the additivity of g on $K(S_0)$ we have $g(u) + g(v_1) = g(u_1) + g(v)$ and $g(u) - g(v) = g(u_1) - g(v_1)$. That $g: V(S_0) \to V(J(S))$ is linear and bijective is now easily verified. (c) Clearly $\sigma(s,t) > 0$. If $\sigma(s,t) = 0$, then there exist sequences $\lambda_i \to 0$, $\lambda_i > 0$, $s_i, t_i \in S$ such that $\langle \lambda_i, s_i, s \rangle = \langle \lambda_i, t_i, t \rangle$. Applying an $f \in \mathfrak{E}(S)$ gives

$$\lambda_i f(s_i) + (1 - \lambda_i)f(s) = \lambda_i f(t_i) + (1 - \lambda_i)f(t). \tag{4.19}$$

Since $0 \leq f(s_i), f(t_i) \leq 1$, taking the limit in (4.19) as $i \to \infty$ we have $f(s) = f(t)$. Since S is total, $s = t$. Define $\sigma': J(S) \times J(S) \to \mathbb{R}$ by $\sigma'(J(s), J(t)) = \sigma(s,t)$. If we can show that σ' is a metric on $J(S)$ it will follow that σ is a metric on S. Since $\frac{1}{2}p + \frac{1}{2}q = \frac{1}{2}q + \frac{1}{2}p$ for $p, q \in J(S)$ we have

$$\sigma'(p,q) = \inf\{0 \leq \lambda \leq 1 : \lambda p_1 + (1 - \lambda)p = \lambda p_2 + (1 - \lambda)q, p_1, p_2 \in J(S)\}.$$

It is clear that $\sigma'(p,q) = \sigma'(q,p)$ and that $\sigma(p,p) \neq 0$ for every $p, q \in J(S)$. For the triangle inequality, it is clear that $\sigma'(p,q) \leq \sigma'(p,r) + \sigma'(r,q)$ if $p = r$ or $q = r$, so we can exclude these cases. Suppose $0 < \lambda_1 < 1$ and $0 < \lambda_2 < 1$ satisfy

$$\lambda_1 p_1 + (1 - \lambda_1)p = \lambda_1 r_1 + (1 - \lambda_1)r,$$
$$\lambda_2 r_2 + (1 - \lambda_2)r = \lambda_2 q_1 + (1 - \lambda_2)q.$$

Letting $\lambda_3 = \lambda_1 + \lambda_2 - 2\lambda_1\lambda_2$,

$$p_2 = \lambda_1(1 - \lambda_2)\lambda_3^{-1}p_1 + \lambda_2(1 - \lambda_1)\lambda_3^{-1}r_2 \in J(S),$$
$$q_2 = \lambda_1(1 - \lambda_2)\lambda_3^{-1}r_1 + \lambda_2(1 - \lambda_1)\lambda_3^{-1}q_1 \in J(S),$$

and $\lambda_0 = \lambda_3(1 - \lambda_1\lambda_2)^{-1}$, a straightforward computation gives

$$\lambda_0 p_2 + (1 - \lambda_0)p = \lambda_0 q_2 + (1 - \lambda_0)q.$$

It is easily shown that $\lambda_0 \leq \lambda_1 + \lambda_2$ and hence

$$\sigma'(p,q) \leq \sigma'(p,r) + \sigma'(r,q). \qquad \square$$

It turns out to be more convenient to make a change of scale and define the distance function $\rho(s,t) = \sigma(s,t)[1 - \sigma(s,t)]^{-1}$ on S. If S is a total convex structure, then it is easily seen that ρ is a metric on S which retains the important properties of σ. In particular, Theorem 4.10(c) and Lemma 4.1(c) hold with σ replaced by ρ. We call ρ the *intrinsic metric* of S.

Let $K = K(J(S))$ be the cone generated by $J(S)$. If $\bar{J}: S_+ \to S^{**}$ is the unique affine extension of J to S_+ given by $\bar{J}((\alpha,s)) = \alpha J(s)$, then $K = \bar{J}(S_+)$. Let $X = V(J(S))$ be the subspace of S^{**} generated by $J(S)$. It is clear that $X = K - K$; that is, K is a generating cone for X. We call X the *state space* for the system described by S. Any $u \in X$ admits at least one representation of the form $u = \alpha s - \beta t$, $\alpha, \beta \geq 0$, $s, t \in J(S)$. We define

$$\|u\| = \inf\{\max(\alpha,\beta) : u = \alpha s - \beta t, \ \alpha, \beta \geq 0, \ s, t \in J(S)\}.$$

It is shown that $\|\cdot\|$ is a norm in the next theorem.

Theorem 4.11. *Let S be a total convex structure.* (a) $\|\cdot\|$ *extends ρ in the sense that* $\|J(s) - J(t)\| = \rho(s,t)$ *for all $s, t \in S$. Moreover $\|J(s)\| = 1$ for all $s \in S$.* (b) $(X, \|\cdot\|)$ *is a real Banach space.* (c) *If $F \in Af(S)$, then $JFJ^{-1}: J(S) \to J(S)$ has a unique linear extension $\hat{F}: X \to X$ and $\|\hat{F}\| \leq 1$. If $f \in S^*$, then $fJ^{-1}: J(S) \to \mathbb{R}$ has a unique linear extension $\hat{f}: X \to \mathbb{R}$.*

PROOF. For simplicity of notation we identify S and $J(S)$. First notice that S is *normalized*; that is, if $s \in S$, then $\alpha s \notin S$ for $\alpha \neq 1$. Indeed, if $\alpha s \in S$, then $1 = (\alpha s)(\tau) = \alpha$.

(a) Let $s, t \in S$. If

$$s - t = \alpha s_1 - \beta t_1, \qquad \alpha, \beta \geq 0, \quad s_1, t_1 \in S, \qquad (4.20)$$

then $s + \beta t_1 = t + \alpha s_1$, so

$$(1+\beta)\left[(1+\beta)^{-1}s + \beta(1+\beta)^{-1}t_1\right] = (1+\alpha)\left[(1+\alpha)^{-1}t + \alpha(1+\alpha)^{-1}s_1\right].$$

Now since $s_2 = (1+\alpha)^{-1}t + \alpha(1+\alpha)^{-1}s_1 \in S$, $(1+\alpha)(1+\beta)^{-1}s_2 \in S$, and S is normalized, we have $(1+\alpha)(1+\beta)^{-1} = 1$ so $\alpha = \beta$. Thus all representations of $s - t$ of the form (4.20) must reduce to $s - t = \alpha(s_1 - t_1)$ for $\alpha \geq 0$, $s_1, t_1 \in S$. Hence

$$\sigma(s,t) = \inf\{0 \leq \lambda \leq 1 : \lambda s_1 + (1-\lambda)s = \lambda t_1 + (1-\lambda)t, \ s_1, t_1 \in S\}$$

$$= \inf\{0 \leq \lambda < 1 : s - t = \lambda(1-\lambda)^{-1}(t_1 - s_1), \ s_1, t_1 \in S\}$$

$$= \inf\{\alpha(1+\alpha)^{-1}, \ \alpha \geq 0 : s - t = \alpha(t_1 - s_1), \ s_1, t_1 \in S\}$$

$$= \inf\{\alpha \geq 0 : s - t = \alpha(t_1 - s_1)\}\left[\inf\{\alpha \geq 0 : s - t = \alpha(t_1 - s_1) + 1\}\right]^{-1}$$

$$= \|s - t\|\left[\|s - t\| + 1\right]^{-1}.$$

97

4. The Operational Approach

Thus $\rho(s,t)=\sigma(s,t)[1-\sigma(s,t)]^{-1}=\|s-t\|$. To show that $\|s\|=1$ for $s\in S$, suppose $s=\alpha s_1-\beta s_2$, $\alpha,\beta\geq 0$, $s_1,s_2\in S$. Operating on τ gives $\alpha-\beta=1$. Hence $\alpha=1+\beta\geq 1$, so $\|s\|\geq 1$. But the representation $s=s$ gives $\|s\|\leq 1$, so $\|s\|=1$. (b) Let

$$D=\{\alpha s-\beta t:0\leq\alpha,\beta\leq 1,s,t\in S\}.$$

It is straightforward to show that D is a convex, balanced, absorbing [225] subset of X containing 0. For $p\in X$, let

$$m(p)=\inf\{\lambda>0:p\in\lambda D\}$$

be the Minkowski functional [225] for D in X. It is well known [225] that m is a seminorm on X. We now show that $m(u)=\|u\|$ for every $u\in X$. By definition

$$m(u)=\inf\{\lambda>0:u=\lambda\alpha s-\lambda\beta t, 0\leq\alpha, \beta\leq 1, s,t\in S\}.$$

Now if $u=\lambda\alpha s-\lambda\beta t$, $0\leq\alpha$, $\beta\leq 1$, then $u=(\lambda\alpha)s-(\lambda\beta)t$, where $\max(\lambda\alpha,\lambda\beta)\leq\lambda$ so $\|u\|\leq m(u)$. Conversely, if $u=\alpha s-\beta t$, $\alpha,\beta>0$, $s,t\in S$, then

$$u=\max(\alpha,\beta)\{[\alpha\max(\alpha,\beta)^{-1}]s-[\beta\max(\alpha,\beta)^{-1}]t\},$$

so $m(u)\leq\|u\|$. It follows that $\|u\|=m(u)$ and $\|u\|$ is a seminorm. To show that $\|\cdot\|$ is a norm, suppose $\|u\|=0$ for $u\in X$. Suppose $u=\alpha s-\beta t$, $\alpha,\beta\geq 0$, $s,t\in S$. Then

$$0=\|u\|=\|\alpha s-\beta t\|\geq|\,\|\alpha s\|-\|\beta t\|\,|$$
$$=|\,\alpha\|s\|-\beta\|t\|\,|=|\alpha-\beta|.$$

Hence $\alpha=\beta$ and $u=\alpha(s-t)$. If $\alpha=0$, then $u=0$. Suppose $\alpha\neq 0$. Since

$$0=\|u\|=\alpha\|s-t\|=\alpha\rho(s,t)$$

we have $\rho(s,t)=0$ and $s=t$. Hence $u=0$. To show that $(X,\|\cdot\|)$ is complete, let u_n be a Cauchy sequence in X. We may assume that $\|u_{n+1}-u_n\|<2^{-n}$, $n=1,2,\ldots$ (if not, a subsequence satisfies this inequality and if we can show that a subsequence converges, then the original Cauchy sequence converges). Now we can write $u_{n+1}-u_n=\alpha_n s_n-\beta_n t_n$, where $0\leq\alpha_n$, $\beta_n<2^{-n}$, $s_n,t_n\in S$. We can assume that $\alpha_1,\beta_1>0$. Let $a_n=\Sigma_{i=1}^n\alpha_i$, $b_n=\Sigma_{i=1}^n\beta_i$. Then

$$\left\{\sum_{i=1}^n a_n^{-1}\alpha_i s_i\right\}\quad\text{and}\quad\left\{\sum_{i=1}^n b_n^{-1}\beta_i t_i\right\},\quad n=1,2,\ldots,$$

are Cauchy sequences in S. Indeed, it is clear that $\{a_n\}$ is a Cauchy sequence and we have

$$\rho\left(\sum_{i=1}^{n+k} a_{n+k}^{-1}\alpha_i s_i, \sum_{i=1}^{n} a_n^{-1}\alpha_i s_i\right) = \left\|(a_{n+k}^{-1} - a_n^{-1})\sum_{i=1}^{n}\alpha_i s_i + a_{n+k}^{-1}\sum_{i=n+1}^{n+k}\alpha_i s_i\right\|$$

$$\leq (a_n^{-1} - a_{n+k}^{-1})\sum_{i=1}^{n}\alpha_i + a_{n+k}^{-1}\sum_{i=n+1}^{n+k}\alpha_i$$

$$= 1 - a_n a_{n+k}^{-1} + a_{n+k}^{-1}(a_{n+k} - a_n)$$

$$= 2(1 - a_n a_{n+k}^{-1}),$$

where the last term approaches 0 as $n, k \to \infty$. Since S is complete, by definition there exist elements $s, t \in S$ such that $\lim \sum_{i=1}^{n} a_n^{-1}\alpha_i s_i = s$ and $\lim \sum_{i=1}^{n} b_n^{-1}\beta_i t_i = t$. Suppose that $\lim a_n = a$ and $\lim b_n = b$. We claim that $\lim u_n = u_1 + as - bt$. Indeed,

$$\|u_{n+1} - u_1 - as + bt\| = \|u_{n+1} - u_n + \cdots + u_2 - u_1 - as + bt\|$$

$$\leq |a|\left\|\sum_{i=1}^{n} a^{-1}\alpha_i s_i - s\right\| + |b|\left\|\sum_{i=1}^{n} b^{-1}\beta_i t_i - t\right\|$$

$$\leq |a|\left[\left\|\sum_{i=1}^{n}(a^{-1}\alpha_i - a_n^{-1}\alpha_i)s_i\right\| + \left\|\sum_{i=1}^{n} a_n^{-1}\alpha_i s_i - s\right\|\right]$$

$$+ |b|\left[\left\|\sum_{i=1}^{n}(b^{-1}\beta_i - b_n^{-1}\beta_i)t_i\right\| + \left\|\sum_{i=1}^{n} b_n^{-1}\beta_i t_i - t\right\|\right]$$

$$\leq |a|\left[1 - a^{-1}a_n + \left\|\sum_{i=1}^{n} a_n^{-1}\alpha_i s_i - s\right\|\right]$$

$$+ |b|\left[1 - b^{-1}b_n + \left\|\sum_{i=1}^{n} b_n^{-1}\beta_i t_i - t\right\|\right] \to 0$$

as $n \to \infty$. (c) Define $\hat{F}(0) = 0$. If $0 \neq u \in X$ and $u = \alpha s - \beta t$, $\alpha, \beta \geq 0$, $s, t \in S$, define $\hat{F}(u) = \alpha F(s) - \beta F(t)$. To show that $\hat{F}$ is well defined suppose that also $u = \alpha_1 s_1 - \beta_1 t_1$, $\alpha_1, \beta_1 \geq 0$, $s_1, t_1 \in S$. Notice that $\alpha_1 + \beta, \alpha + \beta_1 > 0$. Since $\alpha_1 s_1 + \beta t = \alpha s + \beta_1 t_1$, applying both sides to τ gives $\alpha + \beta_1 = \alpha_1 + \beta$. Hence

$$\alpha_1(\alpha_1 + \beta)^{-1}s_1 + \beta(\alpha_1 + \beta)^{-1}t = \alpha(\alpha + \beta_1)^{-1}s + \beta_1(\alpha + \beta_1)^{-1}t_1.$$

Since F is affine we have

$$\alpha_1(\alpha_1 + \beta)^{-1}F(s_1) + \beta(\alpha_1 + \beta)^{-1}F(t)$$

$$= \alpha(\alpha + \beta_1)^{-1}F(s) + \beta_1(\alpha + \beta_1)^{-1}F(t_1).$$

Hence

$$\alpha_1 F(s_1) - \beta_1 F(t_1) = \alpha F(s) - \beta F(t)$$

and $\hat{F}$ is well defined. It is easy to show that $\hat{F}$ is a linear operator on X and it follows that $\hat{F}$ is the unique linear extension of F to X. To show that $\hat{F}$ is a contraction we have for all $u \in X$

$$\|\hat{F}(u)\| = \inf\{\max(\alpha,\beta) : \hat{F}(u) = \alpha s - \beta t, \ \alpha,\beta \geq 0, \ s,t \in S\}$$
$$\leq \inf\{\max(\alpha,\beta) : u = \alpha s - \beta t, \ \alpha,\beta \geq 0, \ s,t \in S\}$$
$$= \|u\|.$$

The statement for $f \in S^*$ is proved in a similar way. $\qquad\square$

It follows from Theorem 4.11(c) that the affine functional τ on $J(S)$ has a unique linear extension (also denoted by τ) to X. The next theorem gives some of the important properties of τ and more information about the cone K.

Theorem 4.12. *Let S be a total convex structure. (a) τ is a continuous linear functional on X satisfying $\tau(p) = \|p\|$ for every $p \in K$. (b) K is a closed generating cone in X. (c) K is a normal, strict b-cone in X; that is, (i) if $u,v \in X$ and $u - v \in K$, then $\|u\| \geq \|v\|$, (ii) for any $\gamma > 1$, $u \in X$, there exist $u_1, u_2 \in K$ such that $u = u_1 - u_2$ and $\|u_1\|, \|u_2\| \leq \gamma \|u\|$.*

PROOF. (a) If $p = \alpha s \in K$, $\alpha \geq 0$, $s \in J(S)$, then $\tau(p) = \alpha = \|\alpha s\| = \|p\|$. To show that τ is continuous on X, let $u_i \in X$ and $u_i \to 0$. Then $u_i = \alpha_i s_i - \beta_i t_i$, $\alpha_i, \beta_i \geq 0$, $s_i, t_i \in J(S)$, $\alpha_i, \beta_i \to 0$. Hence

$$|\tau(u_i)| = |\alpha_i - \beta_i| \leq |\alpha_i| + |\beta_i| \to 0.$$

(b) We have already shown that K is generating. To show that K is closed, suppose $\alpha_i s_i \in K$, $\alpha_i \geq 0$, $s_i \in J(S)$, and $\alpha_i s_i \to u \in X$. Since τ is continuous, $\alpha_i = \tau(\alpha_i s_i) \to \tau(u)$. If $\tau(u) = 0$, then $\|\alpha_i s_i\| = \alpha_i \to 0$ so $u = 0 \in K$. If $\tau(u) \neq 0$ then

$$\|s_i - s_j\| \leq \tau(u)^{-1}\big[\|\tau(u)s_i - u\| + \|u - \tau(u)s_j\|\big]$$
$$\leq \tau(u)^{-1}\big[\|\tau(u)s_i - \alpha_i s_i\| + \|\alpha_i s_i - u\| + \|u - \alpha_j s_j\| + \|\alpha_j s_j - \tau(u)s_j\|\big]$$
$$= \tau(u)^{-1}\big[|\tau(u) - \alpha_i| + \|\alpha_i s_i - u\| + \|u - \alpha_j s_j\| + |\alpha_j - \tau(u)|\big].$$

The last expression approaches 0 as $i,j \to \infty$. Since $J(S)$ is complete, there exists an $s \in J(S)$ such that $s_i \to s$. Hence $\alpha_i s_i \to \tau(u)s$ and $u = \tau(u)s \in K$. (c) To prove (i), since $u - v \in K$ there exists an $\alpha_0 \geq 0$ and $s_0 \in J(S)$ such that $u - v = \alpha_0 s_0$. We may assume that $\alpha_0 > 0$. If $v = \alpha s - \beta t$, $\alpha,\beta \geq 0$, $s,t \in J(S)$,

then

$$u = v + \alpha_0 s_0 = (\alpha_0 + \alpha) \left[\alpha_0 (\alpha + \alpha_0)^{-1} s_0 + \alpha (\alpha + \alpha_0)^{-1} s \right] - \beta t.$$

Since $\max(\alpha_0 + \alpha, \beta) \geq \max(\alpha, \beta)$ we have $\|u\| \geq \|v\|$. To prove (ii), by definition of $\|u\|$ there exist $\alpha, \beta \leq \gamma \|u\|$, $s, t \in S$ such that $u = \alpha s - \beta t$. Letting $u_1 = \alpha s$, $u_2 = \beta t$, we have $u_1, u_2 \in K$, $u = u_1 - u_2$, and

$$\|u_1\| = \|\alpha s\| = \alpha \leq \gamma \|u\|.$$

Similarly, $\|u_2\| \leq \gamma \|u\|$. $\qquad\qquad\square$

Another norm on X which is used is the norm

$$\|u\|_1 = \inf\{\alpha + \beta : u = \alpha s - \beta t; \alpha, \beta \geq 0, s, t \in J(S)\}.$$

It is clear that $\|u\| \leq \|u\|_1 \leq 2\|u\|$ for all $u \in X$ so the two norms are equivalent. Moreover, it is clear that $\|s\|_1 = 1$ for all $s \in J(S)$ and

$$\|u\|_1 = \inf\{\|p\|_1 + \|q\|_1 : u = p - q, p, q \in K\}.$$

Finally, $\|s - t\|_1 = 2\rho(s, t)$ for all $s, t \in J(S)$, since from the proof of Theorem 5.2(a)

$$\|s - t\|_1 = \inf\{2\alpha : s - t = \alpha(s_1 - t_1), \alpha \geq 0, s_1, t_1 \in J(S)\} = 2\|s - t\| = 2\rho(s, t).$$

Because of these relationships, most of the previous results of this section hold for $\|\cdot\|_1$. In particular $(X, \|\cdot\|_1)$ is a Banach space with closed generating cone K and continuous linear functional τ satisfying $\tau(p) = \|p\|_1$ for every $p \in K$. In fact $(X, \|\cdot\|_1, K, \tau)$ is the state space of Davies and Lewis [54, 55].

To summarize, if S is a total convex structure, then (S, ρ) is a complete metric space which is isometrically isomorphic to an essentially unique convex subset $J(S)$ of a real Banach space $(X, \|\cdot\|)$. Moreover, S_+ is isomorphic to a closed generating normal strict b-cone K in X and under this isomorphism τ extends uniquely to a continuous linear functional (also denoted by τ) satisfying $\tau(p) = \|p\|$ for every $p \in K$. If x is an observable on S, then by Theorem 4.11(c), x is isomorphic to a map $\tilde{x}: \mathfrak{B}(\mathbb{R}) \to X^*$. It easily follows that $\tilde{x}$ satisfies the conditions:

(i) $0 \leq \tilde{x}(E) \leq \tilde{x}(\mathbb{R})$ for all $E \in \mathfrak{B}(\mathbb{R})$;
(ii) $\tilde{x}(\mathbb{R}) = \tau$;
(iii) for any countable family $\{E_i\}$ of pairwise disjoint sets in $\mathfrak{B}(\mathbb{R})$, $\tilde{x}(\cup E_i) = \Sigma \tilde{x}(E_i)$, where the sum converges in the weak*-topology of X^*.

Moreover, an instrument $\mathcal{I}$ on S is isomorphic to a map $\tilde{\mathcal{I}}$ from $\mathfrak{B}(\mathbb{R})$ into the set of positive linear operators on X which satisfies

4. The Operational Approach

(a) $\tau[\tilde{\mathcal{I}}(\mathbb{R})v] = \tau(v)$ for every $v \in X$;

(b) for any countable family $\{E_i\}$ of pairwise disjoint sets in $\mathfrak{B}(\mathbb{R})$, $\tilde{\mathcal{I}}(\cup E_i) = \Sigma \tilde{\mathcal{I}}(E_i)$, where the sum converges in the strong operator topology.

These are the usual definitions of observables and instruments of Davies and Lewis [54, 55].

4.6. Hilbert Space Quantum Mechanics

In this section we continue our study of Hilbert space quantum mechanics begun in Section 4.3 and give examples which illustrate some of our previously defined concepts. Let $\mathcal{H}$ be a complex separable Hilbert space and let S be the set of density operators on $\mathcal{H}$. If $D_1 \neq D_2 \in S$ then there exists a $\psi \in \mathcal{H}$ such that $\langle D_1 \psi, \psi \rangle \neq \langle D_2 \psi, \psi \rangle$. If P_ψ is the one-dimensional projection onto the subspace spanned by ψ then

$$P_\psi(D_1) = \tau(P_\psi D_1) = \langle D_1 \psi, \psi \rangle \neq \langle D_2 \psi, \psi \rangle = P_\psi(D_2).$$

Since $P_\psi \in \mathcal{E}(S)$ we see that S is total. By Theorem 4.10, S is isomorphic to a convex subset of a unique (up to isomorphism) real linear space. In fact, in this case S is already a convex subset of a natural real linear space, namely the set of self-adjoint trace class operators X on $\mathcal{H}$. The space X becomes a Banach space under the trace norm $\|A\| = \mathrm{tr}|A|$, $A \in X$. This norm turns out to be the norm $\|\cdot\|_1$ introduced in Section 4.5. In this case S_+ is the cone of positive trace class operators on $\mathcal{H}$, $\mathcal{E}(S)$ is the set of self-adjoint operators A such that $0 \leq A \leq I$, $\mathcal{E}(S_+)$ is the set of bounded positive self-adjoint operators, and the set of all bounded self-adjoint operators $\mathcal{E}(S_+) - \mathcal{E}(S_+)$ on $\mathcal{H}$ is the Banach space dual X^* of X.

Let x be the observable $x(E) = \Sigma\{Q_i : a_i \in E\}$, $E \in \mathfrak{B}(\mathbb{R})$, defined in Section 4.3, and let $\mathcal{I}_x$ be the instrument $\mathcal{I}_x(E)w = \Sigma\{Q_i w Q_i : a_i \in E\}$, $E \in \mathfrak{B}(\mathbb{R})$, $w \in X$. We have seen in Section 4.4 that the dual instrument is $\mathcal{I}_x^*(E)A = \Sigma\{Q_i A Q_i : a_i \in E\}$, $E \in \mathfrak{B}(\mathbb{R})$, $A \in X^*$. We now show that $\mathcal{I}_x^*$ is, in general, not the conditional expectation with respect to $\mathcal{I}_x$ and any state. We call $\mathcal{I}_x$ *atomic* if each of the projections Q_i is one-dimensional. Suppose $\mathcal{I}_x$ is not atomic and hence at least one of the projections, say Q_1 has dimension greater than one. Then there exists a projection P satisfying $0 < P < Q_1$. Now $\mathcal{I}_x^*(\{a_1\})P = P$. Since $S_{\mathcal{I}_x}^* = \overline{\mathrm{span}}\{Q_i : i = 1, \dots\}$ and $P \notin S_{\mathcal{I}_x}^*$ we see that $\mathcal{I}_x^*$ is not the conditional expectation with respect to $\mathcal{I}_x$ and any state. However, if $\mathcal{I}_x$ is atomic, then $\mathcal{I}_x^*$ is the conditional expectation with respect to $\mathcal{I}_x$ and any state. Indeed, if Q_i is one-dimensional, $i =$

$1, 2, \ldots$, it is easy to show that $Q_i B Q_i = \mathrm{tr}(Q_i B) Q_i$ for all $B \in \mathcal{L}(\mathcal{H})$. Hence for all $A \in X^*$, $w \in X$, $E \in \mathcal{B}(\mathbb{R})$ we have

$$A[\mathcal{I}_x(E)w] = \mathrm{tr}[\mathcal{I}_x(E)wA]$$

$$= \mathrm{tr}\left[\sum \{Q_i w Q_i A : a_i \in E\}\right] = \mathrm{tr}\left[\sum \{\mathrm{tr}(Q_i w) Q_i A : a_i \in E\}\right]$$

$$= \sum \{\mathrm{tr}(Q_i A)\, \mathrm{tr}(Q_i w) : a_i \in E\} = \int_E h_A \tau[\mathcal{I}_x(d\lambda)w],$$

where h_A is the Borel function $h_A = \sum_{i=1} \tau(Q_i A) \chi_{\{a_i\}}$. Hence $\mathcal{I}_x$ is nuclear and by Corollary 4.9 our statement is proved.

It is not hard to show that if w_0 is effective with respect to $\mathcal{I}$, then

$$\varepsilon_{\mathcal{I}_x, w_0}(E)A = \sum_{a_i \in E} \frac{1}{\mathrm{tr}(Q_i w_0)} \mathrm{tr}(Q_i w_0 Q_i A) Q_i. \tag{4.21}$$

Moreover, $\varepsilon_{\mathcal{I}_x, w_0}$ is the dual of the instrument

$$\mathcal{I}_1(E)w = \sum_{a_i \in E} \frac{\mathrm{tr}(Q_i w)}{\mathrm{tr}(Q_i w_0)} Q_i w_0 Q_i \tag{4.22}$$

and $\mathcal{I}_1$ also corresponds to the observable x.

We now consider an example of a quantum stochastic process. Let $\mathcal{I}_t$, $t \in [0, \infty)$, be the quantum stochastic process defined by

$$\mathcal{I}_t(E)w = \sum \{Q_{i,t} w Q_{i,t} : a_i \in E\},$$

where $a_i \in \mathbb{R}$ are distinct and $Q_{i,t}$ are projection operators that are mutually orthogonal for fixed $t \in [0, \infty)$, $i = 1, 2, \ldots$. If each of the $Q_{i,t}$, $t \in [0, \infty)$, $i = 1, 2, \ldots$, are one-dimensional, then the finite-dimensional distributions satisfy

$$\mu_{t_1, \ldots, t_n}(\{a_{i_1}\} \times \cdots \times \{a_{i_n}\}) = \tau[\mathcal{I}_{t_n}(\{a_{i_n}\}) \cdots \mathcal{I}_{t_1}(\{a_{i_1}\})w]$$

$$= \mathrm{tr}[Q_{i_n, t_n} \cdots Q_{i_1, t_1} w Q_{i_1, t_1} \cdots Q_{i_n, t_n}]$$

$$= \mathrm{tr}[Q_{i_n, t_n} \cdots Q_{i_2, t_2} \mathrm{tr}(Q_{i_1, t_1} w) Q_{i_1, t_1} Q_{i_2, t_2} \cdots Q_{i_n, t_n}]$$

$$= \mathrm{tr}(Q_{i_1, t_1} w)\, \mathrm{tr}(Q_{i_n, t_n} \cdots Q_{i_2, t_2} Q_{i_1, t_1} Q_{i_2, t_2} \cdots Q_{i_n, t_n})$$

$$= \mathrm{tr}(Q_{i_1, t_1} w)\, \mathrm{tr}(Q_{i_2, t_2} Q_{i_1, t_1}) \cdots \mathrm{tr}(Q_{i_n, t_n} Q_{i_{n-1}, t_{n-1}}).$$

4. The Operational Approach

Since

$$W(\{a_{i_j}\},t_j|\{a_{i_{j-1}}\},t_{j-1}) = \frac{\mu_{t_{j-1},t_j}(\{a_{i_{j-1}}\}\times\{a_{i_j}\})}{\mu_{t_{j-1}}(\{a_{i_{j-1}}\})} = \frac{\mathrm{tr}\left[\mathcal{I}_{t_j}(\{a_{i_j}\})\mathcal{I}_{t_{j-1}}(\{a_{i_{j-1}}\})w\right]}{\mathrm{tr}\left[\mathcal{I}_{t_{j-1}}(\{a_{i_{j-1}}\})w\right]}$$

$$= \frac{\mathrm{tr}\left(Q_{i_j,t_j}Q_{i_{j-1},t_{j-1}}wQ_{i_{j-1},t_{j-1}}Q_{i_j,t_j}\right)}{\mathrm{tr}\left(Q_{i_{j-1},t_{j-1}}wQ_{i_{j-1},t_{j-1}}\right)}$$

$$= \frac{\mathrm{tr}\left[Q_{i_j,t_j}\,\mathrm{tr}\left(Q_{i_{j-1},t_{j-1}}w\right)Q_{i_{j-1},t_{j-1}}Q_{i_j,t_j}\right]}{\mathrm{tr}\left(Q_{i_{j-1},t_{j-1}}wQ_{i_{j-1},t_{j-1}}\right)}$$

$$= \mathrm{tr}\left(Q_{i_j,t_j}Q_{i_{j-1},t_{j-1}}Q_{i_j,t_j}\right),$$

we have

$$\mu_{t_1,\ldots,t_n}(\{a_{i_1}\}\times\cdots\times\{a_{i_n}\})$$
$$= \mu_{t_1}(\{a_{i_1}\})W(\{a_{i_2}\},t_2|\{a_{i_1}\},t_1)\cdots W(\{a_{i_n}\},t_n|\{a_{i_{n-1}}\},t_{n-1}).$$

(4.23)

It follows that Eq. (4.23) holds with the $\{a_{i_j}\}$'s replaced by $E_i \in \mathcal{B}(\mathbb{R})$ and hence $\mathcal{I}_t$ is Markovian.

If the $\mathcal{I}_t$ are not atomic, then the process is non-Markovian in general. Suppose $\mathcal{I}_t$ is not necessarily atomic, w_0 is effective and let $\mathcal{I}'_t$ be the quantum stochastic process defined by

$$\mathcal{I}'_t(E)w = \sum_{a_i \in E} \mathrm{tr}(Q_{i,t}w)\frac{Q_{i,t}w_0Q_{i,t}}{\mathrm{tr}(Q_{i,t}w_0)},$$

$E \in \mathcal{B}(\mathbb{R})$, $w \in S_+$, $t \in [0,\infty)$. Then $\varepsilon_{\mathcal{I}_t,w_0}$ is the dual instrument of $\mathcal{I}'_t$. We now show that $\mathcal{I}'_t$ is Markovian in the state w_0. In fact, a simple computation gives

$$\mu'_{t_1,\ldots,t_n}(\{a_{i_1}\}\times\cdots\times\{a_{i_n}\})$$
$$= \tau\left[\mathcal{I}'_{t_n}(\{a_{i_n}\})\cdots\mathcal{I}'_{t_1}(\{a_{i_1}\})w_0\right]$$
$$= \mathrm{tr}(Q_{i_1,t_1}w_0)\frac{\mathrm{tr}(Q_{i_2,t_2}Q_{i_1,t_1}w_0Q_{i_1,t_1}Q_{i_2,t_2})}{\mathrm{tr}(Q_{i_1,t_1}w_0)}$$
$$\cdots\mathrm{tr}(Q_{i_n,t_n}Q_{i_{n-1},t_{n-1}}w_0Q_{i_{n-1},t_{n-1}}Q_{i_n,t_n})$$
$$= \mu_{t_1}(\{a_{i_1}\})W(\{a_{i_2}\},t_2|\{a_{i_1}\},t_1)\cdots W(\{a_{i_n}\},t_n|\{a_{i_{n-1}}\},t_{n-1}).$$

(4.24)

It thus follows that $\mathcal{I}'_t$ is dual Markovian even in the nonatomic case.

Recall that in this case of Hilbert space quantum mechanics an observable is a POV measure $x: \mathcal{B}(\mathbb{R}) \to X^*$. One can define integrals relative to POV measures in a manner similar to that for PV measures. As a technical device enabling us to work on compact spaces, let $\mathbb{R}_\infty$ be the one-point compactification of $\mathbb{R}$. Then $\mathcal{B}(\mathbb{R})$ and $\mathcal{B}(\mathbb{R}_\infty)$ are Borel isomorphic and every observable x on $\mathcal{B}(\mathbb{R})$ has a unique extension to $\mathcal{B}(\mathbb{R}_\infty)$, also denoted by x, satisfying $x(\{\infty\}) = 0$. Denote the set of continuous real-valued functions on $\mathbb{R}_\infty$ by $C(\mathbb{R}_\infty)$. If $f \in C(\mathbb{R}_\infty)$, then $f(x) \equiv \int f(\lambda) x(d\lambda)$ is a self-adjoint operator on $\mathcal{H}$. The following lemma gives an equivalent way of defining observables.

Lemma 4.13 (Davies [54]). *If x is an observable, then the map $f \mapsto f(x)$ is a positive linear map from $C(\mathbb{R}_\infty)$ to X^* such that $1(x) = I$. Conversely, if $\bar{x}: C(\mathbb{R}_\infty) \to X^*$ is a positive linear map such that $\bar{x}(1) = I$, then there exists a unique observable x such that $\bar{x}(f) = f(x)$ for all $f \in C(\mathbb{R}_\infty)$.*

PROOF. That $f \mapsto f(x)$ is a positive linear map from $C(\mathbb{R}_\infty)$ to X^* such that $1(x) = I$ follows from the definition of the integral. Conversely, suppose $\bar{x}$ is given as above. Then for all $\phi, \psi \in \mathcal{H}$, the map $f \mapsto \langle \bar{x}(f)\phi, \psi \rangle$ is a bounded linear functional on $C(\mathbb{R}_\infty)$, so by the Riesz representation theorem [227] there exists a complex measure $\mu_{\phi,\psi}$ on $\mathcal{B}(\mathbb{R}_\infty)$ such that $\langle \bar{x}(f)\phi, \psi \rangle = \int f(\lambda) \mu_{\phi,\psi}(d\lambda)$ for all $f \in C(\mathbb{R}_\infty)$. The map $(\phi, \psi) \mapsto \mu_{\phi,\psi}(E)$ is bounded and bilinear, so for any $E \in \mathcal{B}(\mathbb{R}_\infty)$ there exists a bounded operator $x(E)$ on $\mathcal{H}$ such that $\mu_{\phi,\psi}(E) = \langle x(E)\phi, \psi \rangle$. That $x(\cdot)$ is an observable is now a straightforward verification. $\square$

We now give an important example of an observable which is a POV measure but is not a PV measure. Let $\mathcal{H} = L^2(\mathbb{R})$. The usual position observable is the PV measure $Q(\cdot)$ on $\mathcal{B}(\mathbb{R})$ defined by $[Q(E)\psi](\lambda) = \chi_E(\lambda)\psi(\lambda)$. Now let f be a probability density on $\mathbb{R}$; that is, f is a nonnegative measurable function on $\mathbb{R}$ such that $\int f(\lambda) d\lambda = 1$. Then for any bounded measurable function g on $\mathbb{R}$, the convolution

$$(f \circ g)(\lambda) = \int f(\alpha) g(\lambda - \alpha) d\alpha$$

is a bounded continuous function on $\mathbb{R}$. It is straightforward to verify that

$$Q_f(E) = \int (f \circ \chi_E)(\lambda) Q(d\lambda) \tag{4.25}$$

defines an observable. This observable is not a PV measure, in general. We call Q_f an *approximate position observable*, the approximation depending on the function f. The usual position observable is the limiting case as f

approaches the Dirac delta function. An approximate momentum observable is defined analogously. The next theorem shows that position and momentum can be approximately measured simultaneously.

Theorem 4.14 (Davies [54]). *If* $\mathcal{H} = L^2(\mathbb{R})$, *there exists an observable* x *on* $\mathcal{B}(\mathbb{R}^2)$ *whose marginal observables* $x_1(E) = x(E \times \mathbb{R})$, $x_2(E) = x(\mathbb{R} \times E)$ *are respectively approximate position and momentum observables.*

PROOF. Let $(P\psi)(\lambda) = -i\psi'(\lambda)$, $(Q\psi)(\lambda) = \lambda\psi(\lambda)$ be the usual momentum and position operators. Let $\xi \in L^2(\mathbb{R})$ be a unit vector in the domain of P and Q satisfying $\langle P\xi, \xi \rangle = \langle Q\xi, \xi \rangle = 0$. If we define $\xi_{\alpha\beta}(\lambda) = e^{i\beta\lambda}\xi(\lambda - \alpha)$, then an elementary computation shows that

$$\langle P\xi_{\alpha\beta}, \xi_{\alpha\beta} \rangle = \beta, \langle Q\xi_{\alpha\beta}, \xi_{\alpha\beta} \rangle = \alpha. \tag{4.26}$$

If $w \in X$, define the continuous function $w(\alpha, \beta)$ by $w(\alpha, \beta) = (2\pi)^{-1}\langle w\xi_{\alpha,\beta}, \xi_{\alpha,\beta} \rangle$. We claim that for any $s \in S$, $s(\alpha, \beta)$ is a probability density on $\mathbb{R}^2$.

It is clear that $s(\alpha, \beta)$ is nonnegative. To show that $\int s(\alpha, \beta) \, d\alpha \, d\beta = 1$, it is sufficient to assume that s is a one-dimensional projection P_ψ, $\|\psi\| = 1$, since every s is a convex combination of states of this form. In this case

$$\int s(\alpha, \beta) \, d\alpha \, d\beta = \int \left| (2\pi)^{-1/2} \langle \psi, \xi_{\alpha\beta} \rangle \right|^2 d\beta \, d\alpha.$$

Now

$$(2\pi)^{-1/2} \langle \psi, \xi_{\alpha\beta} \rangle = (2\pi)^{-1/2} \int e^{-i\beta\lambda} \left[\psi(\lambda)\bar{\xi}(\lambda - \alpha) \right] d\lambda.$$

Hence by the Plancherel theorem

$$\int \left| (2\pi)^{-1/2} \langle \psi, \xi_{\alpha\beta} \rangle \right|^2 d\beta \, d\alpha = \int |\psi(\lambda)\bar{\xi}(\lambda - \alpha)|^2 \, d\lambda \, d\alpha$$

$$= \int |\psi(\lambda)|^2 |\xi(\alpha)|^2 \, d\lambda \, d\alpha = 1.$$

For $E \in \mathcal{B}(\mathbb{R}^2)$, the map $w \mapsto \int_E w(\alpha, \beta) \, d\alpha \, d\beta$ is a positive linear functional on X. Hence, there exists a bounded positive linear operator $x(E)$ on $\mathcal{H}$ such that

$$\text{tr}[wx(E)] = \int_E w(\alpha, \beta) \, d\alpha \, d\beta \tag{4.27}$$

for every $w \in X$. It is straightforward to check that x is an observable.

We now find the marginal observables. If $\psi \in \mathcal{H}$ and $E \in \mathcal{B}(\mathbb{R})$, then

$$\langle x_1(E)\psi, \psi \rangle = \int_E \int_R |\langle \psi, \xi_{\alpha\beta} \rangle|^2 \, d\beta \, d\alpha$$

$$= \int_E \int_R |\psi(\lambda)\bar{\xi}(\lambda - \alpha)|^2 \, d\lambda \, d\alpha$$

$$= \int_{\mathbb{R}^2} |\psi(\lambda)|^2 \chi_E(\lambda)|\xi(\lambda - \alpha)|^2 \, d\lambda \, d\alpha$$

$$= \int_{\mathbb{R}} |\psi(\lambda)|^2 (\chi_E \circ |\xi|^2)(\alpha) \, d\alpha.$$

Hence,

$$x_1(E) = \int (\chi_E \circ |\xi|^2)(\lambda) Q(d\lambda).$$

In a similar way we obtain

$$x_2(E) = \int (\chi_E \circ |\hat{\xi}|^2)(\lambda) P(d\lambda), \tag{4.28}$$

where $\hat{\xi}$ is the Fourier transform of ξ. $\qquad\Box$

We now consider the composition of instruments in Hilbert space quantum mechanics. In this situation, an instrument is a positive map-valued measure $\mathcal{I}: \mathcal{B}(\mathbb{R}) \to \mathcal{L}^+(X)$ such that $\text{tr}[\mathcal{I}(\mathbb{R})w] = \text{tr}(w)$ for every $w \in X$. Just as Lemma 4.13 gives an equivalent definition of observables, there is a similar equivalent definition of instruments.

Lemma 4.15. *If $\mathcal{I}$ is an instrument, then the map $\bar{\mathcal{I}}(f, w) = \int f(\lambda) \mathcal{I}(d\lambda)w$ is a positive bilinear map from $C(\mathbb{R}_\infty) \times X$ to X such that $\text{tr}[\bar{\mathcal{I}}(1, w)] = \text{tr}(w)$ for all $w \in X$. Conversely, if $\bar{\mathcal{I}}: C(\mathbb{R}_\infty) \times X \to X$ is a positive bilinear map satisfying $\text{tr}[\bar{\mathcal{I}}(1, w)] = \text{tr}(w)$ for all $w \in X$, then there exists a unique instrument $\mathcal{I}$ such that $\bar{\mathcal{I}}(f, w) = \int f(\lambda) \mathcal{I}(d\lambda)w$ for every $(f, w) \in C(\mathbb{R}_\infty) \times X$.*

PROOF. Similar to the proof of Lemma 4.13. $\qquad\Box$

We shall need the following measure-theoretic lemma whose proof may be found in [54].

Lemma 4.16. *Let $B: C(\mathbb{R}_\infty) \times C(\mathbb{R}_\infty) \to \mathbb{R}$ be a bilinear function which is positive [i.e., $f_1, f_2 \geq 0$ implies $B(f_1, f_2) \geq 0$]. There exists a measure μ on $\mathcal{B}(\mathbb{R}_\infty^2)$ such that*

$$B(f_1, f_2) = \int_{\mathbb{R}_\infty^2} f_1(\lambda_1) f_2(\lambda_2) \, d\mu(\lambda_1, \lambda_2)$$

for every $f_1, f_2 \in \mathcal{B}(\mathbb{R}_\infty)$.

Our last theorem shows that the composition of instruments always exists in Hilbert space quantum mechanics.

Theorem 4.17. *If $\mathcal{I}_1$ and $\mathcal{I}_2$ are instruments, there exists a unique instrument $\mathcal{I}$ on $\mathcal{B}(\mathbb{R}^2)$ such that for all $w \in X$, $E_1, E_2 \in \mathcal{B}(\mathbb{R})$, we have*

$$\mathcal{I}(E_1 \times E_2)w = \mathcal{I}_1(E_1)\mathcal{I}_2(E_2)w \qquad (4.29)$$

PROOF. Using the bilinear form definition of an instrument in Lemma 4.15, Eq. (4.29) is equivalent to

$$\bar{\mathcal{I}}(f_1 f_2, w) = \bar{\mathcal{I}}_1\big[f_1, \bar{\mathcal{I}}_2(f_2, w)\big]$$

for every $f_1, f_2 \in C(\mathbb{R}_\infty)$. If $w \in S_+$ and A is a positive compact operator, then

$$B(f_1, f_2) = \mathrm{tr}\big[A\bar{\mathcal{I}}_1(f_1, \bar{\mathcal{I}}_2(f_2, w))\big]$$

is a positive bilinear form, so by Lemma 4.16 there exists a measure $\mu_{A,w}$ on $\mathcal{B}(\mathbb{R}^2_\infty)$ such that

$$\mathrm{tr}\big[A\bar{\mathcal{I}}_1(f_1, \bar{\mathcal{I}}_2(f_2, w))\big] = \int_{\mathbb{R}^2_\infty} f_1(\lambda_1)f_2(\lambda_2)\, d\mu_{A,w}(\lambda_1, \lambda_2)$$

for every $f_1, f_2 \in C(\mathbb{R}_\infty)$. If $E \in \mathcal{B}(\mathbb{R}^2_\infty)$, then $A \mapsto \mu_{A,w}(E)$ is a positive linear functional on the set of self-adjoint compact operators $C_s(\mathcal{H})$. Since the dual of $C_s(\mathcal{H})$ is X [67], there exists an element of S_+ which we call $\mathcal{I}(E)w$ such that $\mu_{A,w}(E) = \mathrm{tr}[A\mathcal{I}(E)w]$ for every $A \in C_s(\mathcal{H})$. Since $w \mapsto \mu_{A,w}(E)$ is a positive linear functional for every positive compact operator A, $w \mapsto \mathcal{I}(E)w$ is a positive linear map which we call $\mathcal{I}(E)$. It is straightforward to verify that $\mathcal{I}$ is an instrument satisfying (4.29). $\qquad \square$

Davies and Lewis [55] have shown that Theorem 4.17 holds for any total convex structure.

4.7. Notes and References

Operational quantum mechanics (or the convexity approach to quantum mechanics) was first systematically explored by Ludwig [170, 171] and Mielnik [187–189]. Its present form was developed in the scholarly work of Davies, Edwards, and Lewis [51–55, 69, 70] although many investigators have made important contributions [49, 50, 119, 129, 130, 162, 215, 216]. The operational approach has not only been important for foundational studies, but it has also proved useful in such practical fields as quantum information theory [132, 133, 136] and quantum optics [54, 158].

The convex structure framework is more general than the usual approach of Davies, Edwards, and Lewis. For more details we refer the reader to [108, 111]. The examples in Section 4.3 and the material on conditional expectations in Section 4.4 are based on the work of Cycon and Hellwig [48]. The results on total convex structures in Section 4.5 are due to the author [107]. The results in Section 4.6 are mainly due to Davies and may be found in [54]. Further investigations in quantum stochastic processes may be found in [1, 2, 58–60, 82–85, 89, 120, 165, 233, 251].

4.8. Exercises

1. If $F \in Af(S)$, show that F^* is linear.

2. If $F, G \in Af(S)$, show that $(FG)^* = G^* F^*$.

3. Prove Lemma 4.1(b).

4. If x is an observable and $F \in Af(S)$, prove that $E \mapsto F^*[x(E)]$ is an observable.

5. Show that S_+ satisfies C1 and C2.

6. Let $s \in S$ be fixed and define $\mathcal{G}(E) : S_+ \to S_+$ by $\mathcal{G}(E)(\alpha, t) = (\alpha x(E), s)$. Show that $\mathcal{G}(E) \in Af(S_+)$.

7. Show that $\mathcal{G}$ defined in Eq. (4.5) is an instrument which determines the observable x.

8. Show that $Af(S_+)$ and $\mathcal{O}(S)$ are semigroups under composition.

9. Show that $\mathcal{K} = \mathcal{G} \circ \mathcal{J}$ in Eq. (4.6).

10. Prove that $\mathcal{G} \circ \mathcal{J}$ is unique if it exists.

11. Show that $f \in S_1'$ is a proposition if and only if f is a characteristic function.

12. Show that $x_g(E) = \chi_{g^{-1}(E)}$ defined in Section 4.3 is an observable.

13. Show that $\mathcal{G}_g$ defined in Eq. (4.7) is an instrument.

14. Prove that x_1 defined in Eq. (4.8) is a POV measure.

15. In Section 4.3 show that the map $F_P(w) = PwP$ is an operation where $w \in S_+$ and P is a projection.

16. Show that $\mathcal{G}_x$ defined in Eq. (4.9) is an instrument.

17. If $z_{x,y}$ is defined as in Section 4.3, prove that $z_{x,y}$ is a PV measure if and only if x and y are compatible.

18. Prove that (4.16) and (4.17) are equivalent.

109

4. The Operational Approach

19. Prove that $x(E)s = x(E)t$ for every observable x and $E \in \mathcal{B}(\mathbb{R})$ if and only if $f(s) = f(t)$ for every $f \in \mathcal{E}(S)$.

20. Let S be a total convex structure and define $\rho(s,t) = \sigma(s,t)[1 - \sigma(s,t)]^{-1}$ for all $s, t \in S$. Prove that ρ is a metric and Theorem 4.10(c) and Lemma 4.1(c) hold with σ replaced by ρ.

21. Show that $\tilde{x}$ and $\tilde{\mathcal{G}}$ satisfy the conditions given in the last paragraph of Section 4.5.

22. If Q is a one-dimensional projection, prove that $QBQ = \mathrm{tr}(QB)Q$ for all $B \in \mathcal{L}(\mathcal{H})$.

23. Prove Eq. (4.21).

24. Show that $\varepsilon_{\mathcal{G}_{x}, w_0}$ is the dual of the instrument $\mathcal{G}_1$ in Eq. (4.22) and that $\mathcal{G}_1$ corresponds to the observable x.

25. Prove Eq. (4.24).

26. Complete the details of Lemma 4.13.

27. Prove that Q_f defined in (4.25) is an observable.

28. Show that Q_f defined in (4.25) need not be a PV measure.

29. Verify Eq. (4.26).

30. Show that x defined by (4.27) is an observable.

31. Verify Eq. (4.28).

32. Prove Lemma 4.15.

33. Complete the proof of Theorem 4.17.

5

Mathematical Interlude

This chapter forms a mathematical interlude in which we introduce concepts and results that will be needed in Chapters 6 and 7. We shall mainly be concerned with the space of tempered distributions $\mathcal{S}'(\mathbb{R}^n)$, orthogonally additive functionals on inner product spaces, and C^*-algebras. Tempered distributions will be needed for our study of random fields, orthogonal additivity will be useful in our discussion of random fields with strongly independent values, and C^*-algebras will be important for our development of quantum field theory.

5.1. Tempered Distributions

Let V be a complex linear space and let $\|\cdot\|_i$ be a family of seminorms on V, $i = 1, 2, \ldots$. We assume that the family $\|\cdot\|_i$ is *separating*; that is, $\|v\|_i = 0$ for every i implies that $v = 0$. We endow V with the weakest topology for which the seminorms $\|\cdot\|_i$ are continuous. Thus, a sequence $v_n \in V$ converges to $v \in V$ in this topology if and only if $\lim_{n \to \infty} \|v_n - v\|_i = 0$ for $i = 1, 2, \ldots$. A neighborhood basis of 0 in this topology is given by sets of the form

$$\{v \in V : \|v\|_{i_1}, \ldots, \|v\|_{i_n} < \varepsilon\}.$$

If V is complete in the above topology, we call V a *Fréchet space*. The following lemma will be useful (completeness is not needed for the proof).

Lemma 5.1. *Let $(V_1, \|\cdot\|_i)$ and $(V_2, |\cdot|_i)$ be Fréchet spaces. A linear map $T: V_1 \to V_2$ is continuous if and only if for any positive integer i, there are positive integers $j_1, \ldots, j_n$ and a $K > 0$ with*

$$|Tv|_i \le K(\|v\|_{j_1} + \cdots + \|v\|_{j_n})$$

for all $v \in V_1$.

PROOF. Sufficiency is clear. For necessity, suppose that T is continuous and i is given. Then $\{v \in V_1 : |Tv|_i < 1\}$ is open. Hence, there exist $j_1, \ldots, j_n$ and $\varepsilon > 0$ such that

$$\{v \in V_1 : \|v\|_{j_1}, \ldots, \|v\|_{j_n} < \varepsilon\} \subseteq \{v \in V_1 : |Tv|_i < 1\}.$$

Therefore,

$$\{v \in V_1 : \|v\|_{j_1} + \cdots + \|v\|_{j_n} < \varepsilon\} \subseteq \{v \in V_1 : |Tv|_i < 1\}.$$

Let $K = 2/\varepsilon$ and suppose there exists a $v \in V_1$ such that

$$|Tv|_i > K(\|v\|_{j_1} + \cdots + \|v\|_{j_n}) > 0.$$

Defining

$$u = (\varepsilon/2)(\|v\|_{j_1} + \cdots + \|v\|_{j_n})^{-1} v$$

we have $\|u\|_{j_1} + \cdots + \|u\|_{j_n} < \varepsilon$. But, $|Tu|_i > 1$, which is a contradiction. □

A Fréchet space that will be especially important to us is the space of rapidly decreasing functions $\mathbb{S}(\mathbb{R}^n)$. To define this space we need some notation. Let I_+^n denote the set of n-tuples of nonnegative integers $\alpha = (\alpha_1, \ldots, \alpha_n)$ and let $I_+ = I_+^1$. For $\alpha = (\alpha_1, \ldots, \alpha_n) \in I_+^n$ we use the notation $|\alpha| = \Sigma \alpha_i$. We frequently denote an element $(x_1, \ldots, x_n) \in \mathbb{R}^n$ by x. If $x = (x_1, \ldots, x_n) \in \mathbb{R}^n$ and $\alpha = (\alpha_1, \ldots, \alpha_n) \in I_+^n$ we define $x^\alpha = x_1^{\alpha_1} \cdots x_n^{\alpha_n}$. Moreover, D^α will denote the operator

$$\frac{\partial^{|\alpha|}}{\partial x_1^{\alpha_1} \cdots \partial x_n^{\alpha_n}}.$$

The space of rapidly decreasing functions $\mathbb{S}(\mathbb{R}^n)$ is the set of infinitely differentiable complex-valued functions $f: \mathbb{R}^n \to C$ for which

$$\|f\|_{\alpha, \beta} \equiv \sup_{x \in \mathbb{R}^n} |x^\alpha D^\beta f(x)| < \infty$$

for every $\alpha, \beta \in I_+^n$. We frequently use the notation $\mathbb{S} = \mathbb{S}(\mathbb{R})$. It is clear that $\|\cdot\|_{\alpha,\beta}, \alpha, \beta \in I_+^n$, form a separating, countable set of seminorms on $\mathbb{S}(\mathbb{R}^n)$. To show that $\mathbb{S}(\mathbb{R}^n)$ is a Fréchet space we need only show completeness.

Lemma 5.2. $\mathcal{S}(\mathbb{R}^n)$ *is a Fréchet space.*

PROOF. Suppose that $f_k \in \mathcal{S}(\mathbb{R}^n)$ is Cauchy in each $\|\cdot\|_{\alpha\beta}$. Then $x^\alpha D^\beta f_k \to g_{\alpha,\beta} \in C(\mathbb{R}^n)$ uniformly as $k \to \infty$ since $C(\mathbb{R}^n)$ is complete. If we can show that $g = g_{0,0}$ is C^∞ and $g_{\alpha,\beta} = x^\alpha D^\beta g$, then it will follow that $g \in \mathcal{S}(\mathbb{R}^n)$ and $f_k \to g$ in the topology of $\mathcal{S}(\mathbb{R}^n)$. But an elementary result of real analysis states that if $h_k \to h$ and $h'_k \to h_0$ uniformly, then h' exists and $h' = h_0$. $\square$

One of the reasons $\mathcal{S}(\mathbb{R}^n)$ is useful is that it is large enough to be complete in a manageable topology, yet small enough to be closed under differentiation and multiplication by polynomials. Since differentiation corresponds to momentum operators and multiplication by x_i corresponds to position operators, $\mathcal{S}(\mathbb{R}^n)$ can be used as a common invariant domain on which these operators are continuous. Moreover, it is not hard to show that $\mathcal{S}(\mathbb{R}^n)$ is dense in $L^2(\mathbb{R}^n)$. Examples of functions in $\mathcal{S}(\mathbb{R}^n)$ are those of the form $p(x)\exp(-\Sigma x_i^2)$, where $p(x)$ is a polynomial in $x_1, \dots, x_n$.

The topological dual of $\mathcal{S}(\mathbb{R}^n)$, denoted by $\mathcal{S}'(\mathbb{R}^n)$, is called the space of *tempered distributions*. Thus $\mathcal{S}'(\mathbb{R}^n)$ is the set of continuous complex linear functionals on $\mathcal{S}(\mathbb{R}^n)$. We endow $\mathcal{S}'(\mathbb{R}^n)$ with the weak*-topology $\sigma(\mathcal{S}', \mathcal{S})$; that is, $F_\gamma \to F$ in $\sigma(\mathcal{S}', \mathcal{S})$ if $F_\gamma(f) \to F(f)$ for every $f \in \mathcal{S}(\mathbb{R}^n)$. Under the $\sigma(\mathcal{S}', \mathcal{S})$-topology $\mathcal{S}'(\mathbb{R}^n)$ becomes a topological linear space. The next lemma is a consequence of Lemma 5.1.

Lemma 5.3. *A linear functional F on $\mathcal{S}(\mathbb{R}^n)$ is in $\mathcal{S}'(\mathbb{R}^n)$ if and only if there exist $(\alpha_1, \beta_1), \dots, (\alpha_n, \beta_n), K > 0$ such that*

$$|Ff| \leq K\big(\|f\|_{\alpha_1,\beta_1} + \cdots + \|f\|_{\alpha_n,\beta_n}\big)$$

for all $f \in \mathcal{S}(\mathbb{R}^n)$.

An important class of elements of $\mathcal{S}'(\mathbb{R}^n)$ are the functions in $L^1(\mathbb{R}^n)$. For $f \in L^1(\mathbb{R}^n)$, define the linear functional $F_f : \mathcal{S}(\mathbb{R}^n) \to C$ by $F_f(\phi) = \int f(x)\phi(x)\,dx$. If we denote the $L^1(\mathbb{R}^n)$ norm by $\|f\|_1 = \int |f(x)|\,dx$, then clearly

$$|F_f(\phi)| \leq \|f\|_1 \|\phi\|_{0,0}. \tag{5.1}$$

It follows that $F_f \in \mathcal{S}'(\mathbb{R}^n)$. If we define $i : L^1(\mathbb{R}^n) \to \mathcal{S}'(\mathbb{R}^n)$ by $i(f) = F_f$, then i gives a natural imbedding of $L^1(\mathbb{R}^n)$ into $\mathcal{S}'(\mathbb{R}^n)$. We frequently use the notation $f(\phi) = F_f(\phi)$ for $f \in L^1(\mathbb{R}^n), \phi \in \mathcal{S}(\mathbb{R}^n)$.

We can also define $F_f \in \mathcal{S}'(\mathbb{R}^n)$ for some measurable functions f which are not in $L^1(\mathbb{R}^n)$. If the measurable function f is *polynomial bounded*, that is, there exists a $K > 0$ and a polynomial $p(x)$ such that $|f(x)| < K|p(x)|$ for

$x \in \mathbb{R}^n$, then $F_f(\phi) = \int f(x)\phi(x)\,dx$ is a tempered distribution. To see this, let us suppose for simplicity that $n = 1$, the general case being similar. For $\phi \in \tilde{S}$ we have $\sup|x^2 p(x)\phi(x)| < \infty$. Hence, there exists a $K_1 > 0$ such that $|p(x)\phi(x)| < K_1 x^{-2}$ for all $x \neq 0$. Hence,

$$\int |f(x)\phi(x)|\,dx$$

$$\leq \|\phi\|_{0,0} \int_{[-1,1]} |f(x)|\,dx + K \int_{|x|>1} |\phi(x)p(x)|\,dx$$

$$\leq K\|\phi\|_{0,0} \int_{[-1,1]} |p(x)|\,dx + KK_1 \int_{|x|>1} \frac{1}{x^2}\,dx < \infty.$$

Another example of an element of $S'(\mathbb{R}^n)$ is the delta function. For $a \in \mathbb{R}^n$, let δ_a be the linear functional $\delta_a(\phi) = \phi(a), \phi \in S(\mathbb{R}^n)$. Since $|\delta_a(\phi)| \leq \|\phi\|_{0,0}, \delta_a \in S'(\mathbb{R}^n)$. There is no function $g(x)$ such that $\delta_a(\phi) = \int g(x)\phi(x)\,dx$ for every $\phi \in S(\mathbb{R}^n)$. For example, consider δ_0 on S and suppose that $\delta_0(\phi) = \int g(x)\phi(x)\,dx$ for every $\phi \in S$. It is clear that g is locally integrable; that is, $\int_{[-a,a]} |g(x)|\,dx < \infty$ for every $a \geq 0$. Now let

$$\phi_b(x) = \begin{cases} \exp[b^2/(x^2 - b^2)] & \text{if } -b < x < b \\ 0 & \text{if } |x| \geq b. \end{cases} \tag{5.2}$$

It is not hard to show that $\phi_b \in S$ for every $b > 0$. We then have $\int g(x)\phi_b(x)\,dx = e^{-1}$ for every $b > 0$. But by the dominated convergence theorem, the integral approaches 0 as $b \to 0$ which is a contradiction. Even though δ_a cannot be represented by a function, there is a measure, namely the probability measure μ_a concentrated at $a \in \mathbb{R}^n$ such that $\delta_a(\phi) = \int \phi(x)\,d\mu_a(x)$ for all $\phi \in S(\mathbb{R}^n)$.

Although δ_a is not of the form $F_f, f \in S(\mathbb{R}^n)$, δ_a is a limit of such tempered distributions in the $\sigma(S', S)$-topology. For example, let

$$f_n(x) = \begin{cases} c_n \exp[-(1 - n^2 x^2)^{-1}] & \text{for } |x| < n \\ 0 & \text{for } |x| \geq n, \end{cases}$$

where

$$c_n^{-1} = \int \exp[-(1 - n^2 x^2)]\,dx$$

for $n = 1, 2, \ldots$. Notice that $f_n = c_n \phi_{n^{-1}}$, where ϕ_b is defined by Eq. (5.2). Hence, $f_n \in S(\mathbb{R})$. It is straightforward to show that

$$\lim_{n \to \infty} \int f_n(x)\phi(x)\,dx = \phi(0) \tag{5.3}$$

for all $\phi \in S(\mathbb{R})$. Hence $F_{f_n} \to \delta_0$ in the $\sigma(S', S)$-topology. Part of the next result shows that any tempered distribution is a weak limit of F_f's, $f \in S(\mathbb{R})$.

Theorem 5.4. *The imbedding* $i : S(\mathbb{R}^n) \to S'(\mathbb{R}^n)$ *is a continuous linear injection of* $S(\mathbb{R}^n)$ *onto a dense subspace of* $S'(\mathbb{R}^n)$.

PROOF. We prove this result for $n = 1$, the more general case being similar. It is clear that i is linear. To show that i is injective, suppose that $F_f = F_g, f, g \in S$. Since S is dense in $L^2(\mathbb{R})$, it follows that $\int f(x)\phi(x)dx = \int g(x)\phi(x)dx$ for all $\phi \in L^2(\mathbb{R})$. Since $f, g \in L^2(\mathbb{R})$ we conclude that $f = g$ almost everywhere. Since f and g are continuous, $f = g$. If we show that $\|\cdot\|_1$ is a continuous norm on S, then it will follow from (5.1) that i is continuous. Suppose $f_k \to 0$ in S. Then for any $\varepsilon > 0$ there exists an N such that $k \geq N$ implies that $\|f_k\|_{0,0}, \|f_k\|_{2,0} < \varepsilon/4$. Hence for $k \geq N$ we have

$$\int |f_k(x)| dx$$

$$= \int_{|x| < 1} |f_k(x)| dx + \int_{|x| \geq 1} |f_k(x)| dx < \frac{\varepsilon}{2} + \frac{\varepsilon}{4} \int_{|x| \geq 1} x^{-2} dx = \varepsilon.$$

Hence, $\|\cdot\|_1$ is continuous on S.

We now show that $i(S)$ is dense in S'. For $\phi \in S$, let ϕ_y be the function in S defined by $\phi_y(x) = \phi(x - y)$. For $F \in S', \phi \in S$, define the function F^ϕ by $F^\phi(y) = F(\phi_y)$. We now show that $F^\phi \in S$. It is easy to show that F^ϕ is continuous. To show that F^ϕ is differentiable, suppose that $y_n \to y_0, y_n \neq y_0$. It easily follows that

$$\lim_{n \to \infty} \frac{\phi_{y_n} - \phi_{y_0}}{y_n - y_0} = \frac{d}{dy} \phi_y \big|_{y = y_0} = -\phi'(x - y_0)$$

in the topology of S. Hence,

$$\lim_{y \to y_0} \frac{F^\phi(y) - F^\phi(y_0)}{y - y_0} = \lim_{y \to y_0} F\left[\frac{\phi_y - \phi_{y_0}}{y - y_0} \right] = -F[\phi'(x - y_0)].$$

Similarly,

$$D^\beta F^\phi(y) = (-1)^\beta F\left[\phi^{(\beta)}(x - y) \right]$$

for all $\beta \in I_+$, so $F^\phi \in C^\infty$. Since $F \in S'$ we conclude from Lemma 5.3 that there exist $(\alpha_1, \beta_1), \ldots, (\alpha_n, \beta_n), K > 0$ such that

$$|F\phi| \leq K \big(\|\phi\|_{\alpha_1, \beta_1} + \cdots + \|\phi\|_{\alpha_n, \beta_n} \big)$$

115

for all $\phi \in \mathcal{S}$. Hence,

$$\|F^\phi\|_{\alpha,\beta}$$

$$= \sup_y |y^\alpha D^\beta F^\phi(y)| = \sup_y |F[y^\alpha \phi^{(\beta)}(x-y)]|$$

$$\leq K(\|\phi\|_{\alpha_1+\alpha,\beta_1+\beta} + \cdots + \|\phi\|_{\alpha_n+\alpha,\beta_n+\beta}).$$

Therefore, $F^\phi \in \mathcal{S}$. Let $\phi_n \in \mathcal{S}$ with $F_{\phi_n} \to \delta_0$ in the $\sigma(\mathcal{S}', \mathcal{S})$-topology (the ϕ_n exist by the paragraph preceding this theorem). If $\phi \in \mathcal{S}$ it is easy to show that

$$\int \phi_{ny}(x)\phi(y)\,dy = \int \phi_n(x-y)\phi(y)\,dy \in \mathcal{S}.$$

Hence,

$$F^{\phi_n}(\phi)$$

$$= \int F^{\phi_n}(y)\phi(y)\,dy = \int F(\phi_{ny})\phi(y)\,dy$$

$$= F\int \phi_{ny}(x)\phi(y)\,dy = F\int \phi_n(x-y)\phi(y)\,dy \to F(\phi)$$

as $n \to \infty$. Thus, $F^{\phi_n} \to F$ in the $\sigma(\mathcal{S}', \mathcal{S})$-topology and $i(\mathcal{S})$ is dense in $\mathcal{S}'$. $\square$

If $S: \mathcal{S}(\mathbb{R}^n) \to \mathcal{S}(\mathbb{R}^n)$ is a continuous linear operator, we define the *adjoint* $S': \mathcal{S}'(\mathbb{R}^n) \to \mathcal{S}'(\mathbb{R}^n)$ of S to be the linear operator defined by $(S'F)(\phi) = F(S\phi)$ for all $F \in \mathcal{S}'(\mathbb{R}^n), \phi \in \mathcal{S}(\mathbb{R}^n)$. The adjoint S' is continuous in the $\sigma(\mathcal{S}', \mathcal{S})$-topology since if $F_\gamma \to F$, then

$$(S'F_\gamma)(\phi) = F_\gamma(S\phi) \to F(S\phi) = (S'F)(\phi).$$

Let $T: \mathcal{S}(\mathbb{R}^n) \to \mathcal{S}(\mathbb{R}^n)$ be a continuous linear operator. Because of the natural imbedding $i: \mathcal{S}(\mathbb{R}^n) \to \mathcal{S}'(\mathbb{R}^n)$, $\mathcal{S}(\mathbb{R}^n)$ can be thought of as a dense subspace of $\mathcal{S}'(\mathbb{R}^n)$. This suggests that we can extend T to a continuous linear operator on $\mathcal{S}'(\mathbb{R}^n)$. In fact, since $i[\mathcal{S}(\mathbb{R}^n)]$ is dense in $\mathcal{S}'(\mathbb{R}^n)$, there is a unique continuous linear extension of $i \circ T: \mathcal{S}(\mathbb{R}^n) \to \mathcal{S}'(\mathbb{R}^n)$ to $\mathcal{S}'(\mathbb{R}^n)$. Frequently, this extension can be found as follows. Suppose there exists a continuous linear operator $S: \mathcal{S}(\mathbb{R}^n) \to \mathcal{S}(\mathbb{R}^n)$ such that the restriction $S'|\mathcal{S}(\mathbb{R}^n) = T$. Then $S': \mathcal{S}'(\mathbb{R}^n) \to \mathcal{S}'(\mathbb{R}^n)$ is the unique continuous linear extension of T to $\mathcal{S}'(\mathbb{R}^n)$. We call this the *extension principle*.

As our first application of the extension principle, let $T: \mathcal{S}(\mathbb{R}^n) \to \mathcal{S}(\mathbb{R}^n)$ be the continuous linear operator $T = D^\beta$. For $f \in \mathcal{S}(\mathbb{R}^n)$, integrating by parts gives

$$(T'f)(\phi) = f(T\phi) = f(D^\beta \phi) = \int f(x) D^\beta \phi(x)\,dx$$

$$= (-1)^{|\beta|} \int D^\beta f(x)\phi(x)\,dx = (-1)^{|\beta|}(D^\beta f)(\phi).$$

Hence $(-1)^{|\beta|} T' | \mathbb{S}(\mathbb{R}^n) = D^\beta$. We thus define $D^\beta : \mathbb{S}'(\mathbb{R}^n) \to \mathbb{S}'(\mathbb{R}^n)$ to be the continuous linear operator $(-1)^{|\beta|} T'$. Hence,

$$(D^\beta F)(\phi) = F[(-1)^{|\beta|} D^\beta \phi]$$

for every $F \in \mathbb{S}'(\mathbb{R}^n), \phi \in \mathbb{S}(\mathbb{R}^n)$. For example,

$$(D^\beta \delta_a)(\phi) = \delta_a[(-1)^{|\beta|} D^\beta \phi] = (-1)^{|\beta|} D^\beta \phi(a)$$

for every $\phi \in \mathbb{S}(\mathbb{R}^n)$.

We now show that $\delta_0 \in \mathbb{S}'$ is the second derivative of a polynomial bounded continuous function. Let $f(x) = x$ for $x \geq 0$ and $f(x) = 0$ for $x < 0$. Then f is a polynomial bounded continuous function which is not differentiable in the classical sense. Integrating by parts twice gives

$$\frac{(d^2 f)}{dx^2}(\phi)$$
$$= f(\phi'') = \int_0^\infty x \phi''(x) \, dx = -\int_0^\infty \phi'(x) \, dx$$
$$= \phi(0) = \delta_0(\phi).$$

Hence, $\delta_0 = D^2 f$. This is typical of tempered distributions as the next theorem shows.

Theorem 5.5. *If* $F \in \mathbb{S}'(\mathbb{R}^n)$, *then* $F = D^\beta f$ *for some polynomial bounded continuous function* f.

PROOF. We give the proof for the case $n = 1$, the general case being similar. Since $F \in \mathbb{S}'$, by Lemma 5.1 there exists a $K > 0$ and an integer $n > 0$ such that

$$|F(\phi)| \leq K \sum_{i,j=0}^{n} \left\| x^i \frac{d^j}{dx^j} \phi \right\|_\infty$$

for every $\phi \in \mathbb{S}$, where $\| \cdot \|_\infty$ is the supremum norm. Let $C_n(\mathbb{R})$ be the Banach space of continuous functions f with $\| x^i f(x) \|_\infty < \infty$ for $i = 1, \ldots, n$ and norm $\| f \| = \sum_{i=0}^{n} \| x^i f(x) \|_\infty$. For $\phi \in \mathbb{S}$, let

$$h(\phi) = (\phi, \phi', \ldots, \phi^{(n)}) \in C_n(\mathbb{R}) \oplus \cdots \oplus C_n(\mathbb{R}).$$

Then h is a continuous injection of $\mathbb{S}$ onto $h(\mathbb{S})$. Define $\bar{F} : h(\mathbb{S}) \to C$ by $\bar{F}(\phi, \phi', \ldots, \phi^{(n)}) = F(\phi)$. Then $\bar{F}$ is linear and, moreover, $\bar{F}$ is continuous

117

since

$$|\bar{F}(\phi,\phi',\ldots,\phi^{(n)})| = |F(\phi)| \le K \sum_{i,j=0}^{n} \|x^i \frac{d^j}{dx^j}\phi\|_\infty$$

$$= K\left[\sum_{i=0}^{n} \|x^i\phi(x)\|_\infty + \cdots + \sum_{i=0}^{n} \|x^i\phi^{(n)}(x)\|_\infty\right]$$

$$= K\|(\phi,\phi',\ldots,\phi^{(n)})\|.$$

By the Hahn–Banach theorem, there is a continuous linear extension $\tilde{F}$ of $\bar{F}$ to $C_n(\mathbb{R}) \oplus \cdots \oplus C_n(\mathbb{R})$. Define $\tilde{F}_i : C_n(\mathbb{R}) \to C$ by

$$\tilde{F}_i\phi = \tilde{F}(0,\ldots,0,\phi,0,\ldots,0), \tag{5.4}$$

where ϕ appears in the ith entry on the right-hand side of (5.4), $i = 0, 1, \ldots, n$. Then $\tilde{F}_i$ is continuous and

$$\tilde{F}(f_0,\ldots,f_n)$$
$$= \tilde{F}[(f_0,0,\ldots,0) + \cdots + (0,\ldots,0,f_n)]$$
$$= \tilde{F}_0 f_0 + \cdots + \tilde{F}_n f_n.$$

By the Hahn–Banach theorem there exists a continuous linear extension $\hat{F}_i$ of $\tilde{F}_i$ to the Banach space $C_0(\mathbb{R})$ of continuous functions which vanish at infinity. By the Riesz representation theorem there exist measures μ_i such that $\hat{F}_i(f) = \int f d\mu_i$ for every $f \in C_0(\mathbb{R}), i = 0, \ldots, n$. Hence, for any $\phi \in \mathcal{S}$ we have

$$F(\phi) = \bar{F}(\phi,\phi',\ldots,\phi^{(n)}) = \tilde{F}_0\phi + \cdots + \tilde{F}_n\phi^{(n)}$$

$$= \sum_{i=0}^{n} \int \phi^{(i)} d\mu_i. \tag{5.5}$$

Define $u_i(x) = \mu_i((0,x])$ and $v_i(x) = \int_0^x u_i(y)\,dy, i = 0, 1, \ldots, n$. By the definition of the Riemann–Stieltjes integral and integration by parts we have

$$\int \phi\,d\mu_i = \int \phi\,du_i = -\int \phi' u_i\,dx = \int \phi'' v_i\,dx, \tag{5.6}$$

$i = 0, 1, \ldots, n$. Now v_i is continuous and from (5.6) it follows that v_i is polynomial bounded, $i = 0, 1, \ldots, n$. Hence, by (5.5) and (5.6)

$$F(\phi) = \sum_{i=0}^{n} v_i(\phi^{(i+2)}) = \sum_{i=0}^{n} (-1)^i v_i^{(i+2)}(\phi), \tag{5.7}$$

Let $(Jg)(x) = \int_0^x g(y)\,dy$. Then $(Jf)(x)$ is a polynomial bounded continuous function if f is such a function. Let

$$f = \sum_{i=0}^{n} (-1)^i J^{n-i} v_i.$$

Then f is a polynomial bounded continuous function and from (5.7) we have

$$F(\phi) = \sum_{i=0}^{n} (-1)^i (D^{i+2} v_i)(\phi)$$

$$= \sum_{i=0}^{n} (-1)^i (D^{n+2} J^{n-i} v_i)(\phi) = (D^{n+2} f)(\phi)$$

for every $\phi \in S$. $\qquad\qquad\square$

As a second application of the extension principle, let $C_b^{\infty}(\mathbb{R}^n)$ be the set of infinitely differentiable functions on $\mathbb{R}^n$ which together with their derivatives are polynomial bounded. If $g \in C_b^{\infty}(\mathbb{R}^n)$, then it is clear that the map $(T\phi)(x) = g(x)\phi(x)$ is a continuous linear operator on $S(\mathbb{R}^n)$. The adjoint T' satisfies

$$(T'f)(\phi) = f(T\phi) = f(g\phi) = \int f(x)g(x)\phi(x)\,dx = (gf)(\phi) = Tf(\phi)$$

for all f, $\phi \in S(\mathbb{R}^n)$. Hence $T = T'$ on $S(\mathbb{R}^n)$ and so we define $(gF)(\phi) = F(g\phi)$ for all $F \in S'(\mathbb{R}^n)$. It is easy to show that

$$(gF)' = g'F + gF' \tag{5.8}$$

for all $g \in C_b^{\infty}(\mathbb{R}^n), F \in S'(\mathbb{R}^n)$.

A third application of the extension principle is the Fourier transform. For $\phi \in S(\mathbb{R}^n)$, we define the *Fourier transform* $\mathcal{F}\phi$ by

$$(\mathcal{F}\phi)(y) = (2\pi)^{-n/2} \int \phi(x) e^{ix\cdot y}\,dx$$

where $x \cdot y = x_1 y_1 + \cdots + x_n y_n$. It is clear that $\mathcal{F}$ is linear and it is not hard to show that $\mathcal{F}\phi \in S(\mathbb{R}^n)$. In fact,

$$(D^{\alpha} \mathcal{F}\phi)(y) = (2\pi)^{-n/2} \int (ix)^{\alpha} \phi(x) e^{ix\cdot y}\,dx, \tag{5.9}$$

where $x^{\alpha} = x_1^{\alpha_1} \cdots x_n^{\alpha_n}$. Moreover, $\mathcal{F}: S(\mathbb{R}^n) \to S(\mathbb{R}^n)$ is continuous. The *inverse Fourier transform* $\tilde{\mathcal{F}}$ is given by

$$(\tilde{\mathcal{F}}\phi)(y) = (2\pi)^{-n/2} \int \phi(x) e^{-ix\cdot y}\,dx$$

and again $\tilde{\mathcal{F}}: S(\mathbb{R}^n) \to S(\mathbb{R}^n)$ is a continuous linear operator. A standard result of Fourier analysis states that

$$\mathcal{F}\tilde{\mathcal{F}} = \tilde{\mathcal{F}}\mathcal{F} = I \qquad (5.10)$$

and hence $\mathcal{F}$ is a continuous automorphism on $\mathcal{S}(\mathbb{R}^n)$. The adjoint $\mathcal{F}'$ satisfies

$$(\mathcal{F}'f)(\phi)(y) = f(\mathcal{F}\phi)(y)$$

$$= (2\pi)^{-n/2} \int \int f(y)\phi(x)e^{ix\cdot y}\,dx\,dy = (\mathcal{F}f)(\phi)(y)$$

for every $f, \phi \in \mathcal{S}(\mathbb{R}^n)$. Hence, $\mathcal{F} = \mathcal{F}'$ on $\mathcal{S}(\mathbb{R}^n)$ and so we define $\mathcal{F}F = \mathcal{F}'F$ for every $F \in \mathcal{S}'(\mathbb{R}^n)$; that is, $(\mathcal{F}F)(\phi) = F(\mathcal{F}\phi)$. Similarly, we define $(\tilde{\mathcal{F}}F)(\phi) = F(\tilde{\mathcal{F}}\phi)$. Equation (5.10) holds for $\mathcal{F} : \mathcal{S}'(\mathbb{R}^n) \to \mathcal{S}'(\mathbb{R}^n)$ and hence, $\mathcal{F}$ is a continuous automorphism on $\mathcal{S}'(\mathbb{R}^n)$.

As an example, let us find the Fourier transform of δ_a. For $\phi \in \mathcal{S}(\mathbb{R}^n)$ we have

$$(\mathcal{F}\delta_a)\phi = \delta_a(\mathcal{F}\phi) = (2\pi)^{-n/2}\delta_a\left[\int \phi(x)e^{ix\cdot y}\,dx \right]$$

$$= (2\pi)^{-n/2} \int \phi(x)e^{ix\cdot a}\,dx = \left[(2\pi)^{-n/2}e^{ix\cdot a}\right]\phi.$$

Hence $\mathcal{F}\delta_a = (2\pi)^{-n/2}e^{ix\cdot a}$. Moreover, it follows that $\tilde{\mathcal{F}}(e^{ix\cdot a}) = (2\pi)^{n/2}\delta_a$.

If $a \in \mathbb{R}^n$, then we define the translation operator $T_a : \mathcal{S}(\mathbb{R}^n) \to \mathcal{S}(\mathbb{R}^n)$ by $(T_a\phi)(x) = \phi(x - a)$. It follows from the extension principle that T_a has a unique continuous extension to $\mathcal{S}'(\mathbb{R}^n)$ given by $(T_aF)(\phi) = F(T_{-a}\phi)$.

The *support* of $\phi \in \mathcal{S}(\mathbb{R}^n)$, denoted $\operatorname{supp}\phi$, is the closure of the set $\{x \in \mathbb{R}^n : \phi(x) \neq 0\}$. Let Ω be an open set of $\mathbb{R}^n$. We say that $F \in \mathcal{S}'(\mathbb{R}^n)$ *vanishes in* Ω if $F(\phi) = 0$ whenever $\operatorname{supp}\phi \subseteq \Omega$. The *support* of F, denoted $\operatorname{supp}F$, is the complement of the largest open set on which F vanishes. For example, $\operatorname{supp}D^\alpha\delta_0 = \{0\}$ for every $\alpha \in I_+^n$. Conversely, we have the following result (for a proof see [87, 135, 236]).

Theorem 5.6. *If* $\operatorname{supp}F = \{0\}$, *then there exist constants* c_α *and a nonnegative integer* m *such that* $F = \sum_{|\alpha| \leq m} c_\alpha D^\alpha \delta_0$.

For $\phi \in \mathcal{S}(\mathbb{R}^n)$ and $\psi \in \mathcal{S}(\mathbb{R}^m)$, we define $\phi \otimes \psi : \mathbb{R}^{n+m} \to \mathbb{C}$ by

$$\phi \otimes \psi(x_1, \ldots, x_{n+m}) = \phi(x_1, \ldots, x_n)\psi(x_{n+1}, \ldots, x_{n+m}).$$

It is clear that $\phi \otimes \psi \in \mathcal{S}(\mathbb{R}^{n+m})$. We now have the important kernel or nuclear theorem (for a proof, see [87, 135, 236]).

Theorem 5.7. *Let* $B(\phi, \psi)$ *be a separately continuous bilinear functional on* $\mathcal{S}(\mathbb{R}^n) \times \mathcal{S}(\mathbb{R}^m)$. *Then there exists a unique* $F \in \mathcal{S}'(\mathbb{R}^{n+m})$ *with* $B(\phi, \psi) = F(\phi \otimes \psi)$ *for every* $\phi \in \mathcal{S}(\mathbb{R}^n), \psi \in \mathcal{S}(\mathbb{R}^m)$.

Applying Theorems 5.5 and 5.7 we obtain the following result.

Corollary 5.8. *Let $B(\phi,\psi)$ be a separately continuous bilinear functional on $S(\mathbb{R}^n) \times S(\mathbb{R}^m)$. Then there exists a polynomial bounded continuous function $f: R^{n+m} \to C$ and $\alpha, \beta \in I^n_+$ such that*

$$B(\phi,\psi) = \int \int f(x,y) D^\alpha \phi(x) D^\beta \psi(y) \, dx \, dy,$$

where $x \in \mathbb{R}^n, y \in \mathbb{R}^m$.

We shall have occasion to use a certain "product" of tempered distributions. If $F \in S'(\mathbb{R}^n)$ and $G \in S'(\mathbb{R}^m)$ we define $F \otimes G \in S'(\mathbb{R}^{n+m})$ by

$$F \otimes G[\phi \otimes \psi] = F(\phi) G(\psi) \tag{5.11}$$

for all $\phi \in S(\mathbb{R}^n), \psi \in S(\mathbb{R}^m)$. Since linear combinations of elements $\phi \otimes \psi$ are dense in $S(\mathbb{R}^{n+m})$, (5.11) defines $F \otimes G$ uniquely.

5.2. Orthogonal Additivity

Let V be a real inner product space with dim $V \geq 2$, inner product $\langle \cdot, \cdot \rangle$ and norm $\|x\| = \langle x,x \rangle^{\frac{1}{2}}$. (Although we consider real spaces for simplicity, many of our results carry over to complex spaces.) If $x \perp y$ (i.e., $\langle x,y \rangle = 0$), then clearly $\|x+y\|^2 = \|x\|^2 + \|y\|^2$. In fact, this might be thought of as a generalized Pythagoras' theorem. This result states that the function $f(x) = \|x\|^2$ is *orthogonally additive* in the sense that $f(x+y) = f(x) + f(y)$ whenever $x \perp y$. In this section we are mainly concerned with a converse. That is, if $f: V \to \mathbb{R}$ is orthogonally additive, is $f(x) = c\|x\|^2$ for some $c \in \mathbb{R}$? Obviously, the answer to this question as it stands is no, since any linear functional is orthogonally additive, for example. Thus, to get a converse we must require some additional conditions for f. We shall show that if f is orthogonally additive and satisfies condition 1 below or conditions 2 and 3 below, then $f(x) = c\|x\|^2$ for some $c \in \mathbb{R}$:

1. $f(x) \geq 0$ for every $x \in V$ (*nonnegativity*);
2. $f(-x) = f(x)$ for every $x \in V$ (*evenness*);
3. if $\lambda_i \to \lambda$, then $f(\lambda_i x) \to f(\lambda x)$ for every $x \in V$ (*hemi-continuity*).

We say that $f: V \to \mathbb{R}$ is *odd* if $f(-x) = -f(x)$ for every $v \in V$.

Theorem 5.9. *Let $f: V \to \mathbb{R}$ be orthogonally additive and hemi-continuous.*

(a) *If f is odd, then f is linear.*
(b) *If f is even, then $f(x) = c\|x\|^2$ for some $c \in \mathbb{R}$.*
(c) *In general, $f(x) = c\|x\|^2 + f_1(x)$, where $c \in \mathbb{R}$ and f_1 is linear.*

PROOF. (a) Let $0 \neq x \in V$. Choose a $y \in V$ such that $y \perp x$ and $\|y\| = \|x\|$. Then $x + y \perp x - y$ and hence

$$f(2x) = f[(x+y) + (x-y)] = f(x+y) + f(x-y) = 2f(x).$$

Replacing x by $x/2$ gives $f(x/2) = f(x)/2$. Induction gives $f(2^n x) = 2^n f(x)$ for every integer n. Moreover,

$$f(3x) - f(y) = f(3x - y) = f[(x+y) + 2(x-y)]$$
$$= f(x+y) + f[2(x-y)] = 3f(x) - f(y).$$

Again, induction gives $f(3^m x) = 3^m f(x)$ for every integer m and hence, $f(2^n 3^m x) = 2^n 3^m f(x)$ for all integers n, m. Since rational numbers of the form $2^n 3^m$ are dense in $\mathbb{R}$, hemi-continuity and oddness gives $f(\alpha x) = \alpha f(x)$ for every $\alpha \in \mathbb{R}$. It follows that $f(\alpha x + \beta y) = \alpha f(x) + \beta f(y)$ for every $\alpha, \beta \in \mathbb{R}$. Thus f is linear on the span of x and y. It follows that f is linear on V.

(b) For $0 \neq x \in V$, choose $y \in V$ as in the proof of (a). Then

$$f(2x) = f(x+y) + f(x-y) = 2f(x) + 2f(y).$$

Similarly, $f(2y) = 2f(x) + 2f(y)$. Replacing x by $x/2$ gives $f(x) = f(y)$ so $f(2x) = 4f(x)$. Replacing x by $x/4$ gives $f(x/2) = 4^{-1} f(x)$ and by induction we obtain $f(2^n x) = 2^{2n} f(x)$ for integer n. Moreover,

$$f(3x) + f(y) = f(3x - y) = f[(x+y) + 2(x-y)]$$
$$= f(x+y) + f[2(x-y)] = 10f(x).$$

Induction gives $f(3^m x) = 3^{2m} f(x)$ and hence $f(2^n 3^m x) = 2^{2n} 3^{2m} f(x)$ for all integers n, m. As in the proof of (a), hemi-continuity and evenness gives $f(\alpha x) = \alpha^2 f(x)$ for $\alpha \in \mathbb{R}$. Now let u be any vector in the span of x and y satisfying $\|u\| = \|x\|$. Then u has the form $u = \alpha x + \beta y$ where $\alpha^2 + \beta^2 = 1$. Hence

$$f(u) = f(\alpha x + \beta y) = \alpha^2 f(x) + \beta^2 f(x) = f(x).$$

It follows that if $f(x) = c$ then for any unit vector u, we have $f(u) = c$. Hence for any $0 \neq v \in V$, we have

$$f(v) = f[\|v\|(v/\|v\|)] = c\|v\|^2.$$

(c) Let $f_2(x) = \frac{1}{2}[f(x) + f(-x)]$, $f_1(x) = \frac{1}{2}[f(x) - f(-x)]$. Then $f = f_1 + f_2$, f_1 is odd, hemi-continuous, and f_2 is even, hemi-continuous. By parts (a) and (b) we have that f_1 is linear and $f_2(x) = c\|x\|^2$ for some $c \in \mathbb{R}$. $\qquad \square$

Notice that Theorem 5.9 does not hold if $\dim V = 1$. Indeed, in this case $x \perp y$ if and only if x or y equals 0. It follows that $f : V \to \mathbb{R}$ is orthogonally additive if and only if $f(0) = 0$. Moreover, in this case hemi-continuity is the same as continuity. Of course, there are odd continuous functions f with

$f(0) = 0$ which are not linear, and even continuous functions g with $g(0) = 0$ which are not of the form $g(x) = cx^2$.

We now show that orthogonality determines the inner product up to a multiplicative constant.

Corollary 5.10. *Let* $\langle \cdot, \cdot \rangle_1$ *be another inner product on* V. *If* $x \perp y$ *implies* $x \perp_1 y$, *then there exists a* $c > 0$ *such that* $\langle u, v \rangle_1 = c \langle u, v \rangle$ *for every* $u, v \in V$.

PROOF. Let $g(w) = \|w\|_1^2$. If $x \perp y$, then $x \perp_1 y$ so $g(x + y) = g(x) + g(y)$. Hence, g is orthogonally additive. Clearly, g is hemi-continuous and even. Hence, by Theorem 5.9(b) there exists a $c > 0$ with $\|w\|_1^2 = g(w) = c\|w\|^2$ for every $w \in V$. Applying the polarization identity gives

$$\langle u, v \rangle_1 = \left[\|u + v\|_1^2 - \|u - v\|_1^2 \right] / 4$$
$$= c \left[\|u + v\|^2 - \|u - v\|^2 \right] / 4 = c \langle u, v \rangle$$

for every $u, v \in V$. $\qquad \square$

We next prove a generalization of the Riesz lemma.

Corollary 5.11. *If* $f : V \to \mathbb{R}$ *is orthogonally additive and satisfies* $|f(x)| \leq M\|x\|$ *for every* $x \in V$, *then* f *is a continuous linear functional. If, in addition,* V *is a Hilbert space, then* $f(x) = \langle x, x_0 \rangle$ *for some* $x_0 \in V$.

PROOF. We can assume that $M > 0$. Clearly f is continuous at 0. Let $0 \neq x \in V$. We first show that $\beta \to 1$ implies that $f(\beta x) \to f(x)$. Let $\beta > 1, y \perp x, \|y\| = 1$, and $u = x + (\beta - 1)^{\frac{1}{2}} \|x\| y$. Then $(u - x) \perp x$ and $(u - \beta x) \perp u$. Now

$$f(u) = f(u - x + x) = f(u - x) + f(x)$$

so $f(u) - f(x) = f(u - x)$, and similarly, $f(\beta x) - f(u) = f(\beta x - u)$. Hence

$$|f(x) - f(\beta x)|$$
$$\leq |f(x) - f(u)| + |f(u) - f(\beta x)|$$
$$= |f(u - x)| + |f(\beta x - u)| \leq M(\|u - x\| + \|\beta x - x\|)$$
$$= M\left[(\beta - 1)^{\frac{1}{2}} \|x\| + \|(\beta - 1)x - (\beta - 1)^{\frac{1}{2}} \|x\| y\| \right]$$
$$\leq M\|x\| \left[2(\beta - 1)^{\frac{1}{2}} + (\beta - 1) \right].$$

Now let $0 < \beta < 1$, $y \perp x, \|y\| = 1$, and $u = \beta x + (1 - \beta)^{\frac{1}{2}} \beta^{\frac{1}{2}} \|x\| y$. Then $(u - \beta x) \perp \beta x$ and $(x - u) \perp u$. Again, since $f(u) - f(\beta x) = f(u - \beta x)$ and

5. Mathematical Interlude

$f(x) - f(u) = f(x - u)$, we have

$$|f(x) - f(\beta x)| \le M\big[\,\|x - u\| + \|u - \beta x\|\,\big]$$
$$= M\big[\,\|(1-\beta)x - (1-\beta)^{\frac{1}{2}}\beta^{\frac{1}{2}}\|x\|\,y\| + (1-\beta)^{\frac{1}{2}}\beta^{\frac{1}{2}}\|x\|\,\big]$$
$$\le M\|x\|\big[\,(1-\beta) + 2(1-\beta)^{\frac{1}{2}}\beta^{\frac{1}{2}}\,\big].$$

It follows that $f(\beta x) \to f(x)$ as $\beta \to 1$. Now suppose that $\lambda_i \to \lambda \ne 0$. Then $\lambda_i/\lambda \to 1$ and hence

$$f(\lambda_i x) = f\big[\,(\lambda_i/\lambda)\lambda x\,\big] \to f(\lambda x).$$

Thus, f is hemi-continuous and by Theorem 5.9(c) there exists a linear functional f_1 and a $c \in \mathbb{R}$ such that $f(x) = c\|x\|^2 + f_1(x)$. Hence,

$$|f_1(x)| \le |f(x)| + |c|\,\|x\|^2 \le M\|x\| + |c|\,\|x\|^2.$$

It follows that f_1 is a continuous linear functional. Similarly,

$$|c|\,\|x\|^2 \le M\|x\| + |f_1(x)| \le (M + \|f_1\|)\|x\|$$

and

$$|c|\,\|x\| \le M + \|f_1\|.$$

This implies that $c = 0$. $\qquad\qquad\qquad\qquad\qquad\qquad\qquad\qquad\square$

We now consider a generalization of orthogonal additivity. A function $f: V \to \mathbb{R}$ is *orthogonally increasing* if $x \perp y$ implies that $f(x+y) \ge f(x)$. For example, if $g: \mathbb{R}^+ \to \mathbb{R}$, where $R^+ = \{\lambda \in \mathbb{R}: \lambda \ge 0\}$, is any nondecreasing function, then $f(x) = g(\|x\|)$ is orthogonally increasing since $x \perp y$ implies that

$$f(x+y) = g(\|x+y\|) \ge g(\|x\|) = f(x).$$

We now show that every orthogonally increasing function is essentially of this form.

Theorem 5.12. *If f is orthogonally increasing, then there exists a countable number of spheres $S_1, S_2, \ldots$ such that f is norm continuous at w if and only if $w \notin \bigcup S_i$. Moreover, there is a nondecreasing function $g: \mathbb{R}^+ \to \mathbb{R}$ such that $f(w) = g(\|w\|)$ for every $w \notin \bigcup S_i$.*

PROOF. Let $\alpha \ge 1$. We shall show that $f(\alpha y) \ge f(y)$ for all $y \in V$. Since the inequality clearly holds if $y = 0$, suppose that $y \ne 0$. Let $x \perp y$ and $\|x\| = (\alpha - 1)^{\frac{1}{2}}\|y\|$. A direct computation gives $(y + x) \perp [(\alpha - 1)y - x]$. Hence

$$f(\alpha y) = f\big[\,y + x + (\alpha - 1)y - x\,\big] \ge f(y + x) \ge f(y).$$

We now show that if $\|y\| > \|x\|$, then $f(y) \geq f(x)$. By definition $f(y) \geq f(0)$ for every $y \in V$ so we can assume that $x \neq 0$. Let V_2 be any two-dimensional subspace containing x and y. Since V_2 is isomorphic to $\mathbb{R}^2$ with the usual Euclidean inner product, we can assume that $V_2 = \mathbb{R}^2$. Let $\theta =$ angle(y, x). Applying L'Hospital's rule to $\lim_{h \to 0} \log \cos h\theta / h$ we conclude that $\lim_{n \to \infty} [\cos(\theta / n)]^n = 1$. Hence there exists an n such that $\alpha = \|x\|^{-1} \|y\| (\cos \theta / n)^n \geq 1$. Define elements $y_0 = \alpha x, y_1, y_2, \ldots, y_n = y$ as follows:

$$\text{angle}(y_i, y_{i-1}) = \theta / n, \qquad \|y_{i-1}\| = \|y_i\| \cos(\theta / n), \qquad i = 1, \ldots, n.$$

Notice that $(y_i - y_{i-1}) \perp y_{i-1}, 1 \leq i \leq n$. Hence,

$$f(y) = f[y_{n-1} + (y_n - y_{n-1})] \geq f(y_{n-1}).$$

Continuing this process gives

$$f(y) \geq f(y_{n-1}) \geq f(y_{n-2}) \geq \cdots \geq f(\alpha x) \geq f(x).$$

Now let $x \neq 0$. We show next that if $f(\alpha_i x) \to f(x)$ as $\alpha_i \to 1$, then f is norm continuous at x. Let x_i be a sequence converging in norm to x. Clearly, we can assume that $x_i \not\perp x$. Let $s_i = \langle x_i, x \rangle \|x\|^{-2} x$ and let $r_i = \|x_i\|^2 \langle x, x_i \rangle^{-1} x$. Then $(r_i - x_i) \perp x_i, (x_i - s_i) \perp s_i, i = 1, 2, \ldots$. Hence,

$$f(r_i) = f(r_i - x_i + x_i) \geq f(x_i) = f(x_i - s_i + s_i) \geq f(s_i).$$

Since the coefficients of x in s_i and r_i converge to 1 it follows that

$$\lim_{i \to \infty} f(r_i) = \lim_{i \to \infty} f(s_i) = f(x),$$

so $\lim f(x_i) = f(x)$.

Let $x \neq 0$ and $S = \{\lambda x : \lambda > 0\}$. We now show that f is norm continuous at every point of S except for a countable set $\hat{S} \subseteq S$. By the above, f restricted to S is nondecreasing (regarded as a function of λ). Hence f restricted to S is continuous except for a countable set $\hat{S} \subseteq S$. We now show that f itself (not just restricted to S) is norm continuous on $S - \hat{S}$. Let $y \in S - \hat{S}$. Then $f(\alpha_i y) \to f(y)$ as $\alpha_i \to 1$, so by the previous paragraph f is norm continuous at y.

Suppose that f is norm continuous at x and $\|y\| = \|x\|$. We shall show that $f(y) = f(x)$ and f is norm continuous at y. We assume that $x \neq 0$, since the result holds trivially otherwise. If $\lambda > 1$, then $\|\lambda x\| > \|y\|$ so by the above, i.e., $f(\lambda x) \geq f(y)$. Letting $\lambda \to 1$ we conclude that $f(x) \geq f(y)$. Similarly, $f(x) \leq f(y)$, so $f(y) = f(x)$. Let $y_i \to y$. As $\|y\| = \|x\| > 0$, we can find a sequence $a_i \in \mathbb{R}$ such that $a_i > 0$ and $\|y_i\| - a_i > 0$. Let

$$x_i = (\|y_i\| + a_i) x / \|y\|, \qquad z_i = (\|y_i\| - a_i) x / \|y\|.$$

Then $\|x_i\| > \|y_i\| > \|z_i\|$ so by the above, $f(x_i) \geq f(y_i) \geq f(z_i)$. Now $x_i \to x, z_i \to x$ so by the continuity of f at x we have $f(y_i) \to f(x) = f(y)$. Hence, f is norm continuous at y.

It now follows that f is norm continuous in V except for a countable number of spheres S_i centered at 0.

Let $x_0 \in V$ be a fixed unit vector. Define $g : \mathbb{R}^+ \to \mathbb{R}$ by $g(\alpha) = f(\alpha x_0)$. Then g is nondecreasing and if $w \notin \cup S_i$, we have

$$f(w) = f(\|w\|x_0) = g(\|w\|),$$

which completes the proof. $\qquad\qquad\qquad\qquad\qquad\qquad\qquad\qquad\qquad\square$

Corollary 5.13. *If $f : V \to \mathbb{R}^+$ is orthogonally additive, then there exists a $c \geq 0$ such that $f(x) = c\|x\|^2$ for every $x \in V$.*

PROOF. If $x \perp y$, then $f(x+y) = f(x) + f(y) \geq f(x)$, so f is orthogonally increasing. Hence, by Theorem 5.12, f is norm continuous except on a countable number of spheres. Let $x \in V, \alpha \in \mathbb{R}$. There exists a $y \in V$ such that $y \perp x$ and f is norm continuous at $\alpha x + y$. If $\alpha_i \to \alpha$, then $\alpha_i x + y \to \alpha x + y$ so

$$f(\alpha_i x) + f(y) = f(\alpha_i x + y) \to f(\alpha x + y) = f(\alpha x) + f(y).$$

Hence, $f(\alpha_i x) \to f(\alpha x)$ and f is hemi-continuous everywhere. Hence, by Theorem 5.9(c), $f(x) = c\|x\|^2 + f_1(x)$, where $c \in \mathbb{R}$ and f_1 is linear. For $\lambda > 0$,

$$c\lambda^2 \|x\|^2 - \lambda f_1(x) = f(-\lambda x) \geq 0.$$

Hence, $f_1(x) \leq c\lambda \|x\|^2$ for every $\lambda > 0$. Letting $\lambda \to 0$ gives $f_1(x) \leq 0$ for all $x \in V$. It follows that $f_1 = 0$. $\qquad\qquad\qquad\qquad\qquad\qquad\square$

Corollary 5.14. *If $f : V \to \mathbb{R}$ is orthogonally additive and $f(x) \geq -M\|x\|^2$ for every $x \in V$ and some $M \geq 0$, then there exists a $c \in \mathbb{R}$ such that $f(x) = c\|x\|^2$.*

PROOF. If $g(x) = f(x) + M\|x\|^2$, then $g : V \to \mathbb{R}^+$ is orthogonally additive. Hence, there exist a $c \geq 0$ such that $g(x) = c\|x\|^2$. Thus, $f(x) = (c - M)\|x\|^2$. $\qquad\qquad\qquad\qquad\qquad\qquad\qquad\qquad\qquad\qquad\qquad\qquad\square$

Corollary 5.15. *If $f : V \to \mathbb{R}$ is orthogonally additive and $f(x) \geq f(x_0)$ for all $x \in V$ and some fixed $x_0 \in V$, then there exists a $c \geq 0$ such that $f(x) = c\|x - x_0\|^2 + f(x_0)$. Moreover, if f is not identically 0, then x_0 is unique.*

PROOF. Let $g(x) = f(x + x_0) - f(x_0)$. Then $g : V \to \mathbb{R}^+$. Let $x \perp y$ and write $x_0 = x_1 + x_2 + x_3$ where x_1 is a multiple of x, x_2 is a multiple of y, and

$x_3 \perp x, x_3 \perp y$. Then

$$g(x+y) = f(x+x_1) + f(y+x_2+x_3) - f(x_0)$$
$$= f(x+x_1+x_2) - f(x_2) + f(y+x_0) - f(x_1) - f(x_0)$$
$$= f(x+x_0) - f(x_3) - f(x_2) - f(x_0) + f(y+x_0) - f(x_1)$$
$$= f(x+x_0) + f(y+y_0) - 2f(x_0) = g(x) + g(y).$$

Hence, $g(x) = c\|x\|^2$ for some $c \geq 0$ and $f(x+x_0) = c\|x\|^2 + f(x_0)$. Hence, $f(x) = c\|x - x_0\|^2 + f(x_0)$. Suppose that f is not identically 0 and $f(x) \geq f(y_0)$ for all $x \in V$. Then $f(y_0) = f(x_0)$ and $f(y_0) = c\|y_0 - x_0\|^2 + f(x_0)$. Since $c \neq 0$, we have $\|y_0 - x_0\| = 0$ so $y_0 = x_0$. $\qquad \square$

We have seen that the function $f_0(x) = c\|x\|^2, c \geq 0$, is characterized by the conditions: $f_0(x) \geq 0, f_0(x) + f_0(y) = f_0(x+y)$ whenever $x \perp y$. We now solve another functional equation using Theorem 5.12 and give an application of the solution.

Let $f_1 : V \to \mathbb{R}$ be the function $f_1(x) = (1 + \|x\|^2)^{-1}$ and define

$$w(x,y) = (2 + \|x\|^2 + \|y\|^2)^{-1}[(1 + \|y\|^2)x + (1 + \|x\|^2)y].$$

Then it is not hard to verify that f_1 satisfies the conditions:

$$f_1(x) \geq 0, \qquad f_1(x) + f_1(y) = f_1[w(x,y)] \tag{5.12}$$

whenever $\langle x,y \rangle = -1$. We shall show that these conditions characterize f_1 to within a constant. First, it is easy to see that $\langle x,y \rangle = -1$ if and only if $y = -\|x\|^{-2}x + z$ where $z \perp x$. It is clear that $w(x,y) = w(y,x)$ and that $w(x,y)$ lies on the straight line $\overline{xy}$ from x to y. Moreover, if $\langle x,y \rangle = -1$, then $w(x,y) \perp w(x,y) - x$ and $w(x,y) \perp w(x,y) - y$ so the line $\overline{Ow(x,y)}$ is perpendicular to the line $\overline{xy}$.

Theorem 5.16. *A function* $f : V \to \mathbb{R}^+$ *satisfies* $f(x) + f(y) = f[w(x,y)]$ *whenever* $\langle x,y \rangle = -1$ *if and only if* $f(x) = f(0)(1 + \|x\|^2)^{-1}$.

PROOF. Sufficiency is straightforward. For necessity we first show that $f(x+y) \leq f(y)$ if $x \perp y$. Clearly the inequality holds if $x = 0$. If $y = 0, x \neq 0$, then $f(x+y) = f(x) = f(0) - f(-\|x\|^{-2}x) \leq f(0) = f(y)$. Now suppose x, $y \neq 0, x \perp y$. Let z be the point on the straight line $\overline{y(x+y)}$ satisfying $\langle z, x+y \rangle = -1$. Then it is easily checked that $w(z, x+y) = y$ and hence $f(x+y) \leq f(x+y) + f(z) = f(y)$. Hence, f is orthogonally decreasing and by the result analogous to Theorem 5.12 we have that f is continuous except on a countable number of spheres in V. Suppose $\langle x,y \rangle = -1$ and f is continuous at x. We now show f is continuous at y. Let r_i be a sequence converging to y. Since $y \neq 0$ we can assume $r_i \neq 0, i = 1, 2, \ldots$. Let $z =$

$(1-\lambda)y+\lambda x$ for some $0<\lambda<1$. We can assume $r_i\neq z, i=1,2,\ldots$. Let $r_i'=(1-\lambda_i)r_i+\lambda_i z$ and $s_i=(1-\mu_i)r_i+\mu_i z$ where

$$\lambda_i=-(1+\langle r_i,z\rangle)(\|z\|^2-\langle r_i,z\rangle)^{-1}$$

and

$$\mu_i=-(1+\|r_i\|^2)(\langle r_i,z\rangle-\|r_i\|^2)^{-1}.$$

Now $\langle r_i,s_i\rangle=\langle z,r_i'\rangle=-1$. Also we see that $s_i\to x$. Indeed,

$$\lim_{i\to\infty}s_i=(\langle z,y\rangle+1)(\langle y,z\rangle-\|y\|^2)^{-1}y-(1+\|y\|^2)(\langle z,y\rangle-\|y\|^2)^{-1}z$$

$$=(\lambda-1)\lambda^{-1}y+\lambda^{-1}z=x.$$

Similarly,

$$\lim_{i\to\infty}r_i'=(\|z\|^2+1)(\|z\|^2-\langle z,y\rangle)^{-1}y-(1+\langle z,y\rangle)(\|z\|^2-\langle y,z\rangle)^{-1}z\equiv z_1.$$

It is easily checked that $\langle z,z_1\rangle=-1$ and since z and z_1 lie on the line $\overline{xy}$ we have $w(z,z_1)=w(x,y)$. It follows that $f(x)+f(y)=f(z_1)+f(z)$. Also since r_i,s_i,z,r_i' are on the line $\overline{zr_i}$ we have

$$f(r_i)+f(s_i)=f(r_i')+f(z)=f(r_i')+f(x)+f(y)-f(z_1).$$

Hence

$$|f(r_i)-f(y)|\leq|f(x)-f(s_i)|+|f(r_i')-f(x)|+|f(x)-f(z_1)|.$$

Let $\varepsilon>0$ be given. Since f is continuous at x there is a $\delta>0$ such that $|x-w|<\delta$ implies $|f(x)-f(w)|<\varepsilon$. Clearly as $\lambda\to0$, $z\to y$ and $z_1\to x$. Let λ be such that $|x-z_1|<\delta/2$. Then for i sufficiently large

$$|r_i'-x|\leq|r_i'-z_1|+|z_1-x|<\delta$$

and hence, $|f(r_i')-f(x)|<\varepsilon$. It follows that if i is sufficiently large, $|f(r_i)-f(y)|<3\varepsilon$ so f is continuous at y. Now let $0\neq x_0\in V$. Since f is continuous on a dense set, there is a vector $0\neq x$ not in the one-dimensional subspace generated by x_0 at which f is continuous. Now there is a vector z such that $\langle z,x_0\rangle=\langle z,x\rangle=-1$. By the above, f is continuous at z and again f is continuous at x_0. Hence f is continuous everywhere on $V-\{0\}$. It follows from the result analogous to Theorem 5.12 that there exists a nonincreasing functon $g:\mathbb{R}^+\to\mathbb{R}^+$ which is continuous on $\mathbb{R}^+-\{0\}$ such that $f(x)=g(\|x\|)$. Let $\lambda>0$ and $\mu\geq1/\lambda$. Then there exist $x,y\in V$ such that $\|x\|=\lambda,\|y\|=\mu$ and $\langle x,y\rangle=-1$. Since $f(x)+f(y)=f(w(x,y))$ we have

$$g(\lambda)+g(\mu)=g(\|w(x,y)\|)$$
$$=g\left[(\mu^2\lambda^2-1)^{\frac{1}{2}}(2+\lambda^2+\mu^2)^{-1/2}\right].$$

Make the change of variables $\alpha = (1+\lambda^2)^{-1}, \beta = (1+\mu^2)^{-1}$, and introduce the function $h(v) = g[(1-v)^{1/2}v^{-1/2}], 0 < v < 1$. Then $h(\alpha) + h(\beta) = h(\alpha + \beta)$ for the relevant range of values for α and β. Since h is continuous, it follows that $h(v) = cv$ for some $c \geq 0$. Hence $g(\lambda) = h[(1+\lambda^2)^{-1}] = c(1+\lambda^2)^{-1}$ for $\lambda > 0$. Therefore $f(x) = g(\|x\|) = c(1 + \|x\|^2)^{-1}$ for $x \neq 0$. If $\|x_0\| = 1$ then $f(x_0) + f(-x_0) = f(0) = c$. Thus $c = f(0)$ and $f(x) = f(0)(1 + \|x\|^2)^{-1}$ for all $x \in V$. $\qquad \square$

As a corollary to this theorem we can obtain a result due to Gleason [90]. Gleason's full result is more general than the one we obtain, however his proof is more sophisticated and requires eight pages. Let $\mathcal{H}$ be a Hilbert space and let $\mathcal{P}(\mathcal{H})$ denote the set of orthogonal projections on $\mathcal{H}$. A *state* is a map $m: \mathcal{P}(\mathcal{H}) \to [0,1]$ satisfying $m(I) = 1$, $m(\Sigma_{i=1} P_i) = \Sigma_{i=1} m(P_i)$ if the P_i's are mutually orthogonal and the first sum is in the strong operator topology. We denote the projection onto the one-dimensional subspace generated by a vector $\phi \neq 0$ by P_ϕ. A state m is *atomic* if there is a P_ϕ such that $m(P_\phi) = 1$. We now prove Gleason's theorem for atomic states.

Corollary 5.17. *Let $\mathcal{H}$ be a real separable Hilbert space of dimension ≥ 3 and let m be an atomic state satisfying $m(P_\phi) = 1$ where $\|\phi\| = 1$. If $P \in \mathcal{P}(\mathcal{H})$, then $m(P) = \langle P\phi, \phi \rangle$.*

PROOF. If $\psi \perp \phi$ then $1 \geq m(P_\psi + P_\phi) = m(P_\psi) + 1$ so $m(P_\psi) = 0$. Now assume $0 \neq \psi \not\perp \phi$. Let $0 \neq \phi_1 \perp \phi$ so that ϕ_1, ϕ generate the two-dimensional subspace spanned by ϕ and ψ, and let ϕ_2 be orthogonal to ϕ_1 and ϕ so that ϕ, ϕ_1, ϕ_2 span a three-dimensional subspace M. Let P_1 be the plane in M satisfying $P_1 = \{x \in M : \langle x, \phi \rangle = 1\}$. Now the plane P_1 can be thought of as a two-dimensional real inner product space with origin ϕ and inner product

$$\langle x, y \rangle_1 = \langle x, \phi_1 \rangle \langle y, \phi_1 \rangle + \langle x, \phi_2 \rangle \langle y, \phi_2 \rangle.$$

Define a function $f: P_1 \to \mathbb{R}^+$ by $f(x) = m(P_x)$. If $\langle x, y \rangle_1 = -1, x, y \in P_1$, then

$$\langle x, y \rangle = \langle x, \phi \rangle \langle y, \phi \rangle + \langle x, \phi_1 \rangle \langle y, \phi_1 \rangle + \langle x, \phi_2 \rangle \langle y, \phi_2 \rangle = 0$$

so $x \perp y$. Now $z = -\langle y, w(x,y) \rangle \langle x, w(x,y) \rangle^{-1} x + y$ is a vector in the subspace generated by x and y which is orthogonal to $w(x,y)$. We now show $x \perp \phi$. Indeed,

$$\langle z, \phi \rangle = -\langle y, w(x,y) \rangle \langle x, w(x,y) \rangle^{-1} + 1.$$

But

$$\langle y, w(x,y) \rangle = \left(2 + \|x\|_1^2 + \|y\|_1^2\right)^{-1}\left(1 + \|x\|_1^2\right)\|y\|^2$$
$$= \left(2 + \|x\|_1^2 + \|y\|_1^2\right)^{-1}\|x\|^2\|y\|^2$$

and

$$\langle x, w(x,y) \rangle = \left(2 + \|x\|_1^2 + \|y\|_1^2\right)^{-1}\left(1 + \|y\|_1^2\right)\|x\|^2$$
$$= \left(2 + \|x\|_1^2 + \|y\|_1^2\right)^{-1}\|x\|^2\|y\|^2.$$

We thus have

$$f(x) + f(y) = m(P_x) + m(P_y) = m(P_x + P_y)$$
$$= m(P_{w(x,y)} + P_z) = m(P_{w(x,y)}) + m(P_z)$$
$$= m(P_{w(x,y)}) = f[w(x,y)].$$

Let $\psi_1 = \langle \psi, \phi \rangle^{-1}$ so that $\psi_1 \in P_1$. Applying Theorem 5.16 we have

$$m(P_\psi) = f(\psi_1) = f(\phi)\left(1 + \|\psi_1\|_1^2\right)^{-1} = \|\psi_1\|^{-2}$$
$$= |\langle \psi, \phi \rangle|^2 = \langle P_\psi \phi, \phi \rangle.$$

We have thus shown that $m(P) = \langle P\phi, \phi \rangle$ for all one-dimensional projections P. Now for arbitrary $P \in \mathscr{P}(\mathcal{H})$, there exist mutually orthogonal one-dimensional projections P_i such that $P = \Sigma P_i$ so

$$m(P) = \sum m(P_i) = \sum \langle P_i \phi, \phi \rangle = \left\langle \sum P_i \phi, \phi \right\rangle = \langle P\phi, \phi \rangle$$

and the proof is complete. $\qquad\qquad\qquad\qquad\qquad\qquad\qquad\qquad\square$

There is another type of orthogonal additivity which is of importance. This type has been studied on function spaces such as $C(\Omega)$, $L^P(\Omega, \Sigma, \mu)$ and $\mathcal{S}(\mathbb{R}^n)$. We shall briefly consider the representation of such orthogonally additive functionals on $\mathcal{S}(\mathbb{R}^n)$.

We say that two functions $\phi, \psi \in \mathcal{S}(\mathbb{R}^n)$ are *L-orthogonal*, denoted $\phi \perp_L \psi$, if $\operatorname{supp}\phi \cap \operatorname{supp}\psi = \varnothing$; that is, ϕ and ψ have disjoint support. Thus $\phi \perp_L \psi$ if and only if $\phi(x)\psi(x) = 0$ for all $x \in \mathbb{R}^n$. The L stands for lattice and the terminology stems from the fact that $\phi \perp_L \psi$ if and only if $|\phi| \wedge |\psi| = 0$, where $\wedge$ stands for the minimum of the two functions. Notice that L-orthogonality is stronger than L^2-orthogonality since $\phi \perp_L \psi = 0$ implies that $\int \phi(x)\psi(x)\,dx = 0$, but the converse need not hold.

A map $F: \mathcal{S}(\mathbb{R}^n) \to C$ is *L-additive* if $F(\phi + \psi) = F(\phi) + F(\psi)$ whenever $\phi \perp_L \psi$. We now give an example of an L-additive functional on $\mathcal{S}$. Let $G: C^{m-1} \times \mathbb{R} \to C$ be a bounded Borel function satisfying $G(0, \ldots, 0, x) = 0$

for all $x \in \mathbb{R}$ and let μ be a finite measure on $\mathcal{B}(\mathbb{R})$. Define $F: \mathcal{S} \to C$ by

$$F(\phi) = \int G[\phi(x), D\phi(x), \ldots, D^m\phi(x), x]\,\mu(dx). \tag{5.13}$$

To show that F is L-additive, suppose that $\phi \perp_L \psi$. If $A = \operatorname{supp} \phi$ and $B = \operatorname{supp} \psi$, since $A \cap B = \varnothing$ we have

$$F(\phi + \psi) = \int_A G[\phi(x), D\phi(x), \ldots, D^m\phi(x), x]\,\mu(dx)$$

$$+ \int_B G[\psi(x), D\psi(x), \ldots, D^m\psi(x), x]\,\mu(dx)$$

$$= F(\phi) + F(\psi).$$

The next result shows that L-additive functionals with certain additional requirements are all of the form (5.13). This result is due to M. M. Rao and the proof can be found in [224]. For $F: \mathcal{S}(\mathbb{R}^n) \to C, \phi \in \mathcal{S}(\mathbb{R}^n)$ we define F_ϕ by $F_\phi(\psi) = F(\psi + \phi) - F(\phi)$.

Theorem 5.18. *Let $F: \mathcal{S}(\mathbb{R}^n) \to C$ be a continuous L-additive functional such that F_ϕ is L-additive for every $\phi \in \mathcal{S}(\mathbb{R}^n)$. Then F has the form*

$$F(\phi) = \int G(\phi(x), D\phi(x), \ldots, D^\alpha\phi(x), x)\,d\mu(x)$$

where $|\alpha| \leq m$ for some m, μ is a measure on $\mathcal{B}(\mathbb{R}^n)$ such that $\mu(A) < \infty$ for compact $A \subseteq \mathbb{R}^n$, and the function $G: C^\nu \times \mathbb{R}^n \to C$ has the following properties:

   (i) $G(\cdot, \ldots, x)$ *is continuous for all $x \in \mathbb{R}^n$;*
   (ii) $G(0, \ldots, 0, x) = 0$ *for all $x \in \mathbb{R}^n$;*
   (iii) $G(\lambda_1, \ldots, \lambda_\nu, \cdot)$ *is a Borel function on $\mathbb{R}^n$ for all $\lambda_1, \ldots, \lambda_\nu \in C$.*

Rao has also proved a converse of the above theorem [224].

5.3. *C*-Algebras*

In this section we develop the basic theory of C^*-algebras. Such algebras give the mathematical framework for another approach to axiomatic quantum mechanics, namely the algebraic approach. We shall briefly consider this approach in the present section.

Let $\mathcal{Q}$ be a complex algebra. An *involution* on $\mathcal{Q}$ is a map $*: \mathcal{Q} \to \mathcal{Q}$ satisfying: $A^{**} = A$, $(\alpha A + \beta B)^* = \bar{\alpha}A^* + \bar{\beta}B^*$, and $(AB)^* = B^*A^*$ for every α, $\beta \in C$, $A, B \in \mathcal{Q}$. An algebra $\mathcal{Q}$ with involution $*$ is called a

*-algebra. We call $\mathcal{C}$ a *normed algebra* if $\mathcal{C}$ is equipped with a norm $\|\cdot\|$ satisfying $\|AB\| \le \|A\| \|B\|$ for all $A, B \in \mathcal{C}$. A *Banach *-algebra* is a complete normed *-algebra $\mathcal{C}$ satisfying $\|A^*\| = \|A\|$ for all $A \in \mathcal{C}$. A *C*-algebra* is a complete normed *-algebra satisfying $\|A^*A\| = \|A\|^2$ for all $A \in \mathcal{C}$. A C^*-algebra is a Banach *-algebra. Indeed, if $\mathcal{C}$ is a C^*-algebra, then for all $A \in \mathcal{C}$ we have

$$\|A\|^2 = \|A^*A\| \le \|A^*\| \|A\|.$$

Hence, $\|A\| \le \|A^*\|$ and replacing A by A^* gives $\|A^*\| = \|A\|$ so $\mathcal{C}$ is a Banach *-algebra. We shall always assume that a *-algebra contains an identity I. Since

$$I^*I = (I^*I)^* = I^{**}I$$

we have $I^* = I$, and similarly $0^* = 0$.

There are many examples of C^*-algebras. The most common are the complex numbers C, the continuous functions $C(\Omega)$ on a compact Hausdorff space Ω, and the set of bounded linear operators $\mathcal{L}(\mathcal{K})$ on a Hilbert space $\mathcal{K}$. In these examples, the operations of addition, multiplication, scalar multiplication, and the norms are defined as usual. The involution in the first two examples is given by complex conjugation and in the third example the involution is the adjoint operation. The first two examples are commutative and the third is a noncommutative C^*-algebra.

Let $\mathcal{C}$ be a *-algebra. An element $A \in \mathcal{C}$ is *self-adjoint* if $A = A^*$; A is *positive* if $A = B^*B$ for some $B \in \mathcal{C}$. In the algebraic approach to axiomatic quantum mechanics, the self-adjoint elements of $\mathcal{C}$ represent physical observables. A *positive linear functional* on $\mathcal{C}$ is a linear functional $\omega: \mathcal{C} \to C$ which satisfies $\omega(A^*A) \ge 0$ for all $A \in \mathcal{C}$. A *state* is a positive linear functional ω for which $\omega(I) = 1$. The states represent expectation functionals corresponding to physical states. The following lemma gives some of the important properties of positive linear functionals on Banach *-algebras.

Lemma 5.19. *Let ω be a positive linear functional on a Banach *-algebra $\mathcal{C}$.*

(a) $\overline{\omega(A^*)} = \omega(A)$ *for all $A \in \mathcal{C}$*:

(b) $|\omega(A^*B)|^2 \le \omega(A^*A)\omega(B^*B)$ *for all $A, B \in \mathcal{C}$ (Schwarz's inequality);*

(c) $|\omega(A)| \le \omega(I)\|A\|$ *for all $A \in \mathcal{C}$, and hence is continuous.*

PROOF. (a) Suppose A is self-adjoint. Then

$$\omega(I) + 2\omega(A) + \omega(A^*A) = \omega\big[(I+A)^*(I+A)\big] \ge 0.$$

Since $\omega(A^*A) \ge 0$ and $\omega(I) = \omega(I^*I) \ge 0$, we conclude that $\omega(A)$ is real. If A

is arbitrary, then we can write $A = A_1 + iA_2$, where

$$A_1 = \frac{1}{2}(A + A^*), \qquad A_2 = \frac{1}{2i}(A - A^*)$$

are self-adjoint. Since $\omega(A_1)$ and $\omega(A_2)$ are real we have

$$\omega(A^*) = \omega(A_1 - iA_2) = \omega(A_1) - i\omega(A_2)$$
$$= \overline{\omega(A_1) + i\omega(A_2)} = \overline{\omega(A)}.$$

(b) Applying (a) we have for any $\lambda \in C$

$$\omega(A^*A) - \lambda\omega(A^*B) - \bar{\lambda}\,\overline{\omega(A^*B)} + \lambda\bar{\lambda}\omega(B^*B) = \omega[(A - \lambda B)^*(A - \lambda B)] \geq 0.$$

If $\omega(B^*B) \neq 0$, let $\lambda = \overline{\omega(A^*B)}/\omega(B^*B)$. If $\omega(B^*B) = 0$, let $\lambda = n\overline{\omega(A^*B)}$ and let $n \to \infty$.

(c) Suppose that A is self-adjoint and that $\|A\| \leq 1$. The binomial series

$$(1 - \lambda)^{1/2} = 1 - \frac{1}{2}\lambda - \frac{1}{2!} \cdot \frac{1}{2} \cdot \frac{1}{2}\lambda^2 - \frac{1}{3!} \cdot \frac{1}{2} \cdot \frac{1}{2} \cdot \frac{3}{2}\lambda^3 - \cdots$$

converges absolutely for $|\lambda| \leq 1$. Since $\mathcal{Q}$ is complete and $\|A\| \leq 1$, if we replace λ by A and 1 by I, the resulting series

$$I - \frac{1}{2}A - \frac{1}{2!} \cdot \frac{1}{2} \cdot \frac{1}{2}A^2 - \frac{1}{3!} \cdot \frac{1}{2} \cdot \frac{1}{2} \cdot \frac{3}{2}A^3 - \cdots$$

converges absolutely in $\mathcal{Q}$. Since the involution is continuous, the sum B of this series is self-adjoint. Moreover, it is easily verified that $B^*B = B^2 = I - A$. Hence, $\omega(I - A) = \omega(B^*B) \geq 0$ so $\omega(A) \leq \omega(I)$. Replacing A by $-A$ gives $-\omega(A) \leq \omega(I)$ so $|\omega(A)| \leq \omega(I)$. Now suppose that $A \neq 0$ is an arbitrary self-adjoint element (if $A = 0$ the inequality obviously holds). If we let $A_1 = A/\|A\|$, then A_1 is self-adjoint and $\|A_1\| = 1$. Hence by the above, $|\omega(A_1)| \leq \omega(I)$ so $|\omega(A)| \leq \omega(I)\|A\|$. Finally, suppose that B is an arbitrary element of $\mathcal{Q}$. Then B^*B is self-adjoint so

$$\omega(B^*B) \leq \omega(I)\|B^*B\| \leq \omega(I)\|B^*\|\,\|B\| = \omega(I)\|B\|^2.$$

Setting $A = I$ in Schwarz's inequality (b) gives

$$|\omega(B)|^2 \leq \omega(I)\omega(B^*B) \leq \omega(I)^2\|B\|^2.$$

The inequality now follows. $\qquad\qquad\qquad\qquad\qquad\qquad\qquad\square$

It is frequently useful to represent a Banach *-algebra as a *C**-algebra of operators on a complex Hilbert space. If $\mathcal{Q}$ is a Banach *-algebra, a *representation* of $\mathcal{Q}$ is a *-homomorphism π from $\mathcal{Q}$ into $\mathcal{L}(\mathcal{K})$, where $\mathcal{K}$ is a complex Hilbert space. That is, $\pi: \mathcal{Q} \to \mathcal{L}(\mathcal{K})$ preserves addition, multiplication, scalar multiplication, and involution. A representation $\pi: \mathcal{Q} \to \mathcal{L}(\mathcal{K})$ is *continuous* if π is a continuous map; that is, if $\|\pi(A)\| \leq M\|A\|$ for all $A \in \mathcal{Q}$ and some fixed $M \geq 0$. It is an interesting fact that representations, which are required to preserve only the algebraic operations, are automatically continuous. We thus obtain topological properties free.

Lemma 5.20. *Every representation* π: $\mathcal{Q} \rightarrow \mathcal{L}(\mathcal{H})$ *of a Banach* *-algebra $\mathcal{Q}$ is continuous. In fact,* $\|\pi(A\| \leq \|A\|$ *for every* $A \in \mathcal{Q}$.

PROOF. Let $\phi \in \mathcal{H}$ and define $\omega(A) = \langle \pi(A)\phi, \phi \rangle$, $A \in \mathcal{Q}$. Since

$$\omega(A^*A) = \langle \pi(A^*A)\phi, \phi \rangle = \langle \pi(A)^*\pi(A)\phi, \phi \rangle$$
$$= \langle \pi(A)\phi, \pi(A)\phi \rangle \geq 0 \qquad (5.14)$$

we see that ω is a positive linear functional on $\mathcal{Q}$. Applying Lemma 5.19(c) gives

$$|\langle \pi(A)\phi, \phi \rangle| \leq \|A\| \langle \phi, \phi \rangle.$$

Replacing A by A^*A we obtain

$$\|\pi(A)\phi\|^2 = \langle \pi(A)\phi, \pi(A)\phi \rangle = \langle \pi(A^*A)\phi, \phi \rangle$$
$$\leq \|A^*A\| \langle \phi, \phi \rangle \leq \|A^*\| \|A\| \|\phi\|^2 = \|A\|^2 \|\phi\|^2.$$

Since $\phi \in \mathcal{H}$ was arbitrary, $\|\pi(A)\| \leq \|A\|$. $\square$

We say that two representations π_1: $\mathcal{Q} \rightarrow \mathcal{L}(\mathcal{H}_1)$, π_2: $\mathcal{Q} \rightarrow \mathcal{L}(\mathcal{H}_2)$ are *equivalent* if there exists an isometry U from $\mathcal{H}_1$ onto $\mathcal{H}_2$ such that $\pi_1(A) = U^{-1}\pi_2(A)U$ for every $A \in \mathcal{Q}$. A representation π: $\mathcal{Q} \rightarrow \mathcal{L}(\mathcal{H})$ is *cyclic* if there exists a vector $\phi \in \mathcal{H}$ such that $\{\pi(A)\phi : A \in \mathcal{Q}\}$ is dense in $\mathcal{H}$. We then call ϕ a *cyclic vector*. Although we shall not give the details here, any representation can be decomposed into a direct sum (or integral) of cyclic representations. Thus, cyclic representations are the basic building blocks for any representation and hence the study of cyclic representations is of fundamental importance. It turns out that a cyclic representation can be uniquely described (to within equivalence) by a positive linear functional and conversely, a positive linear functional generates a unique (again to within inequivalence) cyclic representation. This important fact is given by the next theorem and is called the GNS construction (Gelfand and Naimark [194] and Segal [238]).

Theorem 5.21. *Let* $\mathcal{Q}$ *be a Banach* *-algebra. To every cyclic representation π of $\mathcal{Q}$ with cyclic vector ϕ, there corresponds a positive linear functional* $\omega(A) = \langle \pi(A)\phi, \phi \rangle$. *The representation π is determined uniquely to within equivalence by the functional ω. Conversely, to every positive linear functional ω on $\mathcal{Q}$ there corresponds a cyclic representation π with a cyclic vector ϕ such that* $\omega(A) = \langle \pi(A)\phi, \phi \rangle$ *for all* $A \in \mathcal{Q}$.

PROOF. Let $\pi\colon \mathcal{Q}\to\mathcal{L}(\mathcal{H})$ be a cyclic representation with cyclic vector ϕ. If we define $\omega(A)=\langle\pi(A)\phi,\phi\rangle$, then ω is clearly linear and is positive by (5.14). Now suppose that $\pi_1\colon \mathcal{Q}\to\mathcal{L}(\mathcal{H}_1)$ is another cyclic representation with cyclic vector ϕ_1, $\omega_1(A)=\langle\pi_1(A)\phi_1,\phi_1\rangle$, and $\omega_1=\omega$. We shall show that π and π_1 are equivalent. Define the map

$$U\colon \{\pi(A)\phi:\phi\in\mathcal{Q}\}\to\{\pi_1(A)\phi_1:A\in\mathcal{Q}\}$$

by $U\pi(A)\phi=\pi_1(A)\phi_1$. The map U is well defined since if $\pi(A)\phi=\pi(B)\phi$, then

$$\begin{aligned}
\|\pi_1(A)\phi_1-\pi_1(B)\phi_1\|^2 &= \langle\pi_1(A-B)\phi_1,\pi_1(A-B)\phi_1\rangle\\
&= \langle\pi_1[(A-B)^*(A-B)]\phi_1,\phi_1\rangle = \omega_1[(A-B)^*(A-B)]\\
&= \omega[(A-B)^*(A-B)] = \langle\pi[(A-B)^*(A-B)]\phi,\phi\rangle\\
&= \|\pi(A)\phi-\pi(B)\phi\|^2 = 0.
\end{aligned}$$

It is clear that U is linear. We now show that U is isometric.

$$\begin{aligned}
\langle U\pi(A)\phi, U\pi(B)\phi\rangle &= \langle\pi_1(A)\phi_1,\pi_1(B)\phi_1\rangle\\
&= \langle\pi_1(B^*A)\phi_1,\phi_1\rangle = \omega_1(B^*A) = \omega(B^*A)\\
&= \langle\pi(B^*A)\phi,\phi\rangle = \langle\pi(A)\phi,\pi(B)\phi\rangle
\end{aligned}$$

Since $\{\pi(A)\phi:A\in\mathcal{Q}\}\subseteq\mathcal{H}$ and $\{\pi_1(A)\phi_1:A\in\mathcal{Q}\}\subseteq\mathcal{H}_1$ are dense, U is an isometry from a dense subspace in $\mathcal{H}$ onto a dense subspace of $\mathcal{H}_1$. Hence U has a unique extension to an isometry from $\mathcal{H}$ onto $\mathcal{H}_1$. Moreover, for $\psi=\pi(B)\phi\in\mathcal{H}$ we have

$$\begin{aligned}
U^{-1}\pi_1(A)U\psi &= U^{-1}\pi_1(A)\pi_1(B)\phi_1 = U^{-1}\pi_1(AB)\phi_1\\
&= \pi(AB)\phi = \pi(A)\pi(B)\phi = \pi(A)\psi.
\end{aligned}$$

Hence, $U^{-1}\pi_1(A)U=\pi(A)$ so π_1 and π are equivalent.

Conversely, let ω be a positive linear functional on $\mathcal{Q}$. Let $\mathcal{I}=\{A\in\mathcal{Q}:\omega(A^*A)=0\}$. We now show that $\mathcal{I}$ is a left ideal in $\mathcal{Q}$. If $A\in\mathcal{I}$, then clearly $\lambda A\in\mathcal{I}$ for all $\lambda\in C$. If $A_1, A_2\in\mathcal{I}$, then by Schwarz's inequality we have

$$\begin{aligned}
\omega[(A_1+A_2)^*(A_1+A_2)] &= |\omega(A_1^*A_1)+\omega(A_1^*A_2)+\omega(A_2^*A_1)+\omega(A_2^*A_2)|\\
&\le |\omega(A_1^*A_2)|+|\omega(A_2^*A_1)| \le 2\omega(A_1^*A_1)^{1/2}\omega(A_2^*A_2)^{1/2}\\
&= 0.
\end{aligned}$$

Finally, if $A\in\mathcal{Q}$ and $B\in\mathcal{Q}$, then setting $C=A^*B^*B$ we have by Schwarz's inequality

$$\begin{aligned}
\omega[(BA)^*BA] &= \omega(A^*B^*BA) = |\omega(CA)|\\
&\le \omega(CC^*)^{1/2}\omega(A^*A)^{1/2} = 0.
\end{aligned}$$

The factor space $\mathcal{C}/\mathcal{I}$ consists of the residue classes $[A] \equiv A + \mathcal{I}$, $A \in \mathcal{C}$. It is standard to show that $\mathcal{C}/\mathcal{I}$ is a vector space under the operations $\lambda[A] = [\lambda A]$, $[A] + [B] = [A + B]$. For $[A], [B] \in \mathcal{C}/\mathcal{I}$ we define $\langle [A], [B] \rangle = \omega(B^*A)$. This expression is well defined. In fact, if $[A] = [A_1]$, $[B] = [B_1]$, then $A_1 = A + A_0$, $B_1 = B + B_0$ where $A_0, B_0 \in \mathcal{I}$. Hence,

$$\langle [A_1], [B_1] \rangle = \omega[(B + B_0)^*(A + A_0)]$$

$$= \omega(B^*A) + \omega(B_0{}^*A) + \omega(B^*A_0) + \omega(B_0{}^*A_0).$$

Since the last three terms equal 0 by Schwarz's inequality, we conclude that $\langle [A_1], [B_1] \rangle = \langle [A], [B] \rangle$.

We now show that $\langle \cdot, \cdot \rangle$ is an inner product on $\mathcal{C}/\mathcal{I}$. Linearity in the first argument is clear. That $\langle [A], [B] \rangle = \overline{\langle [B], [A] \rangle}$ follows from Lemma 5.19(a). Positive definiteness holds since $\langle [A], [A] \rangle = \omega(A^*A) \geq 0$ and $\langle [A], [A] \rangle = 0$ implies that $\omega(A^*A) = 0$ so $[A] = \mathcal{I}$, which is the zero element of $\mathcal{C}/\mathcal{I}$.

Let $\mathcal{H}$ be the Hilbert space completion of $\mathcal{C}/\mathcal{I}$ relative to the inner product $\langle \cdot, \cdot \rangle$. We now construct a representation in the space $\mathcal{H}$. For $[B] \in \mathcal{C}/\mathcal{I}$ and $A \in \mathcal{C}$, define $\pi(A)[B] = [AB]$. The map $\pi(A): \mathcal{C}/\mathcal{I} \to \mathcal{C}/\mathcal{I}$ is well defined. Indeed, if $[B] = [C]$ then $C = C + B_0$ where $B_0 \in \mathcal{I}$ and $AC = AB + AB_0$. Since $\mathcal{I}$ is a left ideal, $AB_0 \in \mathcal{I}$ so $[AC] = [AB]$. It is clear that $\pi(A)$ is a linear operator on $\mathcal{C}/\mathcal{I}$ for every $A \in \mathcal{C}$. We now show that $\pi(A)$ is bounded. Define $\omega_1: \mathcal{C} \to \mathbb{C}$ by $\omega_1(C) = \omega(B^*CB)$ for some fixed $B \in \mathcal{C}$. Then ω_1 is clearly linear and moreover, ω_1 is positive since

$$\omega_1(C^*C) = \omega(B^*C^*CB) = \omega[(CB)^*CB] \geq 0.$$

By Lemma 5.19(c) we have $|\omega_1(C)| \leq \omega_1(I) \|C\|$. In particular, for $C = A^*A$ we have

$$\omega_1(A^*A) \leq \omega_1(I) \|A^*A\| \leq \omega_1(I) \|A\|^2$$

$$= \omega(B^*B) \|A\|^2 = \|A\|^2 \|[B]\|^2.$$

Hence,

$$\|\pi(A)[B]\|^2 = \langle \pi(A)[B], \pi(A)[B] \rangle = \langle [AB], [AB] \rangle$$

$$= \omega(B^*A^*AB) = \omega_1(A^*A) \leq \|A\|^2 \|[B]\|^2$$

and $\pi(A)$ is a bounded linear operator with norm

$$\|\pi(A)\| \leq \|A\|. \tag{5.15}$$

It follows that $\pi(A)$ can be uniquely extended to a bounded linear operator on $\mathcal{H}$, which we also denote by $\pi(A)$, and that the norm of the extended operator also satisfies (5.15).

The map $A \mapsto \pi(A)$ is a representation of $\mathcal{C}$. Indeed, it is straightforward to show that π preserves addition, multiplication and scalar multiplication. To show that $\pi(A^*) = \pi(A)^*$, let $[B], [C] \in \mathcal{C}/\mathcal{I}$. Then

$$\langle \pi(A)[B], [C] \rangle = \langle [AB], [C] \rangle = \omega(C^*AB)$$
$$= \omega[(A^*C)^*B] = \langle [B], [A^*C] \rangle = \langle [B], \pi(A^*)[C] \rangle.$$

Hence, $\pi(A)^* = \pi(A^*)$ and π preserves the involution.

We now show that π is cyclic with cyclic vector $[I]$. Indeed,

$$\{\pi(A)[I] : A \in \mathcal{C}\} = \{[A] : A \in \mathcal{C}\} = \mathcal{C}/\mathcal{I}$$

and $\mathcal{C}/\mathcal{I}$ is dense in $\mathcal{H}$.

Finally, for any $A \in \mathcal{C}$ we have

$$\omega(A) = \omega(I^*AI) = \langle [AI], [I] \rangle = \langle \pi(A)[I], [I] \rangle$$

and the proof is complete. $\square$

Of course, the only difference between states and arbitrary positive linear functionals is a normalization constant. Theorem 5.21 could just as well have been phrased in terms of states. The only change needed is that the cyclic vector must be a unit vector. Theorem 5.21 gives the connection between the algebraic approach and Hilbert space quantum mechanics. The philosophy of the algebraic approach is that a quantum mechanical system is described by the C^*-algebra $\mathcal{C}$ generated by the system's bounded observables. One then concludes that Hilbert space quantum mechanics is too restrictive. The Hilbert space that represents $\mathcal{C}$ depends on the particular state of the system and different states give inequivalent representations in general.

For applications to quantum field theory it is important to generalize some of the above concepts to *-algebras. Whereas a C^*-algebra represents the algebra of bounded observables, a *-algebra represents all observables including the unbounded ones. Notice that Lemma 5.19(a) and (b) as well as their proofs go through unchanged for *-algebras.

There is a generalization of the GNS construction for *-algebras. Analogous to the fact that the GNS construction gives a representation of a Banach *-algebra as bounded operators on a Hilbert space, the generalization gives a representation of a *-algebra as unbounded operators (in general) on a Hilbert space. A *representation* π of a *-algebra $\mathcal{C}$ on a Hilbert space $\mathcal{H}$ is a map of $\mathcal{C}$ into a set of linear operators all defined on a common dense domain $D(\pi) \subseteq \mathcal{H}$ satisfying for all $A, B \in \mathcal{C}, \alpha, \beta \in C$:

(i) $\pi(I) = I$, $\pi(\alpha A + \beta B) = \alpha \pi(A) + \beta \pi(B)$;

(ii) $\pi(A)D(\pi) \subseteq D(\pi)$ and $\pi(AB) = \pi(A)\pi(B)$;

(iii) $\langle \pi(A)\phi, \psi \rangle = \langle \phi, \pi(A^*)\psi \rangle$ for every $\phi, \psi \in D(\pi)$.

Property (iii) is equivalent to $\pi(A^*) \subseteq \pi(A)^*$ for every $A \in \mathcal{C}$.

137

If π is a representation of $\mathcal{C}$, the *induced topology* on $D(\pi)$ is the weakest topology for which the maps $\pi(A): D(\pi) \to D(\pi)$ are continuous for every $A \in \mathcal{C}$. The induced topology is generated by the collection of seminorms $\{\phi \mapsto \|\pi(A)\phi\| : A \in \mathcal{C}\}$. We call π a *closed representation* if $D(\pi)$ is complete in the induced topology. A representation of $\mathcal{C}$ is *cyclic* if there exists a vector $\phi \in D(\pi)$ such that $\{\pi(A)\phi : A \in \mathcal{C}\}$ is dense in $D(\pi)$ in the induced topology. We call ϕ a *cyclic vector* for π. The proof of the next theorem follows that of the GNS construction very closely.

Theorem 5.22. *For any positive linear functional ω on a *-algebra $\mathcal{C}$ there exists a closed cyclic representation π with cyclic vector $\phi \in D(\pi)$ such that $\omega(A) = \langle \pi(A)\phi, \phi \rangle$ for every $A \in \mathcal{C}$. Moreover, π is determined by ω up to an equivalence.*

An example of a *-algebra is an algebra of linear operators on a Hilbert space $\mathcal{H}$ with a common dense invariant domain $D \subseteq \mathcal{H}$ such that $D(A^*) \supseteq D$ for every $A \in \mathcal{C}$. We now give another example of a *-algebra which will be useful in Chapter 7. The appearance of $\mathbb{R}^4$ in this example is due to the fact that $\mathbb{R}^4$ represents physical space–time.

Let $\mathcal{S}_0 = C$ and $\mathcal{S}_n = \mathcal{S}(\mathbb{R}^{4n})$, $n = 1, 2, \ldots$. We denote by $\mathcal{S}$ the space of sequences $f = (f_0, f_1, \ldots)$ satisfying:

(i) $f_i \in \mathcal{S}_i$;

(ii) only a finite number of the f_i's are different from zero.

We now define algebraic operations on $\mathcal{S}$. Let $f = (f_0, f_1, \ldots)$, $g = (g_0, g_1, \ldots)$ $\in \mathcal{S}$. We define sums and scalar multiplication component-wise. Multiplication is defined by

$$fg = \left(f_0 g_0, f_0 g_1 + f_1 g_0, \ldots, \sum_{i+j=n} f_i \times g_j, \ldots \right),$$

where $f_i \times g_j$ is a function of $4(i+j)$ variables given by

$$(f_i \times g_j)(x_1, \ldots, x_{i+j}) = f_i(x_1, \ldots, x_i) g_j(x_{i+1}, \ldots, x_{i+j}),$$

where $x_k \in \mathbb{R}^4$. The involution is defined by $f^* = (f_0^*, f_1^*, \ldots)$, where

$$f_0^* = \bar{f}_0, \qquad f_i^*(x_1, \ldots, x_i) = \bar{f}_i(x_i, x_{i-1}, \ldots, x_1), \tag{5.16}$$

and $x_k \in \mathbb{R}^4$.

The proof of the following lemma is straightforward.

Lemma 5.23. *The space $\mathcal{S}$ is a *-algebra under the above operations.*

Notice that the *-algebra $\mathcal{S}$ is noncommutative.

5.4. Notes and References

Laurant Schwartz was the main developer of the theory of distributions [236]. There are many references on this subject but we recommend the following [87, 135, 225]. Schwartz has proved many of the Theorems in Section 5.1 including the important nuclear theorem (Theorem 5.7). Much of the motivation for distributions comes from the Dirac delta function in physics and from finding solutions to partial differential equations. Our discussion follows that of [225] fairly closely.

Theorem 5.9 is due to Sundaresan [258] and the other results on orthogonally additive and orthogonally increasing functionals are due to the author and Strawther [116, 117]. For further discussion of Gleason's theorem see [173, 268]. Our proof of Corollary 5.17 follows some ideas of Piron [212]. Some articles on L-additivity in different settings are [36, 66, 223, 224, 282].

The standard references on C^*-algebras are [64, 194, 226, 234]. Further details concerning the algebraic approach to quantum mechanics may be found in [8, 72, 110, 118, 121, 122, 146, 154, 156, 169, 228, 238, 245, 247, 248]. Some articles on unbounded representations of *-algebras are [217, 261].

5.5. Exercises

1. Prove that $S(\mathbb{R}^n)$ is dense in $L^2(\mathbb{R}^n)$.

2. If $p(x)$ is a polynomial in $x_1, \ldots, x_n$, prove that $p(x)\exp(-\Sigma x_i^2) \in S(\mathbb{R}^n)$.

3. Prove Lemma 5.3.

4. Prove that ϕ_b defined by (5.2) is in S.

5. Verify Eq. (5.3).

6. Supply the missing details in the proof of Theorem 5.4.

7. Show that $D^\beta \delta_a$ cannot be represented by a measure for $\beta \neq 0$.

8. Prove Eq. (5.8).

9. If $g \in C_b^\infty(\mathbb{R}^n)$ and $\phi \in S(\mathbb{R}^n)$, show that $g\phi \in S(\mathbb{R}^n)$.

10. Prove Eq. (5.9)

11. Prove that $\mathcal{F}: S(\mathbb{R}^n) \to S(\mathbb{R}^n)$ and that $\mathcal{F}$ is a continuous linear operator.

12. Prove that the translation operator T_a has a unique continuous extension to $S'(\mathbb{R}^n)$ given by $(T_a F)(\phi) = F(T_{-a}\phi)$.

13. For $F \in S'(\mathbb{R}^n)$ and $G \in S'(\mathbb{R}^m)$, show that $F \otimes G \in S'(\mathbb{R}^{n+m})$.

14. Show that f_1 satisfies (5.12) whenever $\langle x,y\rangle = -1$.

15. Show that $\langle x,y\rangle = -1$ if and only if $y = -\|x\|^2 x + z$ where $z \perp x$.

16. Prove that if $\langle x,y\rangle = -1$, then the line $\overline{Ow(x,y)}$ is perpendicular to the line $\overline{xy}$.

17. Let V be an inner product space, $\dim V \geq 2$. A function $f: V \to \mathbb{R}$ is *radially increasing* if $\alpha > 1$ implies $f(\alpha x) \geq f(x)$ for every $x \in V$, and f is *spherically increasing* if $\|x\| > \|y\|$ implies $f(x) \geq f(y)$. Show that spherically increasing implies radially increasing but the converse need not hold. Show that f is orthogonally increasing if and only if f is spherically increasing.

18. Let $f: V \to \mathbb{R}$ be orthogonally additive. If there is an $x_0 \in V$ such that $f(x_0) = \|x_0\|^2$ and $|f(x)| \leq \|x_0\| \|x\|$ for all $x \in V$, prove that $f(x) = \langle x, x_0\rangle$ for all $x \in V$.

19. Let V be an inner product space, $\dim V \geq 2$. Let $g: V \times V \to \mathbb{R}$ satisfy $g(x,x) \geq 0$, if $x \perp y$ then $g(x,y) = 0$, $g(x+y, x+y) = g(x,x) + g(y,y) + 2g(x,y)$. Prove that there is a $c \geq 0$ such that $g(x,y) = c\langle x,y\rangle$.

20. Show that C, $C(\Omega)$, and $\mathfrak{L}(\mathcal{H})$ are C^*-algebras.

21. Prove that the induced topology is generated by the collection of seminorms $\{\phi \mapsto \|\pi(A)\phi\| : A \in \mathcal{C}\}$.

22. Prove Theorem 5.22.

23. Prove Lemma 5.23.

24. Show that the $*$-algebra $\mathbb{S}$ in Section 5.3 is noncommutative.

25. Let $\pi: \mathcal{C} \to \mathfrak{L}(\mathcal{H})$ be a representation of the Banach $*$-algebra $\mathcal{C}$. We say that π is *irreducible* if $\pi(A)M \subseteq M$ for every $A \in \mathcal{C}$ and a closed subspace M implies that $M = \{0\}$ or $M = \mathcal{H}$. Prove that π is irreducible if and only if $T\pi(A) = \pi(A)T$ for every $A \in \mathcal{C}$ and a $T \in \mathfrak{L}(\mathcal{H})$ implies that $T = \lambda I$ for some $\lambda \in C$.

26. Prove that a representation π of a Banach $*$-algebra is irreducible if and only if every nonzero vector is cyclic for π.

27. A state is *pure* if it is not a nontrivial convex combination of two other states. Let ω be a state on a Banach $*$-algebra $\mathcal{C}$ and let π_ω be the corresponding GNS representation of $\mathcal{C}$. Prove that π_ω is irreducible iff ω is pure.

6

Random Fields

We are frequently interested in describing the manner in which a physical system evolves in time. In many cases the system undergoes random fluctuations which cannot be specified exactly but can be described by random variables. Such random evolutions are given rigorous form in terms of a stochastic process $X(t, \omega)$ where t represents time and $X(t, \cdot)$ is a random variable for every t. Much of classical probability theory is concerned with the study of stochastic processes. As we shall see, stochastic processes are not broad enough to describe certain phenomena and thus necessitate a generalization called random fields. Random fields are not only useful for describing classical systems such as Brownian motion particles, but as we shall see in Chapter 7, they can be utilized in quantum field theory.

6.1. The Wiener Process

The Wiener process is one of the most important examples of a stochastic process. This process also generates a very important random field which we shall study subsequently. We begin with a general discussion of stochastic processes.

Let (Ω, Σ, μ) be a probability space. A *stochastic process* is a map $X: \mathbb{R}^+ \times \Omega \to \mathbb{R}$ such that $\omega \mapsto X(t, \omega)$ is a random variable for all $t \in \mathbb{R}^+$. Stochastic processes in which $\mathbb{R}^+$ is replaced by a set Y are sometimes considered but for now $\mathbb{R}^+$ is general enough for our purposes. (A stochastic process $X: Y \times \Omega \to \mathbb{R}$ is said to be *indexed* by Y.) Usually, t is thought of as time and X represents a random motion or fluctuation evolving in time. For fixed t, the random variable $\omega \mapsto X(t, \omega)$ might represent a position coordinate of a moving particle at time t and for fixed ω the function $t \mapsto X(t, \omega)$ gives a "sample" of the position coordinate as it evolves in time. We call $t \mapsto X(t, \omega)$ a *sample path* for X. Thus each $\omega \in \Omega$ can be used to describe a different sample path, in general.

Let $X_t(\omega) = X(t, \omega), t \in \mathbb{R}^+, \omega \in \Omega$, be a stochastic process. For $t_1 < t_2 \in \mathbb{R}^+$ we call $X_{t_2} - X_{t_1}$ an *increment* of X_t. We say that X_t has *stationary increments* if the distribution of $X_{t_2} - X_{t_1}$ equals the distribution of $X_{t_4} - X_{t_3}$ whenever $t_2 - t_1 = t_4 - t_3$; that is, the distribution of increments depends only on the length of the intervening time. We say that X_t has *independent increments* if for any $t_0 < t_1 < \cdots < t_n \in \mathbb{R}^+$, the increments $X_{t_i} - X_{t_{i-1}}$ are independent, $i = 1, 2, \ldots, n$.

In the early 19th century the botanist Robert Brown noticed that small dust particles undergo continual, irregular motions which he explained by the fact that the particles suffer frequent collisions with the randomly moving air molecules. This phenomenon also usually occurs when small particles are suspended in a fluid; for example, when a drop of ink is placed in a glass of water. Moreover, similar phenomena are observed in systems with noise such as in electronic circuits. Although others certainly observed such motions earlier, Brown mentioned them in some of his papers and the phenomenon now carries the name Brownian motion.

At the beginning of this century, Bachelier and Einstein proposed mathematical descriptions of Brownian motion. Let X_t represent one coordinate of the Brownian particle at time t, which begins at the origin so $X_0 = 0$. Since the underlying molecular motions are only known statistically, X_t is a random variable for every $t \in \mathbb{R}^+$ and thus X_t is a stochastic process. Now an important result of probability theory, the central limit theorem, comes into play (see Section 2.2 for the statement of this theorem). As the displacement of the particle during the time interval $[0, t]$ is the sum of many small, independent contributions, it seems reasonable in light of the central limit theorem that X_t is normally distributed. It is also plausible to assume that displacements suffered by the particle during nonoverlapping time intervals should be independent and hence X_t has independent increments. Moreover, by symmetry, the expectation $E[X_t] = 0$, and if the physical conditions remain unchanged X_t should have stationary increments. Since the distribution of $X_{t+s} - X_t$ is independent of t we have

$$E\left[(X_{t+s} - X_t)^2\right] = g(s).$$

Because of the independent increment and mean zero assumptions we obtain

$$g(s_1 + s_2) = E\left[\left(X_{t+s_1+s_2} - X_{t+s_1} + X_{t+s_1} - X_t\right)^2\right]$$

$$= E\left[\left(X_{t+s_1+s_2} - X_{t+s_1}\right)^2\right] + E\left[\left(X_{t+s_1} - X_t\right)^2\right]$$

$$= g(s_1) + g(s_2).$$

If we make the reasonable assumption that g is continuous, then g must have the form $g(s) = \sigma^2 s$, where σ^2 is a constant. Einstein showed that this model predicts results that are in very close agreement with experiment. In fact, using the constant σ^2 and other measured parameters of a Brownian motion experiment he determined Avogadro's number to a high degree of accuracy.

In 1923, Norbert Wiener published the first completely rigorous results concerning Brownian motion [276]. It is this theory that we now describe. A *Wiener process* with variance σ^2 is a stochastic process $W_t(\omega) = W(t, \omega), t \in \mathbb{R}^+$, which satisfies the following conditions:

(i) $W_0 = 0$ a.e.;
(ii) If $\Delta = [a, b) \subseteq \mathbb{R}^+$ and $W_\Delta = (b - a)^{-1/2}[W_b - W_a]$, then for any sequence of mutually disjoint intervals $\Delta_1, \Delta_2, \ldots$, the random variables $X_{\Delta_1}, X_{\Delta_2}, \ldots$ are independent, normally distributed with mean zero and variance σ^2;
(iii) almost all sample paths of W are continuous.

Condition (iii) is a reasonable physical requirement and (i) is a harmless normalization. Condition (ii) follows from the discussion of the previous paragraph. In particular, the variance condition follows from the form of g in the previous paragraph.

Wiener [276], and later others using different techniques [17, 65], proved that a Wiener process exists and is essentially unique. Also, Wiener showed that $W(t, \omega)$ is nowhere differentiable for almost every ω. It is sometimes useful to find averages of functions over the sample paths of a Wiener process. For example, if $f \in L^2(\mathbb{R})$ one would like to define $\int f(t) dW(t, \omega)$ in a reasonable way. This "integral" is called the *Wiener integral* and the following theorem shows that it exists and has natural properties.

Theorem 6.1. *Let W be the Wiener process with variance σ^2 on (Ω, Σ, μ). There is a unique isometry from $L^2(\mathbb{R}^+, \sigma^2 dt)$ into $L^2(\Omega, \Sigma, \mu)$ denoted $f \mapsto \int f dW$ such that for any $a \leq b \in \mathbb{R}^+$, $\int \chi_{[a,b]} dW = W_b - W_a$. All the* *random variables in the set*

$$\left\{ \int f\,dW : f \in L^2(\mathbb{R}^+, \sigma^2 dt) \right\}$$

are Gaussian (i.e., normally distributed).

PROOF. For $s, s', t, t' \in \mathbb{R}^+$ with $s \le t, s' \le t'$ we first show that

$$E[(W_t - W_s)(W_{t'} - W_{s'})] = \sigma^2 |[s,t] \cap [s',t']|, \tag{6.1}$$

where $|\cdot|$ denotes Lebesgue measure. Indeed, suppose for concreteness that $s \le s' \le t \le t'$. Due to the independence and mean zero condition we have

$$
\begin{aligned}
E[(W_t - W_s)(W_{t'} - W_{s'})] &= E\{[(W_t - W_s) + (W_{s'} - W_s)] \\
&\quad \times [(W_{t'} - W_t) + (W_t - W_{s'})]\} \\
&= E[(W_t - W_{s'})^2] = \sigma^2 |t - s'| = \sigma^2 |[s,t] \cap [s',t']|.
\end{aligned}
$$

The other cases follow in a similar way.

Now let $f = \sum_{i=1}^n c_i \chi_{[a_i, b_i]}$ be a simple function. We define

$$\int f\,dW = \sum_{i=1}^n c_i (W_{b_i} - W_{a_i}).$$

If $g = \sum_{j=1}^m d_j \chi_{[e_j, f_j]}$ is another simple function we have

$$
\begin{aligned}
\left\langle \int f\,dW, \int g\,dW \right\rangle &= \left\langle \sum_{i=1}^n c_i (W_{b_i} - W_{a_i}), \sum_{j=1}^m d_j (W_{f_j} - W_{e_j}) \right\rangle \\
&= \sum_{i=1}^n \sum_{j=1}^m c_i d_j E\left[(W_{b_i} - W_{a_i})(W_{f_j} - W_{e_j}) \right] \\
&= \sum_{i=1}^n \sum_{j=1}^m c_i d_j \sigma^2 |[a_i, b_i] \cap [e_j, f_j]| \\
&= \int f(t)g(t)\sigma^2\,dt = \langle f, g \rangle.
\end{aligned}
$$

Since the simple functions are dense in $L^2(\mathbb{R}^+, \sigma^2 dt)$, the integral extends by continuity to an isometry. Uniqueness and the fact that the random variables are Gaussian are straightforward. $\qquad\square$

6.2. Definitions and Examples

If $a < b$, $a, b \in \mathbb{R}^+$, then $W_b(\omega) - W_a(\omega) = \int \chi_{[a,b]}(t)\,dW(t, \omega)$ gives the displacement of a Brownian particle during the time interval $[a, b]$ for the sample path given by ω. However, in practice, it is impossible to make such a precise measurement. For example, an experimentalist watching a

Brownian particle through a microscope cannot measure the position of the particle precisely at some instant of time. Due to the rapidity of the fluctuations, the best he can do is measure a weighted average $\int \phi \, dW$ of the displacements, where ϕ is some averaging function depending on the measuring apparatus. The closer ϕ is to $\chi_{[a,b]}$ the more accurate will be the measurement. However, no apparatus can turn off and on instantaneously so ϕ will in general be a smooth function approximating $\chi_{[a,b]}$. We thus see that the Wiener integral $\Phi(\phi) = \int \phi \, dW$ describes the actual observation of a Brownian particle more realistically than the Wiener process W itself. As we shall see, the map Φ is an example of a random field.

Let V be a real or complex inner product space with inner product $\langle \cdot, \cdot \rangle$. Unless specified otherwise, we assume that V is equipped with the norm topology determined by the inner product. Let (Ω, Σ, μ) be a probability space and denote the set of complex-valued random variables on this space by $R(\Omega, \Sigma, \mu)$. A map $\Phi: V \to R(\Omega, \Sigma, \mu)$ is called a *random functional*. Hence, a random functional is a stochastic process indexed by V. Two random functionals Φ and Φ_1 are *equivalent* if for any finite set $\{\phi_1, \ldots, \phi_n\} \subseteq V$, the joint distributions of $\Phi(\phi_1), \ldots, \Phi(\phi_n)$ and $\Phi_1(\phi_1), \ldots, \Phi_1(\phi_n)$ are identical. Equivalent random functionals are indistinguishable in so far as their stochastic properties are concerned. A random functional Φ is *linear* if $\Phi(\alpha\phi + \beta\psi) = \alpha\Phi(\phi) + \beta\Phi(\psi)$ for every $\alpha, \beta \in C$ (or $\mathbb{R}$ if V is a real space) and all $\phi, \psi \in V$. We say that Φ is *continuous* if $\phi_i \to \phi$ in V implies that $\Phi(\phi_i) \to \Phi(\phi)$ in probability (i.e., in measure). A continuous linear random functional is a *random field*. We now give some examples.

Example 1. Let $X: \mathbb{R}^n \times \Omega \to C$ be a stochastic process. Then $\Phi(x) = X(x, \cdot)$ is a random functional. If we think of $\mathbb{R}^n$ as a physical space, X could represent a statistical measurement of some physical field in space. The sample path $x \mapsto X(x, \omega_0)$ gives a "sample" of the values of the measurement X at each point of space. The random variable $\omega \mapsto X(x_0, \omega)$ gives the statistical distribution of the values of X at the point $x_0 \in \mathbb{R}^n$.

Example 2. Let $X: \mathbb{R}^n \times \Omega \to C$ be a stochastic process, almost all of whose sample paths are locally integrable. Equip $\mathbb{S}(\mathbb{R}^n)$ with the usual inner product. The map $\Phi: \mathbb{S}(\mathbb{R}^n) \to R(\Omega, \Sigma, \mu)$ given by $[\Phi(\phi)](\omega) = \int \phi(x) X(x, \omega) \, dx$ is a random field. We may think of $\Phi(\phi)$ as a measurement of the physical field "smeared" by the test function ϕ. We say that Φ is *generated* by the stochastic process X.

Example 3. Extend the Wiener process to $\mathbb{R}$ by defining $W(-t, \omega) = W(t, \omega)$, $t \geq 0$, and let $W: \mathbb{R} \times \Omega \to \mathbb{R}$ be this process with variance 1. Then $\Phi: L^2(\mathbb{R}) \to R(\Omega, \Sigma, \mu)$ given by $\Phi(\phi) = \int \phi \, dW$ is a random field. This is an

145

example of the unit Gaussian random field which will be considered in detail later. Notice that Φ is not generated by W in the sense of the definition in Example 2. Formally, we have $\Phi(\phi) = \int \phi(dW/dt)\,dt$ so one might think that Φ is generated by dW/dt if sense can be made of this expression. However, as we have already mentioned, $W(\cdot,\omega)$ is nowhere differentiable for almost every ω so dW/dt does not make sense as a stochastic process. In fact, we shall later show that Φ is not generated by any stochastic process.

Example 4. Let $\mathcal{H}$ be a Hilbert space, $\mathcal{B}$ the Borel σ-algebra on $\mathcal{H}$, and μ a probability measure on $\mathcal{B}$. The map $\Phi\colon \mathcal{H}\to R(\mathcal{H},\mathcal{B},\mu)$ given by $[\Phi(\phi)](\psi) = \langle\phi,\psi\rangle$ is a random field. More generally, let $\mathcal{H}_1$ be a separable Hilbert space and let $f\colon \mathcal{H}\to\mathcal{H}_1$ be a Borel function. If we define $\Phi\colon \mathcal{H}_1\to R(\mathcal{H},\mathcal{B},\mu)$ by $[\Phi(\phi)](\psi) = \langle\phi,f(\psi)\rangle$, then Φ is a random field. A random field $\Phi\colon \mathcal{H}_1\to R(\mathcal{H},\mathcal{B},\mu)$ has this form if and only if $\sum\|[\Phi(\phi_i)](\psi)\|^2 < \infty$ for almost every $\psi\in\mathcal{H}$ and every orthonormal basis $\{\phi_i\}$ of $\mathcal{H}_1$. Indeed, if Φ has the above form, then

$$\sum \left| [\Phi(\phi_i)](\psi)\right|^2 = \sum |\langle\phi_i,f(\psi)\rangle|^2 = \|f(\psi)\|^2 < \infty.$$

Conversely, if Φ satisfies $\sum\|[\Phi(\phi_i)](\psi)\|^2 < \infty$ as above, define $f\colon \mathcal{H}\to\mathcal{H}_1$ by

$$f(\psi) = \sum [\Phi(\phi_i)](\psi)\phi_i,$$

where ϕ_i is an orthonormal basis of $\mathcal{H}_1$. Then for every $\phi\in\mathcal{H}_1$ and $\psi\in\mathcal{H}$ we have

$$[\Phi(\phi)](\psi) = \lim_{n\to\infty} \left[\Phi\left(\sum_{i=1}^n \langle\phi\phi_i\rangle\phi_i\right)\right](\psi)$$

$$= \lim_{n\to\infty} \sum_{i=1}^n [\Phi(\langle\phi,\phi_i\rangle\phi_i)](\psi)$$

$$= \sum [\Phi(\langle\phi,\phi_i\rangle\phi_i)](\psi) = \sum [\Phi(\phi_i)](\psi)\langle\phi,\phi_i\rangle$$

$$= \langle\phi,f(\psi)\rangle.$$

This example will be important later when we consider random fields generated by measures.

We now motivate our definition of a random field. In physics one frequently deals with a physical field such as an electromagnetic field or a nuclear field. Such fields are usually assumed to exist in a Euclidean space $\mathbb{R}^n$; for example, in a configuration space $\mathbb{R}^3$ or in space–time $\mathbb{R}^4$. If a physical field in $\mathbb{R}^n$ is described by a stochastic process $X\colon \mathbb{R}^n\times\Omega\to C$, this corresponds to a "sharp" measurement at each point of the space $\mathbb{R}^n$.

However, in general, it is impossible to make a sharp measurement of a physical field. In classical (nonquantum) physics this is usually due to the inertia of the measuring apparatus. We have already seen this occurring for a Brownian particle. For the quantum case, sharp measurements are not only impossible in practice but also in theory. One reason for this is because of the singularities of quantum fields at points of $\mathbb{R}^n$. For example, an electron would create such a singularity at its point of location. What is actually measured is an average or smeared field over a small region of $\mathbb{R}^n$. This is equivalent to associating a test function on $\mathbb{R}^n$ (say in $S(\mathbb{R}^n)$) with a random variable, thus giving a random functional Φ. It is easy to justify on physical grounds that Φ should be linear and continuous. Hence Φ is a random field.

We now define some important complex-valued functionals which correspond to the random functional $\Phi: V \to R(\Omega, \Sigma, \mu)$. The *mean* functional is the map $M_\Phi(\phi) = E[\Phi(\phi)] = \int \Phi(\phi) \, d\mu$. Of course, $M_\Phi(\phi)$ need not exist for all $\phi \in V$. If M_Φ exists and Φ is a random field, then M_Φ is a linear (not necessarily continuous) functional on V. The *covariance* functional $C_\Phi: V \times V \to C$ is defined by $C_\Phi(\phi, \psi) = E\{[\Phi(\phi) - M_\Phi(\phi)][\Phi(\psi) - M_\Phi(\psi)]^*\}$, where $*$ denotes complex conjugation. Again, $C_\Phi(\phi, \psi)$ need not exist for all $\phi, \psi \in V$. If C_Φ exists and Φ is a random field, then C_Φ is a positive semidefinite, (not necessarily bounded) hermitian bilinear form on V. The *variance* functional $V_\Phi: V \to \mathbb{R}^+$ is defined by $V_\Phi(\phi) = C_\Phi(\phi, \phi)$ and the *correlation* functional $B_\Phi: V \times V \to C$ is defined by

$$B_\Phi(\phi, \psi) = C_\Phi(\phi, \psi) + M_\Phi(\phi) M_\Phi(\psi)^* = E[\Phi(\phi)\Phi(\psi)^*].$$

A random functional Φ is of *second order* if $B_\Phi(\phi, \phi) < \infty$ for all $\phi \in V$ and Φ is *bounded* if there exists a $B \geq 0$ such that $B_\Phi(\phi, \phi)^{1/2} \leq B \|\phi\|$ for all $\phi \in V$. It is easy to see that a bounded random functional is of second order but that the converse need not hold. A random functional Φ is *uncorrelated* if $\langle \phi, \psi \rangle = 0$ implies that

$$E[\Phi(\phi)\Phi(\psi)^*] = E[\Phi(\phi)]E[\Phi(\psi)^*].$$

The *characteristic* functional $L_\Phi: V \to C$ is defined by

$$L_\Phi(\phi) = E[e^{i\Phi(\phi)}] = \int e^{i\Phi(\phi)} \, d\mu.$$

Notice that the characteristic functional exists for all $\phi \in V$ and that if Φ is continuous, then L_Φ is also continuous. The characteristic functional is related to the concept of the characteristic function of a random variable. If f is a random variable on (Ω, Σ, μ), the *characteristic function* $c_f: \mathbb{R} \to C$ of f is defined as

$$c_f(t) = E(e^{itf}) = \int e^{itf(\omega)} \, d\mu(\omega).$$

Moreover, the *joint characteristic* function of random variables $f_1, \ldots, f_n$ is the function $c_{f_1, \ldots, f_n} : \mathbb{R}^n \to C$ defined by

$$c_{f_1, \ldots, f_n}(t_1, \ldots, t_n) = E[\exp(it_1 f_1 + \cdots + it_n f_n)].$$

It can be shown that the joint distribution of $f_1, \ldots, f_n$ is uniquely determined by $c_{f_1, \ldots, f_n}$ [31, 65, 166]. It follows that if Φ is a random field, then L_Φ determines Φ uniquely to within an equivalence.

A function $g: V \to C$ is *positive definite* if for any $\phi_1, \ldots, \phi_n \in V$, $c_1, \ldots, c_n \in C$ ($n < \infty$, arbitrary) we have

$$\sum_{i,j=1}^{n} g(\phi_i - \phi_j) c_i \bar{c}_j \geq 0.$$

We now show that for a random field Φ, L_Φ is positive definite. If $\phi_1, \ldots, \phi_n \in V$, $c_1, \ldots, c_n \in C$, then

$$\sum_{j,k=1}^{n} c_j \bar{c}_k L_\Phi(\phi_j - \phi_k) = \sum_{j,k=1}^{n} c_j \bar{c}_k \int e^{i\Phi(\phi_j - \phi_k)} d\mu$$

$$= \int \sum_{j,k=1}^{n} c_j e^{i\Phi(\phi_j)} \bar{c}_k e^{-i\Phi(\phi_k)} d\mu$$

$$= \int |\sum_j c_j e^{i\Phi(\phi_j)}|^2 d\mu \geq 0.$$

Thus, for a random field Φ, L_Φ is a continuous positive definite functional and $L_\Phi(0) = 1$. Conversely, it can be shown that for any continuous positive definite functional g on V such that $g(0) = 1$ there exists a unique (to within equivalence) random field Φ such that $L_\Phi(\phi) = g(\phi)$ for all $\phi \in V$ [87].

Denote the element of $R(\Omega, \Sigma, \mu)$ which is identically equal to 1 by **1**. We say that the random field Φ is *proper* if Φ is injective and **1** is not in the range of Φ. Postulating that Φ be proper is essentially a matter of convenience. If Φ is not injective then the kernel $K(\Phi)$ of Φ is a closed subspace of V and Φ will be injective on $K(\Phi)^\perp$. Requiring that the range of Φ does not contain **1** eliminates the possibility of having degenerate distributions; that is, distributions concentrated at a single nonzero number. This can always be arranged by working with the zero mean random field $\Phi - M_\Phi$.

To avoid certain pathologies we shall assume in the sequel that $\dim V \geq 2$.

Lemma 6.2. *Let* $\Phi: V \to R(\Omega, \Sigma, \mu)$ *be a second-order random field.*

(a) Φ *is proper if and only if* C_Φ *is an inner product on* V.

(b) *The following statements are equivalent:*

(i) Φ *is proper and uncorrelated*;

(ii) *there exists an $r>0$ such that $C_\Phi(\phi,\psi)=r\langle\phi,\psi\rangle$ for all $\phi,\psi\in V$;*

(iii) *if $C_\Phi(\phi,\psi)=0$ then $\langle\phi,\psi\rangle=0$.*

(c) *If Φ is proper and uncorrelated it is bounded.*

(d) *Suppose V is a Hilbert space and Φ is bounded. Then there exists a unique vector $\phi_\Phi\in V$ and unique bounded nonnegative operators S_Φ, T_Φ on V such that $M_\Phi(\phi)=\langle\phi,\phi_\Phi\rangle, C_\Phi(\phi,\psi)=\langle S_\Phi\phi,\psi\rangle$, and $B_\Phi(\phi,\psi)=\langle T_\Phi\phi,\psi\rangle$ for all $\phi,\psi\in V$. (These are called the **mean vector, covariance operator** and **correlation operator**, respectively, for Φ.)*

PROOF. (a) Suppose Φ is proper and $C_\Phi(\phi,\phi)=0$. It follows that $\Phi(\phi)=M_\Phi(\phi)$ almost everywhere. Since Φ is proper, nonzero constants are not in the range of Φ. Hence $M_\Phi(\phi)=0$. Since Φ is injective, $\phi=0$ and C_Φ is an inner product on V. Conversely, suppose C_Φ is an inner product on V. If $\Phi(\phi)=0$, then $C_\Phi(\phi,\phi)=0$ so $\phi=0$ and Φ is injective. If $\Phi(\phi)=\mathbf{1}$, then $C_\Phi(\phi,\phi)=0$ so $\phi=0$. This gives the contradiction $\Phi(0)=\mathbf{1}$.

(b) (i)$\Rightarrow$(ii) If Φ is proper it follows from (a) that C_Φ is an inner product on V. If Φ is uncorrelated it follows from Corollary 5.10 that there exists an $r>0$ such that $C_\Phi(\phi,\psi)=r\langle\phi,\psi\rangle$ for all $\phi,\psi\in V$. (ii)$\Rightarrow$(iii) Trivial. (iii)$\Rightarrow$(i) If (iii) holds then C_Φ is an inner product on V and applying (a), Φ is proper. Again using Corollary 5.10, there exists an $r>0$ such that $C_\Phi(\phi,\psi)=r\langle\phi,\psi\rangle$. Hence (i) holds.

(c) This follows from (b).

(d) Since

$$|M_\Phi(\phi)| \le \int |\Phi(\phi)|d\mu \le \left[\int |\Phi(\phi)|^2 d\mu\right]^{1/2}$$

$$= B_\Phi(\phi,\phi)^{1/2} \le B\|\phi\|$$

we see that M_Φ is a bounded linear functional on V. By the Riesz theorem there exists a unique vector $\phi_\Phi\in V$ such that $M_\Phi(\phi)=\langle\phi,\phi_\Phi\rangle$. By Schwarz's inequality we have

$$|C_\Phi(\phi,\psi)| \le |C_\Phi(\phi,\phi)|^{1/2}|C_\Phi(\psi,\psi)|^{1/2}$$

$$\le |B_\Phi(\phi,\phi)|^{1/2}|B_\Phi(\psi,\psi)|^{1/2} \le B\|\phi\|\,\|\psi\|.$$

Hence C_Φ is a bounded, positive semidefinite, hermitian bilinear form. The result now follows from a well-known theorem. $\square$

Notice that Example 3 illustrates a proper uncorrelated random field. We now define another important class of random fields. A random field Φ has *strongly independent values* if $\langle\phi,\psi\rangle=0$ implies that $\Phi(\phi)$ and $\Phi(\psi)$

are stochastically independent. If Φ has strongly independent values, it is clear that Φ is uncorrelated.

Lemma 6.3. *A random field Φ has strongly independent values if and only if $\langle \phi, \psi \rangle = 0$ implies $L_\Phi(\phi + \psi) = L_\Phi(\phi) L_\Phi(\psi)$.*

PROOF. Suppose Φ has strongly inaependent values and $\langle \phi, \psi \rangle = 0$. Since $\Phi(\phi)$ and $\Phi(\psi)$ are stochastically independent we have

$$L_\Phi(\phi + \psi) = E\left[e^{i\Phi(\phi + \psi)} \right] = E\left[e^{i\Phi(\phi)} \right] E\left[e^{i\Phi(\psi)} \right]$$

$$= L_\Phi(\phi) L_\Phi(\psi).$$

Conversely, suppose $\langle \phi, \psi \rangle = 0$ implies $L_\Phi(\phi + \psi) = L_\Phi(\phi) L_\Phi(\psi)$. If $\langle \phi, \psi \rangle = 0$, then

$$E\left\{ \exp\left[is(\Phi(\phi) + \Phi(\psi)) \right] \right\} = E\left\{ \exp\left[i\Phi(s\phi + s\psi) \right] \right\}$$

$$= L_\Phi(s\phi + s\psi) = L_\Phi(s\phi) L_\Phi(s\psi)$$

$$= E\left[e^{is\Phi(\phi)} \right] E\left[e^{is\Phi(\psi)} \right].$$

Thus the characteristic function of the random variable $\Phi(\phi) + \Phi(\psi)$ equals the product of the characteristic function of $\Phi(\phi)$ with the characteristic function of $\Phi(\psi)$. It follows from a well-known result in probability theory that $\Phi(\phi)$ and $\Phi(\psi)$ are stochastically independent. $\qquad\square$

6.3. Gaussian Random Fields

A second-order random field Φ whose characteristic functional has the form

$$L_\Phi(\phi) = \exp\left[iM_\Phi(\phi) - V_\Phi(\phi)/2 \right]$$

is called a *Gaussian* random field. A Gaussian random field with given mean and covariance functional can always be constructed [87]. A Gaussian random field Φ_r with covariance function $C_\Phi(\phi, \psi) = r\langle \phi, \psi \rangle$, $r > 0$, is called an *isonormal random field with parameter r*. The *unit* random field Φ_u is the isonormal random field with mean zero and parameter 1. Example 3 of Section 6.2 is a unit random field. Notice that Φ_u is the unique (to within equivalence) random field which satisfies $L_{\Phi_u}(\phi) = e^{-\|\phi\|^2/2}$. It follows from Lemma 6.2(b) that a Gaussian random field is isonormal if and only if it is proper and uncorrelated. If Φ is a bounded Gaussian random field on a Hilbert space, it follows from Lemma 6.2(d) that

$$L_\Phi(\phi) = \exp\left[i\langle \phi, \phi_\Phi \rangle - \langle S_\Phi \phi, \phi \rangle/2 \right].$$

If Φ_r is an isonormal random field, then it is easily seen that $\Phi_1(\phi) =$ $[\Phi_r(\phi) - M_{\Phi_r}(\phi)]/r^{1/2}$ is the unit random field. Phrased differently, any isonormal random field Φ_r can be represented in terms of the unit random field $\Phi_r = r^{1/2}\Phi_u + M_{\Phi_r}$ by a linear change of scale.

In Chapter 7 we shall consider the relationship between random fields and quantum field theory. To appreciate the next theorem we give a little preview.

In quantum field theory, one has a Hilbert space V of one-particle states. The "second quantization" is given by a random field $\Phi: V \to R(\Omega, \Sigma, \mu)$. The Hilbert space of the second quantization is $L^2(\Omega, \Sigma, \mu)$. The unit random field gives the second quantization for noninteracting quantum fields and is equivalent to the Fock representation (see Chapter 7). The next result, which generalizes a theorem due to Urbanik [264] gives a reason why the unit or, more generally, the isonormal random fields generate representations of noninteracting quantum fields. This result states that the isonormal random fields are precisely the proper random fields with strongly independent values. One can reason physically that random fields with strongly independent values correspond to noninteracting quantum fields.

Theorem 6.4. *A random field Φ is isonormal if and only if Φ is proper and has strongly independent values.*

PROOF. Suppose Φ is isonormal. By Lemma 6.2(b), Φ is proper. Since the characteristic functional of Φ is

$$L_\Phi(\phi) = \exp\left[iM_\Phi(\phi) - r\|\phi\|^2/2 \right],$$

it follows that $L_\Phi(\phi + \psi) = L_\Phi(\phi)L_\Phi(\psi)$ whenever $\langle \phi, \psi \rangle = 0$. Hence by Lemma 6.3, Φ has strongly independent values. Conversely, suppose Φ is proper and has strongly independent values. We first show that L_Φ is never zero. Suppose there exists a ϕ such that $L_\Phi(\phi) = 0$. Then $\phi \neq 0$, since $L_\Phi(0) = 1$. Now suppose $\|\psi\| > \|\phi\|$. As in the proof of Theorem 5.12 we can construct a sequence of vectors $\psi_1, \psi_2, \ldots, \psi_n$ such that

$$\phi \perp \psi_1; \qquad \left(\phi + \sum_{i=1}^{j-1} \psi_i\right) \perp \psi_j, \quad j = 2, \ldots, n; \qquad \psi = \phi + \sum_{j=1}^{n} \psi_i.$$

Then by Lemma 6.3

$$L_\Phi(\psi) = L_\Phi\left(\phi + \sum_{i=1}^{n-1} \psi_i + \psi_n\right) = L_\Phi\left(\phi + \sum_{i=1}^{n-1} \psi_i\right) L_\Phi(\psi_n)$$

$$= L_\Phi(\phi)L_\Phi(\psi_1)\ldots L_\Phi(\psi_n) = 0.$$

Now there are vectors ϕ_1, ϕ_2 such that $\phi_1 \perp \phi_2$, $\phi = \phi_1 + \phi_2$, and $\|\phi_1\| = \|\phi_2\| = \|\phi\|/\sqrt{2}$. Then

$$0 = L_\Phi(\phi) = L_\Phi(\phi_1 + \phi_2) = L_\Phi(\phi_1) L_\Phi(\phi_2).$$

Hence $L_\Phi(\phi_1)$ or $L_\Phi(\phi_2)$ equals zero. We therefore see that $L_\Phi(\psi) = 0$ for every ψ which satisfies $\|\psi\| > \|\phi\|/\sqrt{2}$. Continuing this process we conclude that $L_\Phi(\psi) = 0$ for every $\psi \neq 0$. Since L_Φ is continuous, this implies that $L_\Phi(0) = 0$, which is a contradiction. If we write $L_\Phi(\phi) = |L_\Phi(\phi)| \exp[i \arg L_\Phi(\phi)]$, since $L_\Phi(\phi) \neq 0$ we have $\log L_\Phi(\phi) = \log|L_\Phi(\phi)| + i \arg L_\Phi(\phi)$. Now $\log|L_\Phi(\phi)|$ is continuous and orthogonally additive; that is, $\langle \phi, \psi \rangle = 0$ implies

$$\log|L_\Phi(\phi + \psi)| = \log|L_\Phi(\phi)||L_\Phi(\psi)| = \log|L_\Phi(\phi)| + \log|L_\Phi(\psi)|.$$

Also, $\log|L_\Phi(\phi)|$ is even since

$$\log|L_\Phi(-\phi)| = \log|L_\Phi(\phi)^*| = \log|L_\Phi(\phi)|.$$

Hence by Theorem 5.9(b) there exists a $-r_0 \in \mathbb{R}$ such that $\log|L_\Phi(\phi)| = -r_0\|\phi\|^2$. Similarly, $\arg L_\Phi(\phi)$ is continuous, orthogonally additive and odd, so by Theorem 5.9(a) $h(\phi) = \arg L_\Phi(\phi)$ is linear. We hence conclude that $L_\Phi(\phi) = \exp[ih(\phi) - r_0\|\phi\|^2]$. Now the characteristic function of the random variable $\Phi(\phi)$ is given by

$$f(s) = E[e^{is\Phi(\phi)}] = L_\Phi(s\phi) = \exp[ish(\phi) - s^2 r_0 \|\phi\|^2]$$

It is straightforward that the mean and variance of $\Phi(\phi)$ is given by $-if'(0)$ and $-f''(0) + [f'(0)]^2$, respectively. Using these formulas we obtain $M_\Phi(\phi) = h(\phi)$ and $V_\Phi(\phi) = 2r_0\|\phi\|^2$. Hence Φ is Gaussian. Since Φ is proper and uncorrelated it follows from Lemma 6.2(b) that $C_\Phi(\phi, \psi) = r\langle \phi, \psi \rangle, r > 0$. $\square$

For the sufficiency part of the above theorem we did not use the fact that Φ is proper until the very end. We have thus proved the stronger result that if Φ has strongly independent values then Φ is Gaussian and $V_\Phi(\phi) = r\|\phi\|^2$, $r > 0$. As a corollary to this remark we obtain the following generalization of a result due to Kac [148]. (Kac proved this result for finite-dimensional vector spaces.)

Corollary 6.5. *Let $\mathcal{H}$ and $\mathcal{H}_1$ be Hilbert spaces and let μ be a probability measure on the Borel sets $\mathcal{B}$ of $\mathcal{H}$. Let $f: \mathcal{H} \to \mathcal{H}_1$ be a Borel function such that $\psi \mapsto \langle \phi_i, f(\psi) \rangle$, $i = 1, 2$, are stochastically independent whenever ϕ_1, ϕ_2 are orthogonal unit vectors. Then f is Gaussian distributed with radial symmetry.*

PROOF. Define the random field $\Phi: \mathcal{H}_1 \to R(\mathcal{H}, \mathcal{B}, \mu)$ by $[\Phi(\phi)](\psi) = \langle \phi, f(\psi) \rangle$. Then Φ has strongly independent values and the result follows.

$\square$

For the remainder of this section $\mathcal{H}$ will be a real separable Hilbert space, $\mathcal{B}$ its Borel σ-algebra and μ a probability measure on $\mathcal{B}$. Furthermore, Φ will be a random field such that $\Phi(\phi)$ is real valued.

Define the random field $\Phi_0 \colon \mathcal{H} \to R(\mathcal{H}, \mathcal{B}, \mu)$ by $[\Phi_0(\phi)](\psi) = \langle \phi, \psi \rangle$. In this way every Borel probability measure on $\mathcal{B}$ produces a random field. We now ask for the converse. When is a random field on $\mathcal{H}$ produced in this way by a Borel probability measure? We say that a random field $\Phi \colon \mathcal{H} \to R(\Omega, \Sigma, \nu)$ is *generated* by the probability measure μ on $(\mathcal{H}, \mathcal{B})$ if Φ is equivalent to Φ_0 defined above.

Theorem 6.6. *A random field $\Phi \colon \mathcal{H} \to R(\Omega, \Sigma, \nu)$ is generated by a probability measure μ on $(\mathcal{H}, \mathcal{B})$ if and only if for every $\varepsilon > 0$ there exists a density operator D_ε such that for all $\phi \in \mathcal{H}$*

$$1 - \operatorname{Re} L_\Phi(\phi) \leq \langle D_\varepsilon \phi, \phi \rangle + \varepsilon. \tag{6.2}$$

PROOF. If Φ is generated by μ then

$$L_\Phi(\phi) = \int_\Omega e^{i\Phi(\phi)} d\nu = \int_H e^{i\langle \phi, \psi \rangle} d\mu(\psi). \tag{6.3}$$

But the last term in (6.3) is the characteristic functional of the measure μ. Hence by a theorem due to Prohorov [218] (Theorem 2.2 in [163]) density operators satisfying (6.2) exist. Conversely, suppose L_Φ satisfies (6.2). Then by the same theorem of Prohorov, there exists a probability measure μ on $(\mathcal{H}, \mathcal{B})$ whose characteristic functional equals L_Φ. Hence $L_{\Phi_0} = L_\Phi$ and since the characteristic functional determines finite dimensional joint probability distributions, the result follows. $\qquad\square$

We have seen in Lemma 6.2(d) that the covariance operator S_Φ for a bounded random field Φ is a bounded nonnegative operator. The next result characterizes generated Gaussian random fields in terms of S_Φ.

Corollary 6.7. *A Gaussian random field Φ is generated by a probability measure on $(\mathcal{H}, \mathcal{B})$ if and only if S_Φ is a density operator.*

PROOF. It is straightforward to check that if Φ is generated by μ, then μ is a Gaussian measure on $\mathcal{H}$ [163]. Conversely, a Gaussian measure on $\mathcal{H}$ generates a Gaussian random field. The result now follows from another theorem due to Prohorov [218] (Theorem 2.3 in [163]). $\qquad\square$

For an isonormal random field Φ, $S_\Phi = rI$, $r > 0$. If $\dim H = \infty$, then S_Φ is not of trace class. Hence an isonormal random field cannot be generated by a Borel probability measure. We thus have the following result.

Corollary 6.8. *If* $dim\, H = \infty$, *then a proper random field with strongly independent values on* H *is not generated by a Borel probability measure.*

We now relate Gaussian random fields to Gaussian random variables. Let f be a random variable and let c_f be the characteristic function of f. A famous theorem due to Bochner [166] states that $c_f: \mathbb{R} \to C$ is the characteristic function of a random variable if and only if c_f is continuous, positive definite and $c_f(0) = 1$. It is not hard to show that the nth moment $E[f^n]$ of f satisfies

$$E[f^n] = \int f^n d\mu = \int t^n d\mu_f(t) = (-i)^n c_f^{(n)}(0). \tag{6.4}$$

If (Ω, Σ, μ) is a probability space, Σ becomes a ring if the "sum" of two sets $A, B \in \Sigma$ is the symmetric difference $A \triangle B = (A - B) \cap (B - A)$ and their "product" is $A \cap B$. The family

$$\mathcal{I}_\mu = \{A \in \Sigma : \mu(A) = 0\}$$

is an ideal in the ring Σ and μ "lifts" to the quotient ring $\Sigma / \mathcal{I}_\mu$ in the natural way. Two probability spaces (Ω, Σ, μ) and (Ω', Σ', μ') are *isomorphic* if there is an isomorphism of the rings $\Sigma / \mathcal{I}_\mu$ and $\Sigma' / \mathcal{I}_{\mu'}$ taking the measure μ into μ'. If f and f' are random variables on (Ω, Σ, μ) and (Ω', Σ', μ') respectively, and $T: \Sigma / \mathcal{I}_\mu \to \Sigma' / \mathcal{I}_{\mu'}$ is an isomorphism, we say that f and f' *correspond* under the isomorphism T if $T[f^{-1}(A)] = (f')^{-1}(A)$ for every $A \in \mathcal{B}(\mathbb{R})$.

A random variable f is *Gaussian* if

$$c_f(t) = e^{-(1/2)at^2 + imt} \tag{6.5}$$

for $a > 0$. A simple computation using (6.4) shows that m is the mean of f and a is the variance. Moreover, the distribution of f becomes

$$\mu_f(A) = \frac{1}{\sqrt{2\pi a}} \int_A e^{-(1/2)(t-m)^2/a} dt \tag{6.6}$$

for all $A \in \mathcal{B}(\mathbb{R})$, and the moments of f become

$$\int f^{2n+1} d\mu = 0, \quad \int f^{2n} d\mu = \frac{(2n)!}{2^n n!} a^n \tag{6.7}$$

$n = 0, 1, \ldots$. A finite set $f_1, \ldots, f_n$ of random variables is *jointly Gaussian* if the joint characteristic function has the form

$$c_{f_1, \ldots, f_n}(t_1, \ldots, t_n) = \exp\left(-\tfrac{1}{2} \sum a_{ij} t_i t_j + i \sum t_i m_i\right),$$

where $a = [a_{ij}]$ is a real positive definite matrix. In this case m_i is the mean of f_i and

$$a_{ij} = \int (f_i - m_i)(f_j - m_j) \, d\mu$$

is the covariance matrix. The joint distribution becomes

$$\mu_{f_1,\ldots,f_n}(A) = (2\pi)^{-n/2} |\det a|^{-1/2} \int_A \exp\left(-\tfrac{1}{2} \sum b_{ij}(x_i - m_i)(x_j - m_j)\right) d^n x$$

for all $A \in \mathscr{B}(\mathbb{R}^n)$, where $[b_{ij}] = [a_{ij}]^{-1}$. It is not hard to show that $f_1,\ldots,f_n$ are jointly Gaussian if and only if $\sum \alpha_i f_i$ is Gaussian for every $(\alpha_1,\ldots,\alpha_n) \in \mathbb{R}^n$.

We shall usually consider only mean zero Gaussian random variables. The proof of the following lemma is left as a straightforward exercise for the reader.

Lemma 6.9. *If $f_1,\ldots,f_{2n}$ are jointly Gaussian mean zero random variables, then*

$$\int f_1 \ldots f_{2n} \, d\mu = \sum \int f_{i_1} f_{j_1} \, d\mu \cdots \int f_{i_n} f_{j_n} \, d\mu,$$

where the sum is over all $(2n)!/2^n n!$ ways of writing $1,\ldots,2n$ as n distinct unordered pairs $(i_1, j_1), \ldots, (i_n, j_n)$.

Let f be a fixed Gaussian random variable on (Ω, Σ, μ) with mean zero and variance a. Then by (6.7), $f^n \in L^2(\Omega, \Sigma, \mu)$, $n = 0, 1, \ldots$. If $\sum c_n x^n$ is a power series with infinite radius of convergence, then $\sum c_n f^n$ converges pointwise and hence converges in $L^2(\Omega, \Sigma, \mu)$. For such a power series we define

$$\frac{\partial}{\partial f}\left(\sum_{n=0}^{\infty} c_n f^n\right) = \sum_{n=0}^{\infty} (n+1)c_{n+1} f^n. \tag{6.8}$$

For $n = 0, 1, \ldots$, we define the nth *Wick power* $:f^n:$ of f to be the nth degree polynomial in f defined recursively by

$$:f^0: = 1, \qquad \frac{\partial}{\partial f} :f^n: = n:f^{n-1}:, \qquad E[:f^n:] = 0, \tag{6.9}$$

$n = 1, 2, \ldots$. If $\sum c_n f^n$ and $\sum c_n :f^n:$ both converge we define

$$:\sum c_n f^n: = \sum c_n :f^n:.$$

In particular,

$$:e^{\alpha f}: = \sum \frac{\alpha^n}{n!} :f^n:. \tag{6.10}$$

6. Random Variables

From (6.8) and (6.9) we have

$$\frac{\partial}{\partial f} : e^{\alpha f} := \alpha : e^{\alpha f}: \tag{6.11}$$

and

$$E[:e^{\alpha f}:] = 1. \tag{6.12}$$

We now find an expression for $:e^{\alpha f}:$ in terms of f. From (6.10) we know that $:e^{\alpha f}:$ has the form $:e^{\alpha f}:=\sum d_n f^n$. It follows from (6.9) and (6.11) that

$$\sum (n+1) d_{n+1} f^n = \sum \alpha d_n f^n.$$

Hence $d_n = \alpha^n d_0 / n!$, $n = 1, 2, \ldots$, and $:e^{\alpha f}:=d_0 e^{\alpha f}$. Applying (6.12) gives

$$:e^{\alpha f}:= \left(E[e^{\alpha f}] \right)^{-1} e^{\alpha f}. \tag{6.13}$$

Now

$$E[e^{\alpha f}] = \frac{1}{\sqrt{2\pi a}} \int \exp\left(-\frac{t^2}{2a} + \alpha t \right) dt$$

$$= e^{a\alpha^2/2} \frac{1}{\sqrt{2\pi a}} \int \exp\left[-\left(\frac{t}{\sqrt{2a}} - \sqrt{\frac{a}{2}}\, \alpha \right)^2 \right] dt = e^{a\alpha^2/2}.$$

Hence, (6.13) becomes

$$:e^{\alpha f}:= e^{\alpha f - (a\alpha^2/2)}. \tag{6.14}$$

Using (6.14) we can obtain many properties of Wick powers. For example, expanding the right-hand side of (6.14) in a power series and equating coefficients of α with those in (6.10) gives

$$:f^n := \sum_{m=0}^{[n/2]} \frac{n!}{m!(n-2m)!} \left(-\frac{a}{2} \right)^m f^{n-2m}, \tag{6.15}$$

where $[n/2]$ is the integer part of $n/2$. Similarly, we have

$$f^n = \sum_{m=0}^{[n/2]} \frac{n!}{m!(n-2m)!} \left(\frac{a}{2} \right)^m :f^{n-2m}:. \tag{6.16}$$

The following lemma is another example of the usefullness of (6.14).

Lemma 6.10. *Let f and g be jointly Gaussian random variables with mean zero. Then*

$$E[:f^n::g^m:] = \delta_{nm} n! [E(fg)]^n.$$

PROOF. By (6.14), if the variance of g is b, we have

$$:e^{\alpha f}::e^{\beta g}:=\left(e^{\alpha f+\beta g}\right)e^{-(1/2)(a\alpha^2+b\beta^2)}$$

$$=e^{\alpha\beta E[fg]}:e^{\alpha f+\beta g}:.$$

Hence, from (6.12) we have

$$E\left[:e^{\alpha f}::e^{\beta g}:\right]=e^{\alpha\beta E[fg]}.$$

The result follows by expanding the exponentials. □

It follows from Lemma 6.10 that $:f^n:$, $n=0,1,\ldots$, is a sequence of orthogonal polynomials of f in $L^2(\Omega,\Sigma,\mu)$. Thus another way to get the Wick powers is to orthogonalize the sequence $f^0, f^1, \ldots$ in $L^2(\Omega,\Sigma,\mu)$ using the Gram–Schmidt process subject to the normalization $\|:f^n:\|=n!\|f\|^n$.

Now let $f_1,\ldots,f_k$ be jointly Gaussian mean zero random variables. The *Wick product* $:f_1^{n_1}\cdots f_k^{n_k}:$ is the polynomial in $f_1,\ldots,f_k$ of degree $n=n_1+\cdots+n_k$ defined recursively by

$$:f_1^0\cdots f_k^0:=1,$$

$$\frac{\partial}{\partial f_i}:f_1^{n_1}\cdots f_k^{n_k}:=n_i:f_1^{n_1}\cdots f_i^{n_i-1}\cdots f_k^{n_k}:,$$

$$E\left[:f_1^{n_1}\cdots f_k^{n_k}:\right]=0,\qquad n\neq 0.$$

With this definition one can now prove the following binomial formula

$$:(\alpha f_1+\beta f_2)^n:=\sum_{m=0}^n\binom{n}{m}\alpha^m\beta^{n-m}:f_1^n f_2^{n-m}:. \qquad (6.17)$$

Also, the analogous multinomial formula holds. Using the multinomial formula and Lemma 6.10 we have the following corollary.

Corollary 6.11. *Let* $f_1,\ldots,f_n,g_1,\ldots,g_m$ *be jointly Gaussian random variables with mean zero.*

(a) *If* $n\neq m$, *then* $E[:f_1\cdots f_n::g_1\cdots g_m:]=0.$

(b) *If* $E[f_if_j]=\delta_{ij}$, *then*

$$E\left[:f_1^{n_1}\cdots f_k^{n_k}::f_1^{m_1}\cdots f_k^{m_k}:\right]=\delta_{n_1m_1}\cdots\delta_{n_km_k}n_1!\cdots n_k!.$$

A family f_α of random variables on (Ω,Σ,μ) is *full* if Σ is the smallest σ-algebra with respect to which all the f_α's are measurable.

Lemma 6.12. *Let* $\{f_\alpha\}_{\alpha\in I}$ *be a full set of random variables on* (Ω,Σ,μ). *Then*

$$\mathcal{Q}=\left\{F(f_{\alpha_1},\ldots,f_{\alpha_n}):F\in\mathbb{S}(\mathbb{R}^n),\alpha_i\in I,n=1,2,\ldots\right\}$$

is dense in $L^2(\Omega,\Sigma,\mu)$.

PROOF. Since the argument is fairly standard, we shall only outline the proof. Let $\Sigma(f_\alpha : \alpha \in J)$ be the σ-algebra generated by $\{f_\alpha : \alpha \in J\}, J \subseteq I$. Then $\Sigma = \Sigma(f_\alpha : \alpha \in I)$. Let f be measurable with respect to Σ. Since $\Sigma(f)$ is generated by a countable collection of sets (say the inverse images of intervals with rational endpoints), it follows that there exists a $J \subseteq I$, J countable, such that f is measurable with respect to $\Sigma(f_\alpha : \alpha \in J)$. Moreover, if $J = \{\alpha_1, \alpha_2, \dots\}$ we have

$$\Sigma(f_\alpha : \alpha \in J) = \bigcup_{n=1}^{\infty} \Sigma(f_\alpha : \alpha \in \{\alpha_1, \dots, \alpha_n\}).$$

Then there exists a sequence of functions f_n with f_n measurable with respect to

$$\Sigma(f_\alpha : \alpha \in \{\alpha_1, \dots, \alpha_n\}) = \Sigma_n$$

such that $f_n \to f$ a.e. But every function g measurable with respect to Σ_n has the form $F(f_{\alpha_1}, \dots, f_{\alpha_n})$, $F \in \mathcal{B}(\mathbb{R}^n)$. Indeed, if an indicator function χ_Λ is measurable with respect to Σ_n then

$$\Lambda = \left\{ \omega : \left(f_{\alpha_1}(\omega), \dots, f_{\alpha_n}(\omega) \right) \in A \right\}$$

for some $A \in \mathcal{B}(\mathbb{R}^n)$. Hence $\chi_\Lambda = \chi_A(f_1, \dots, f_n)$. Hence, simple functions which are measurable with respect to Σ_n have the above form, and since g is a limit of simple functions there exists a sequence of $\mathcal{B}(\mathbb{R}^n)$ measurable functions F_m such that $F_m(f_{\alpha_1}, \dots, f_{\alpha_n}) \to g$. But F_m converges to a $\mathcal{B}(\mathbb{R}^n)$-measurable function F and hence $g = F(f_{\alpha_1}, \dots, f_{\alpha_n})$. Hence, $f_n = F_n(f_{\alpha_1}, \dots, f_{\alpha_n})$ for some $\mathcal{B}(\mathbb{R}^n)$-measurable function F_n. If $f \in L^2(\Omega, \Sigma, \mu)$, then the f_n can be chosen so that $f_n \in L^2(\Omega, \Sigma_n, \mu)$. One finally shows that the corresponding functions F_n can be approximated by elements of $\mathbb{S}(\mathbb{R}^n)$. $\quad\square$

We say that a random field Φ is *full* if $\{\Phi(\phi) : \phi \in V\}$ is a full set of random variables. Let $\mathcal{K}$ be a Hilbert space and let $\Phi : \mathcal{K} \to R(\Omega, \Sigma, \mu)$ be a full unit Gaussian random field. Let $\Gamma(\mathcal{K}) = L^2(\Omega, \Sigma, \mu)$ and let $\Gamma_n(\mathcal{K})$ be the closed subspace of $\Gamma(\mathcal{K})$ generated by

$$\{ :\Phi(\phi_1) \cdots \Phi(\phi_n) : \phi_i \in \mathcal{K} \}.$$

Part of the next theorem shows that there is only one unit Gaussian random field on $\mathcal{K}$ up to an isomorphism.

Theorem 6.13. (a) $\Gamma_n(\mathcal{K}) \perp \Gamma_m(\mathcal{K})$ *if* $n \neq m$ *and* $\Gamma(\mathcal{K}) = \bigoplus \Gamma_n(\mathcal{K})$. (b) *If* $\Phi : \mathcal{K} \to R(\Omega, \Sigma, \mu)$ *and* $\Phi' : \mathcal{K} \to R(\Omega', \Sigma', \mu')$ *are full unit Gaussian random fields, then* (Ω, Σ, μ) *and* (Ω', Σ', μ') *are isomorphic and* $\Phi(\phi)$ *corresponds to* $\Phi'(\phi)$ *under the isomorphism for every* $\phi \in \mathcal{K}$.

PROOF. (a) That $\Gamma_n(\mathcal{H}) \perp \Gamma_m(\mathcal{H})$ if $n \neq m$ follows from Corollary 6.11(a). Since

$$:e^{i\Phi(\phi)}: = \sum \frac{i^n}{n!} :\Phi(\phi):$$

it follows that $:e^{i\Phi(\phi)}: \in \bigoplus \Gamma_n(\mathcal{H})$ and hence $e^{i\Phi(\phi)} \in \bigoplus \Gamma_n(\mathcal{H})$ for every $\phi \in \mathcal{H}$. Therefore, for every $F \in \mathcal{S}(\mathbb{R}^n)$ and $\phi_1, \ldots, \phi_n \in \mathcal{H}$ we have

$$F[\Phi(\phi_1), \ldots, \Phi(\phi_n)] = (2\pi)^{-n/2} \int \hat{F}(t) \exp\left[\sum t_i \Phi(\phi_i) \right] d^n t$$

is in $\bigoplus \Gamma_n(\mathcal{H})$. Applying Lemma 6.12, such random variables are dense in $\Gamma(\mathcal{H})$, which gives $\Gamma(\mathcal{H}) = \bigoplus \Gamma_n(\mathcal{H})$.

(b) Define $U: \Gamma(\mathcal{H}) \to \Gamma'(\mathcal{H})$ by

$$U: \Phi(\phi_1) \cdots \Phi(\phi_n): = :\Phi'(\phi_1) \cdots \Phi'(\phi_n):.$$

By Lemma 6.12 and part (a), U is a unitary map from $\Gamma(\mathcal{H})$ onto $\Gamma'(\mathcal{H})$. Since

$$U e^{i\Phi(\phi)} U^{-1} = e^{i\Phi'(\phi)}$$

for all $\phi \in \mathcal{H}$, where $e^{i\Phi(\phi)}$ is considered as a multiplication operator, we have

$$U\left[F(\Phi(\phi_1), \ldots, \Phi(\phi_n)) \right] U^{-1} = F\left[\Phi'(\phi_1), \ldots, \Phi'(\phi_n) \right]$$

for every $F \in \mathcal{S}(\mathbb{R}^n)$, $\phi_1, \ldots, \phi_n \in \mathcal{H}$. If $B \in \mathcal{B}(C)$ it follows that

$$U \chi_{\Phi(\phi)^{-1}(B)} U^{-1} = \chi_{\Phi'(\phi)^{-1}(B)}. \tag{6.18}$$

Moreover, if $A \in \Sigma$, there exists a bounded Borel function $F': \Omega' \to C$ such that $U \chi_A U^{-1} = F'$. Since χ_A considered as a multiplication operator is an orthogonal projection, so is F'. Hence, there exists an $A' \in \Sigma'$ such that $U \chi_A U^{-1} = \chi_{A'}$. Define $T: \Sigma/\mathcal{I}_\mu \to \Sigma'/\mathcal{I}_{\mu'}$ by $TA = A'$. It easily follows that T is an isomorphism and by (6.18) $\Phi(\phi)$ corresponds to $\Phi'(\phi)$ under T for all $\phi \in \mathcal{H}$. $\qquad \square$

We now give an explicit construction which produces the full unit Gaussian random field on a separable real Hilbert space $\mathcal{H}$. Let $\{e_n\}$ be an orthonormal basis for $\mathcal{H}$. Let $\mathbb{R}_\infty$ be the one-point compactification of $\mathbb{R}$ and let $\Omega = \prod \mathbb{R}_\infty$ be the Cartesian product of a countably infinite number of copies of $\mathbb{R}_\infty$. Then Ω is a compact Hausdorff space under the Tychonov topology. Let $C(\Omega)$ be the set of continuous functions on Ω and let $T(\Omega) \subseteq C(\Omega)$ be the set of functions $F(x_1, \ldots, x_n)$ in $C(\Omega)$ dependent on only finitely many coordinates. If we endow $C(\Omega)$ with the supremum norm topology, then by the Stone–Weierstrass theorem, $T(\Omega)$ is dense in

$C(\Omega)$. For $F \in T(\Omega)$, define

$$h(F) = (2\pi)^{-n/2} \int F(x_1, \ldots, x_n) \exp\left(-\sum_{i=1}^{n} \frac{x_i^2}{2}\right) d^n x.$$

Then h is a well-defined linear functional on $T(\Omega)$ and $|h(F)| \leq \|F\|_{\infty}$. Since $T(\Omega)$ is dense in $C(\Omega)$, h extends to a unique continuous linear functional on $C(\Omega)$. By the Riesz representation theorem there exists a Borel measure μ on the Borel sets Σ of Ω such that $h(F) = \int F d\mu$ for every $F \in C(\Omega)$. Then (Ω, Σ, μ) is a probability space. Let $\Phi(e_n): \Omega \to \mathbb{R}_\infty$ be multiplication by x_n, $n = 1, 2, \ldots$. Then $\Phi(e_n)$ takes values which are almost everywhere in $\mathbb{R}$, so $\Phi(e_n)$ can be viewed as a random variable. It is not hard to show that $\Phi(e_n) \in L^2(\Omega, \Sigma, \mu)$ and that $\{\Phi(e_n): n = 1, 2, \ldots\}$ is full. If $\phi \in \mathcal{K}$ and $\phi = \Sigma a_n e_n$, then by explicit computation $\Sigma a_n \Phi(e_n)$ converges in $L^2(\Omega, \Sigma, \mu)$ to an element we call $\Phi(\phi)$. It is straightforward to show that Φ becomes the full unit Gaussian random field on $\mathcal{K}$. The above construction goes through for nonseparable $\mathcal{K}$ and thus proves the existence of the unit Gaussian random field on any real Hilbert space.

6.4. Random Fields on $S(\mathbb{R}^n)$

Let $V = S(\mathbb{R}^n)$ be the Schwartz space of rapidly decreasing functions. Other Schwartz spaces can be used, but for concreteness we shall stay with this one. In this section, random fields will be assumed to be continuous in the Schwartz topology. The *derivative* Φ' of a random field Φ on $S(\mathbb{R}^n)$ is defined as $\Phi'(\phi) = -\Phi(\phi')$. Of course, Φ' is a random field. If Φ is Gaussian, then so is Φ'.

We shall assume that stochastic processes (see Example 1, Section 6.2) have almost all sample paths square integrable and, for simplicity, that all random fields and stochastic processes have mean zero. Recall that a random field $\Phi: S(\mathbb{R}^n) \to R(\Omega, \Sigma, \mu)$ is *generated* by a stochastic process (see Example 2, Section 6.2) $W: \mathbb{R}^n \times \Omega \to C$ if $[\Phi(\phi)](\omega) = \int \phi(x) W(x, \omega) dx$ for all $\phi \in S(\mathbb{R}^n)$. A stochastic process $W(x, \omega)$ is of *second order* if $\int |W(x, \omega)|^2 \mu(d\omega) < \infty$ for all $x \in \mathbb{R}^n$. The *correlation function* $C_W: \mathbb{R}^n \times \mathbb{R}^n \to C$ of a second-order stochastic process $W(x, \omega)$ is defined by $C_W(x, y) = \int_\Omega W(x, \omega) W(y, \omega)^* \mu(d\omega)$. Notice that C_W exists for second-order stochastic processes. Since stochastic processes correspond to sharp measurements of physical fields, it is important to know when a random field is generated by a stochastic process. Furthermore, if a random field is generated by a stochastic process it may be studied using the established literature of classical probability theory.

Theorem 6.14. (a) *If Φ is generated by a second-order stochastic process $W(x,\omega)$, then $C_\Phi(\phi,\psi) = \int C_W(x,y)\phi(x)\psi(y)^* \, dx \, dy$. (b) Let Φ be a proper Gaussian random field. If there exists a continuous function $C: \mathbb{R}^n \times \mathbb{R}^n \to C$ such that $C_\Phi(\phi,\psi) = \int C(x,y)\phi(x)\psi(y)^* \, dx \, dy$, then there exists a Gaussian stochastic process $W(x,\omega)$ which generates Φ such that $C_W = C$.*

PROOF. (a) Using Fubini's theorem we have

$$
\begin{aligned}
C_\Phi(\phi,\psi) &= E\big[\Phi(\phi)\Phi(\psi)^*\big] \\
&= \int \left[\int \phi(x) W(x,\omega)\, dx \int \psi(y)^* W(y,\omega)^* \, dy \right] d\mu \\
&= \int \phi(x)\psi(y)^* \left[\int W(x,\omega) W(y,\omega)^* \, d\mu \right] dx\, dy \\
&= \int C_W(x,y)\phi(x)\psi(y)^* \, dx\, dy.
\end{aligned}
$$

(b) Let $x_1,\ldots,x_m \in \mathbb{R}^n$. Since $C(x,y)$ is continuous, it can be shown that there exist linearly independent test functions $\phi_1,\ldots,\phi_m \in \mathbb{S}(\mathbb{R}^n)$ such that $C(x_i,x_j) = C_\Phi(\phi_i,\phi_j)$, $1 \le i$, $j \le m$. Since Φ is proper, it follows that the matrix $B = [C_\Phi(\phi_i,\phi_j)] = [C(x_i,x_j)]$ is positive definite. Let $\Lambda = B^{-1}$ and let $\mu(x_1,\ldots,x_m)$ be the probability measure on $\mathbb{R}^m$ defined by

$$
\mu(x_1,\ldots,x_m)(A) = (\det \Lambda)^{1/2}(2\pi)^{m/2} \int_A e^{-(1/2)\langle \Lambda x, x \rangle} \, dx,
$$

where A is any Borel set in $\mathbb{R}^m$. It can be shown that the collection

$$
\{ \mu(x_1,\ldots,x_m) : (x_1,\ldots,x_m) \in \mathbb{R}^m, \, m = 1,2,\ldots, \}
$$

forms a consistent set of probability measures. By the Kolmogorov theorem [160], there exists a probability space (Ω,Σ,μ) and a collection of random variables $\{f_x : x \in \mathbb{R}^m\}$ on (Ω,Σ,μ) such that the joint distribution of $f_{x_1},\ldots,f_{x_m}$ is $\mu(x_1,\ldots,x_m)$ for all $(x_1,\ldots,x_m) \in \mathbb{R}^m$, $m = 1,2,\ldots$. Let $W(x,\omega)$ be the stochastic process defined by $W(x,\omega) = f_x(\omega)$, $x \in \mathbb{R}^m$. Then $W(x,\omega)$ is a Gaussian stochastic process with covariance function $C(x,y)$. Using a separability argument, W has a version which is continuous almost everywhere. Define the random field $[\Phi_1(\phi)](\omega) = \int \phi(x) W(x,\omega)\, dx$. Then it is straightforward to show that Φ_1 is a Gaussian random field and by (a) we have that $C_{\Phi_1}(\phi,\psi) = \int C(x,y)\phi(x)\psi(y)^* \, dx \, dy$. Since Φ and Φ_1 are both Gaussian with mean zero and the same covariance functional, they must have the same finite dimensional joint distributions and hence are equivalent. $\qquad\square$

6. Random Variables

Example 1. Let Φ be the isonormal random field with

$$C_\Phi(\phi,\psi) = r\langle\phi,\psi\rangle = r\int \phi(x)\psi(x)^* dx.$$

Then C_Φ is not of the form $\int C(x,y)\phi(x)\psi(y)^* dx\,dy$. In fact, $C(x,y)$ would have to be the delta function $r\delta(x-y)$ for equality to hold. It follows from Theorem 6.14 that Φ is not generated by a stochastic process. Thus proper random fields with strongly independent values are not generated by stochastic processes. This is why there is no analog to strongly independent values for stochastic processes.

Example 2. Let $W(t,\omega)$, $t\in[0,\infty)$, be the Wiener process with variance 1 and let Φ be the random field generated by $W(t,\omega)$. Then Φ is a Gaussian random field. By the proof of Theorem 6.1, the covariance function is

$$C_W(s,t) = |[0,s]\cap[0,t]| = \min(s,t).$$

Hence, by Theorem 6.14(a)

$$C_\Phi(\phi,\psi) = \int_0^\infty \int_0^\infty \min(s,t)\phi(s)\psi(t)\,ds\,dt$$

$$= \int_0^\infty [\hat\phi(t) - \hat\phi(\infty)][\hat\psi(t) - \hat\psi(\infty)]\,dt,$$

where $\hat\phi(t) = \int_0^t \phi(s)\,ds$ and $\hat\psi(t) = \int_0^t \psi(s)\,ds$. Since sample paths of W are nowhere differentiable, the derivative of $W(t,\omega)$ does not exist as a stochastic process. However, since the derivative of a random field always exists, Φ' is a well-defined Gaussian random field. The covariance functional of Φ' is

$$C_{\Phi'}(\phi,\psi) = C_\Phi(\phi',\psi') = \int_0^\infty \phi(t)\psi(t)\,dt = \langle\phi,\psi\rangle.$$

Therefore Φ' is the unit random field. We thus see that the unit random field is a rigorous version of white noise (the "derivative" of Brownian motion).

Example 3. In Chapter 7 we shall define Markov random fields and show how they can be used in quantum field theory. There is an important class of random fields that lie between the random fields with strongly independent values and the Markov random fields. A random field Φ on $S(\mathbb{R}^n)$ has *independent values* if $\Phi(\phi)$ and $\Phi(\psi)$ are stochastically independent whenever $\phi(x)\psi(x) = 0$ for all $x\in\mathbb{R}^n$; that is $\operatorname{supp}\phi\cap\operatorname{supp}\psi = \varnothing$. Physically this means that the results of measuring Φ in nonintersecting regions of space (or time or space–time) are independent. Gelfand and Vilenkin [87] have shown that a Gaussian random field Φ on $S(\mathbb{R}^n)$ has independent values if and only if there exist a finite number of continuous functions F_{ij} on $\mathbb{R}^n$ such that

$$C_\Phi(\phi,\psi) = \int \sum_{i,j} F_{ij}(x)\phi^{(i)}(x)\psi^{(j)}(x)\,d^nx.$$

The proof of the following characterization is the same as that of Lemma 6.3.

Lemma 6.15. *A random field on* $S(\mathbb{R}^n)$ *has independent values if and only if* $L_\Phi(\phi + \psi) = L_\Phi(\phi)L_\Phi(\psi)$ *whenever* $supp\,\phi \cap supp\,\psi = \varnothing.$

The next result follows from Theorem 5.18.

Theorem 6.16. *Let* Φ *be a random field on* $S(\mathbb{R}^n)$ *for which* L_Φ *satisfies the conditions of Theorem 5.18. Then* Φ *has independent values if and only if* $L_\Phi(\phi) = e^{M(\phi)}$, *where*

$$M(\phi) = \int G[\phi(x), \ldots, D^\alpha \phi(x), x]\,d\mu(x)$$

and G and μ *satisfy the conditions of Theorem 5.18.*

6.5. Notes and References

Some early papers on Brownian motion are [32, 33, 39, 207, 210, 276]. More details on the Wiener process may be found in [17, 31, 65, 68, 163, 164, 166, 185, 283]. Our discussion of random fields and Gaussian random fields follows [87, 88, 250]. Some of this material can also be found in [109]. Material on Markov fields which will be considered in greater detail in Chapter 7 may be found in [197, 199, 200, 250]. Additional papers on Markov and Wiener processes in quantum mechanics are [1, 2, 41, 43, 51–53, 57–60, 74, 80, 82–85, 89, 120, 151, 165, 196, 198, 233, 251].

6.6. Exercises

1. Complete the proof of Theorem 6.1.

2. Show that Φ defined by $[\Phi(\phi)](\psi) = \langle \phi, f(\psi) \rangle$ in Example 4 of Section 6.2 is a random field.

3. Prove that a bounded random functional is of second order but that the converse need not hold.

4. Prove that the characteristic functional $L_\Phi(\phi)$ exists for all $\phi \in V$ and that if Φ is continuous then L_Φ is also continuous.

5. Show that Φ contains **1** in its range if and only if there are degenerate distributions.

6. Show that if Φ has strongly independent values, then Φ is uncorrelated.

7. Show that Example 3 of Section 6.2 is a unit random field.

8. If Φ_r is an isonormal random field, show that $\Phi_1(\phi) = [\Phi_r(\phi) - M_{\Phi_r}(\phi)]/r^{1/2}$ is the unit random field.

9. If a random variable f has the characteristic function c_f, show that the mean and variance of f are given by $-ic_f'(0)$ and $-c_f''(0) + [c_f'(0)]^2$, respectively. Moreover, $\int f^n d\mu = (-i)^n c_f^{(n)}(0)$.

10. If c_f is given by (6.5), show that m is the mean of f and a is the variance.

11. Prove Eqs. (6.6) and (6.7).

12. Show that $f_1, \ldots, f_n$ are jointly Gaussian if and only if $\Sigma \alpha_i f_i$ is Gaussian for every $(\alpha_1, \ldots, \alpha_n) \in \mathbb{R}^n$.

13. Prove Lemma 6.9.

14. Derive Eqs. (6.15) and (6.16).

15. Fill in the details of the proof of Lemma 6.10.

16. Prove Eq. (6.17).

17. Prove Corollary 6.11.

18. Supply the details for the construction in the last paragraph of Section 6.3.

19. If Φ is a random field, show that the derivative Φ' is a random field.

20. If Φ is a Gaussian random field, show that Φ' is Gaussian.

7

Quantum Field Theory

Quantum systems with a finite number of degrees of freedom have a complete and rigorous mathematical description. No such mathematical framework has yet been found for general systems with an infinite number of degrees of freedom. Since an infinite number of degrees of freedom are necessary to investigate systems involving interactions between particles and fields and interactions in which the creation or annihilation of particles are present, a satisfactory theory for such systems would have immense practical applications. In this chapter we present some of the recent progress which is bringing us closer to the solution of this problem. Much of the material of the previous two chapters will find application here and stochastic methods will prove to be extremely useful.

7.1. Canonical Commutation Relations

Let us begin with a mathematical description of a single spinless, nonrelativistic particle with one degree of freedom. Such a system is traditionally described by the Hilbert space $L^2(\mathbb{R})$. The position operator Q and the momentum operator P play fundamental roles. The position operator Q is the self-adjoint operator with domain

$$D(Q) = \{\phi \in L^2(\mathbb{R}) : x\phi(x) \in L^2(\mathbb{R})\}$$

and defined by $(Q\phi)(x) = x\phi(x)$. The domain of P is the set of all absolutely continuous functions ϕ on $\mathbb{R}$ such that $\phi' \in L^2(\mathbb{R})$ and P is the self-adjoint operator $P = -id/dx$ (we assume $\hbar = 1$). Notice that the

Schwartz space $S = S(\mathbb{R})$ satisfies $S \subseteq D(Q) \cap D(P)$ and that S is a dense subspace of $L^2(\mathbb{R})$ which is left invariant by Q and P. Moreover, it is clear that

$$QP - PQ = iI \tag{7.1}$$

on S. We call (7.1) the *Heisenberg form* of the *canonical commutation relations* (CCR).

There are other forms of the CCR. Since P is self-adjoint, the operator $U(a) = e^{-iaP}$, $a \in \mathbb{R}$, is unitary. In fact, it is well known that for any self-adjoint operator P, the set of operators $\{U(a) : a \in \mathbb{R}\}$ forms a one-parameter strongly continuous unitary group. That is,

(i) $U(a_1)U(a_2) = U(a_1 + a_2)$ for all $a_1, a_2 \in \mathbb{R}$;
(ii) $\lim_{a \to a_0} U(a)\phi = U(a_0)\phi$ for all $\phi \in L^2(\mathbb{R})$.

For this reason, we call $a \mapsto U(a)$ a *unitary representation* of the additive group $\mathbb{R}$. Using Taylor's theorem we see that for all $\phi \in S$

$$e^{-iaP}\phi(x) = \sum \frac{(-iaP)^n}{n!}\phi(x) = \sum \frac{(-a)^n}{n!}\phi^{(n)}(x)$$
$$= \phi(x - a).$$

Extending by continuity we conclude that

$$U(a)\phi(x) = \phi(x - a)$$

for all $\phi \in L^2(\mathbb{R})$. Now $D(Q)$ is invariant under $U(a)$ for every $a \in \mathbb{R}$, and for every $\phi \in D(Q)$ we have

$$U(a)QU(-a)\phi(x) = QU(-a)\phi(x - a)$$
$$= (x - a)U(-a)\phi(x - a) = (x - a)\phi(x).$$

Hence, on $D(Q)$ we have

$$U(a)QU(-a) = Q - aI. \tag{7.2}$$

We call (7.2) the *Schrödinger form* of the CCR.

Since Q is self-adjoint it generates a strongly continuous one-parameter unitary group $V(b)$ given by

$$V(b)\phi(x) = e^{-ibQ}\phi(x) = e^{-ibx}\phi(x).$$

If $\phi \in S$, $a, b \in \mathbb{R}$ we obtain

$$U(a)V(b)\phi(x) = V(b)\phi(x - a) = e^{-ib(x-a)}\phi(x - a)$$
$$= e^{iab}e^{-ibx}\phi(x - a) = e^{iab}V(b)U(a)\phi(x).$$

Extending this equation to $L^2(\mathbb{R})$ by continuity we conclude that

$$U(a)V(b) = e^{iab}V(b)U(a) \tag{7.3}$$

for every $a, b \in \mathbb{R}$. Equation (7.3) is called the *Weyl form* of the CCR.

Let us now approach the subject from an axiomatic point of view as Heisenberg did when he introduced matrix mechanics. For a quantum system with one degree of freedom we assume that there are two fundamental observables, These observables are represented by two self-adjoint operators Q and P which leave invariant a dense subspace $D \subseteq L^2(\mathbb{R})$ and which satisfy the Heisenberg form (7.1) of the CCR on D. Unfortunately, these conditions do not determine Q and P uniquely to within a unitary equivalence. To see this, we have already observed that $Q\phi(x) = x\phi(x)$ and $P\phi(x) = i\phi'(x)$ satisfy (7.1). Now let $H_1 = L^2([0,1])$ and define the self-adjoint operators Q_1 and P_1 as follows:

$$D(Q_1) = \{\phi \in \mathcal{H}_1 : x\phi(x) \in \mathcal{H}_1\}, \qquad Q_1\phi(x) = x\phi(x),$$

$$D(P_1) = \{\phi \in H_1 : \phi(0) = \phi(1), \phi' \in \mathcal{H}_1\}, \qquad P_1\phi(x) = -i\phi'(x). \quad (7.4)$$

Then it is easy to see that

$$Q_1P_1 - P_1Q_1 = iI$$

on the dense subspace of $\mathcal{H}_1$ consisting of the infinitely differentiable functions ϕ satisfying $\phi(0) = \phi(1)$. But the pair (Q,P) is not unitarily equivalent to the pair (Q_1, P_1) since Q and P are unbounded with purely continuous spectrum, while Q_1 is bounded and P_1 has discrete point spectrum. We thus see that there are inequivalent ways of representing the CCR in the Heisenberg form (7.1).

The situation is quite different for the Weyl form (7.3) of the CCR. In this case the von Neumann uniqueness theorem, which we shall consider in detail later, takes effect. This theorem states that all irreducible [i.e., $U(a)M \subseteq M$, $V(b)M \subseteq M$, $a,b \in \mathbb{R}$, for a closed subspace M implies $M = \{0\}$ or $\mathcal{H}$] representations of the Weyl form of the CCR for one degree of freedom are unitarily equivalent. Since any representation is the direct sum of irreducible representations it follows that if $U_1(a)$ and $V_1(b)$ satisfy (7.3), then $U_1(a)$ and $V_1(b)$ must be the direct sum of copies of operators of the form $U(a)\phi(x) = \phi(x - a)$, $V(b)\phi(x) = e^{-ibx}\phi(x)$.

It follows from the above that the Heisenberg and Weyl forms of the CCR are not logically equivalent. It can be shown, however, that the Schrödinger and the Weyl forms are equivalent. The Weyl form (7.3) thus enjoys at least two advantages. It is phrased in terms of bounded operators and it uniquely determines the operators $U(a)$ and $V(b)$ to within an equivalence (in the irreducible case).

We have noted that the Heisenberg form does not have the uniqueness property of the Weyl form. We now show that it cannot have the boundedness property.

Lemma 7.1. *If $QP - PQ = iI$, then at least one of the operators Q or P must be unbounded.*

PROOF. Suppose, on the contrary, that there exist two bounded operators Q and P acting on a Hilbert space and satisfying (7.1). We can assume without loss of generality that P is invertible, for if not we could replace P by $P - \lambda I$, where $\lambda > \|P\|$, without changing the commutation relation. Since

$$PQ - \alpha I = P(QP - \alpha I)P^{-1}$$

for all $\alpha \in C$, we conclude that the spectrum $\sigma(PQ) = \sigma(QP)$. Since $PQ = QP - iI$, we have

$$PQ - i(\alpha - 1)I = QP - i\alpha I$$

for every $\alpha \in C$. Hence $i\alpha \in \sigma(QP)$ if and only if $i(\alpha - 1) \in \sigma(PQ)$. Since $\sigma(PQ) \neq \varnothing$ it follows that there exists an $\alpha \in C$ such that

$$\{i(\alpha + n) : n = 0, 1, 2, \dots\} \subseteq \sigma(PQ).$$

Thus $\sigma(PQ)$ is unbounded. This contradicts the fact that PQ is a bounded operator. $\qquad\square$

All that we have said so far can be easily generalized to systems with a finite number n of degrees of freedom. In this case we have $2n$ operators $Q_1, \dots, Q_n, P_1, \dots, P_n$ on $L^2(\mathbb{R}^n)$ satisfying the Heisenberg form of the CCR

$$Q_\kappa P_j - P_j Q_\kappa = i\delta_{\kappa j} I. \tag{7.5}$$

For $a = (a_1, \dots, a_n), b = (b_1, \dots, b_n) \in \mathbb{R}^n$ we form the unitary operators $U(a)\phi(x) = \phi(x - a), V(b)\phi(x) = e^{-i\langle b, x\rangle}\phi(x), \phi \in L^2(\mathbb{R}^n)$. As before, $U(a)$ and $V(b)$ are unitary representations of the additive group $\mathbb{R}^n$ [i.e., satisfy (i) and (ii)] and, moreover, the Weyl form of the CCR holds:

$$U(a)V(b) = e^{i\langle a, b\rangle}V(b)U(a) \tag{7.6}$$

for every $a, b \in \mathbb{R}^n$. Again, the von Neumann uniqueness theorem states that (7.6) uniquely determines $U(a)$ and $V(b)$ to within an equivalence (in the irreducible case).

7.2. Infinite Degrees of Freedom

We now jump to the infinite number of degrees of freedom case. In this case $\mathbb{R}^n$ must be replaced by an infinite dimensional real inner product space $\mathcal{V}$. A *unitary representation* U of $\mathcal{V}$ on a Hilbert space $\mathcal{H}$ is a map from $\mathcal{V}$ into the set of unitary operators on $\mathcal{H}$ which satisfies:

(i) $U(\phi + \psi) = U(\phi) U(\psi)$ for all $\phi, \psi \in \mathcal{V}$;

(ii) if $\phi_i \to \phi$ in $\mathcal{V}$, then $U(\phi_i)f \to U(\phi)f$ for every $f \in \mathcal{H}$.

A $\mathcal{V}$-*representation* of the CCR on $\mathcal{H}$ is a pair of unitary representations (U, V) of $\mathcal{V}$ on $\mathcal{H}$ such that

$$U(\phi) V(\psi) = e^{i\langle \phi, \psi \rangle} V(\psi) U(\phi) \tag{7.7}$$

for every $\phi, \psi \in \mathcal{V}$. Of course, this is a straightforward generalization of the finite number of degrees of freedom case. But now the von Neumann uniqueness theorem does not hold and we can have many inequivalent $\mathcal{V}$-representations of the CCR. We shall only consider the so-called cyclic representations. A $\mathcal{V}$-representation of the CCR is *cyclic* with cyclic vector $h \in \mathcal{H}$ if

$$\mathcal{H} = \overline{\text{span}} \{ V(\phi)h : \phi \in V \}.$$

This is the definition of cyclicity used in [87] (and elsewhere); other authors (e.g., [72]) define cyclicity in terms of U and V. Care must be taken here, since the definitions are not equivalent. In this section we give a correspondence between cyclic $\mathcal{V}$-representations of the CCR and certain random fields on $\mathcal{V}$. We first need the relevant definitions.

Let $\Phi: \mathcal{V} \to R(\Omega, \Sigma, \mu)$ be a full random field. For $\psi \in \mathcal{V}$, define $\Phi_\psi: \mathcal{V} \to R(\Omega, \Sigma, \mu)$ by $\Phi_\psi(\phi) = \Phi(\phi) - \langle \phi, \psi \rangle$. The random field Φ_ψ corresponds to a translation of Φ by the vector ψ. Let $m(\Omega, \Sigma, \mu)$ be the group of measurable bijections on Ω. If $T \in m(\Omega, \Sigma, \mu)$ we define $\hat{T}: R(\Omega, \Sigma, \mu) \to R(\Omega, \Sigma, \mu)$ by $(\hat{T}f)(\omega) = f(T^{-1}\omega)$. An *action* of $\mathcal{V}$ on (Ω, Σ, μ) is a group homomorphism $T: \mathcal{V} \to m(\Omega, \Sigma, \mu)$ such that $\psi_i \to \psi$ in $\mathcal{V}$ implies that $\hat{T}_{\psi_i} f \to \hat{T}_\psi f$ in probability. An action T is *quasi-invariant* when $\mu(A) = 0$ implies $\mu(T_\psi A) = 0$ for every $\psi \in \mathcal{V}$. If we define the measure $\mu_\psi(A) = \mu(T_\psi A)$ this is equivalent to μ_ψ being absolutely continuous relative to μ for every $\psi \in \mathcal{V}$. If T is an action of $\mathcal{V}$ on (Ω, Σ, μ) and $\Phi: \mathcal{V} \to R(\Omega, \Sigma, \mu)$ is a random field, we denote the random field $\phi \to \hat{T}_\psi[\Phi(\phi)]$ by $\hat{T}_\psi \Phi$ and say that Φ is T-*covariant* if $\Phi_\psi = \hat{T}_\psi \Phi$ for every $\psi \in \mathcal{V}$. A *covariant random field* is a pair (Φ, T) where $\Phi: \mathcal{V} \to R(\Omega, \Sigma, \mu)$ is a random field, T is a quasi-invariant action of $\mathcal{V}$ on (Ω, Σ, μ) and Φ is T-covariant.

One of the most difficult conditions to verify for a covariant random field (Φ, T) is that T is quasi-invariant. Our first result gives two sufficient conditions for T to be quasi-invariant. If F and F_1 are positive definite functions on $\mathcal{V}$, we say that F_1 *dominates* F if there exists an $M > 0$ such that $MF_1 - F$ is positive definite.

Theorem 7.2. *Let* $T: \mathcal{V} \to m(\Omega, \Sigma, \mu)$ *be an action and* $\Phi: \mathcal{V} \to R(\Omega, \Sigma, \mu)$ *a random field. Then the following statements are equivalent.*

1. L_Φ *dominates* $L_{\hat{T}_\psi \Phi}$ *for every* $\psi \in \mathcal{V}$.
2. $\hat{T}_\psi$ *is a bounded operator from* $\mathcal{H} = L^2(\Omega, \Sigma, \mu)$ *to itself for every* $\psi \in \mathcal{V}$.
3. T *is quasi-invariant and* $f_\psi = d\mu_\psi / d\mu \in \mathcal{H}$ *for every* $\psi \in \mathcal{V}$.

PROOF. We first show that (1) and (2) are equivalent. If (1) holds then for every $\psi \in \mathcal{V}$ there exists an $M_\psi > 0$ such that $M_\psi L_\Phi - L_{\hat{T}_\psi \Phi}$ is positive definite. Since Σ is generated by $\{\Phi(\phi): \phi \in \mathcal{V}\}$, it follows that $\text{span}\{e^{i\Phi(\phi)}: \phi \in \mathcal{V}\} = \mathcal{H}$. Let $Y = \text{span}\{e^{i\Phi(\phi)}: \phi \in \mathcal{V}\}$. We now show that the restriction $\hat{T}_\psi | Y$ is a bounded operator from Y to $\mathcal{H}$.

$$\| \hat{T}_\psi | Y \sum \lambda_k e^{i\Phi(\phi_k)} \|^2 = \| \sum \lambda_k e^{i\hat{T}_\psi \Phi(\phi_k)} \|^2$$

$$= \sum_{j,k} \lambda_j \lambda_k^* \int e^{i\hat{T}_\psi \Phi(\phi_j - \phi_k)} \, d\mu$$

$$= \sum_{j,k} \lambda_j \lambda_k^* L_{\hat{T}_\psi \Phi}(\phi_j - \phi_k)$$

$$\leq M_\psi \sum_{j,k} \lambda_j \lambda_k^* L_\Phi(\phi_j - \phi_k) = M_\psi \| \sum \lambda_k e^{i\Phi(\phi_k)} \|^2.$$

Thus $\| \hat{T}_\psi | Y \| \leq M_\psi^{1/2}$ and $\hat{T}_\psi | Y$ is bounded. Since $\bar{Y} = \mathcal{H}$, $\hat{T}_\psi | Y$ has a unique bounded extension $S_\psi: \mathcal{H} \to \mathcal{H}$. We next show that $S_\psi = \hat{T}_\psi$ on $\mathcal{H}$. If $f \in \mathcal{H}$, there exists a sequence $f_i \in Y$ such that $f_i \to f$ in norm. Hence $S_\psi f = \lim S_\psi f_i$. Now there exists a subsequence $f_{i'}$ such that $f_{i'} \to f$ almost everywhere. Then

$$(S_\psi f)(\omega) = \lim(S_\psi f_{i'})(\omega) = \lim(\hat{T}_\psi f_{i'})(\omega)$$

$$= \lim f_{i'}(T_\psi^{-1} \omega) = f(T_\psi^{-1} \omega) = (\hat{T}_\psi f)(\omega).$$

Hence (2) holds. Conversely, suppose (2) holds and $\phi_1, \ldots, \phi_n \in \mathcal{V}$, $\lambda_1, \ldots, \lambda_n \in C$. Then

$$\sum_{j,k} \lambda_j \lambda_k^* L_{\hat{T}_\psi \Phi}(\phi_j - \phi_k) = \| \sum \lambda_k e^{i\hat{T}_\psi \Phi(\phi_k)} \|^2$$

$$= \| \hat{T}_\psi \sum \lambda_k e^{i\Phi(\phi_k)} \|^2 \leq \| \hat{T}_\psi \|^2 \| \sum \lambda_k e^{i\Phi(\phi_k)} \|^2$$

$$= \| \hat{T}_\psi \|^2 \sum_{j,k} \lambda_j \lambda_k^* L_\Phi(\phi_j - \phi_k).$$

Hence (1) holds. We now show that (2) and (3) are equivalent. If (2) holds, then the map $f \mapsto \langle \hat{T}_\psi f, 1 \rangle$ is a bounded linear functional on $\mathcal{H}$. By the Riesz theorem there exists an $f_\psi \in \mathcal{H}$ such that $\langle \hat{T}_\psi f, 1 \rangle = \langle f, f_\psi \rangle$ for all $f \in \mathcal{H}$. Hence for every $f \in \mathcal{H}$ we have

$$\int f(T_\psi^{-1} \omega) \, d\mu(\omega) = \int f(\omega) f_\psi(\omega) \, d\mu(\omega).$$

Letting $f = \chi_A$, for $A \in \Sigma$ we obtain

$$\mu_\psi(A) = \mu(T_\psi A) = \int \chi_A(T_\psi^{-1}\omega) \, d\mu(\omega) = \int_A f_\psi(\omega) \, d\mu(\omega).$$

Hence μ_ψ is absolutely continuous relative to μ and $f_\psi = d\mu_\psi / d\mu \in \mathcal{H}$. Conversely, suppose (3) holds and let $\Sigma \lambda_j \chi_{A_j}$ be a simple function in $\mathcal{H}$, where $A_i \cap A_j = \varnothing$, $i \neq j$. Then

$$\|\hat{T}_\psi \sum \lambda_j \chi_{A_j}\|^2 = \int |\sum \lambda_j \hat{T}_\psi \chi_{A_j}|^2 \, d\mu = \int |\sum \lambda_j \chi_{A_j}(T_\psi^{-1}\omega)|^2 \, d\mu$$

$$= \int |\sum \lambda_j \chi_{A_j}(\omega)|^2 \, d\mu_\psi = \int |\sum \lambda_j \chi_{A_j}|^2 f_\psi \, d\mu$$

$$= \int \sum_{j,k} \lambda_i \lambda_j^* \chi_{A_i} \chi_{A_j} f_\psi \, d\mu = \sum |\lambda_j|^2 \int \chi_{A_j} f_\psi \, d\mu$$

$$\leq \sum |\lambda_j|^2 \|f_\psi\|^2 \mu(A_j) = \|f_\psi\|^2 \| \sum \lambda_j \chi_{A_j}\|^2.$$

Hence the restriction of $\hat{T}_\psi$ to the subspace of simple functions is bounded. Since the simple functions are dense in $\mathcal{H}$, this restriction has a unique bounded extension to $\mathcal{H}$. By an argument similar to that used above we conclude that (2) holds. $\qquad \square$

An inner product space $\mathcal{V}$ is *completely separable* if there exists a countable orthonormal set $\{x_i\}$ in $\mathcal{V}$ such that $\{x_i\}$ is a basis for the completion $\overline{\mathcal{V}}$ of $\mathcal{V}$. Clearly, a completely separable inner product space is separable. However, the converse need not hold.

Theorem 7.3. (a) Let (Φ, T) be a covariant random field from an inner product space $\mathcal{V}$ to a probability space (Ω, Σ, μ). Then there exists a random functional $\Psi: \mathcal{V} \to R(\Omega, \Sigma, \mu)$ such that (U_0, V_0) defined by Eqs. (7.8) and (7.9) forms a cyclic $\mathcal{V}$-representation of the CCR on $\mathcal{H} = L^2(\Omega, \Sigma, \mu)$:

$$[V_0(\phi)f](\omega) = e^{-i\Phi(\phi)(\omega)} f(\omega), \tag{7.8}$$

$$[U_0(\psi)f](\omega) = [\Psi(\psi)](\omega)(\hat{T}_\psi f)(\omega). \tag{7.9}$$

(b) Conversely, let $\mathcal{V}$ be a completely separable inner product space. If (U, V) is a cyclic $\mathcal{V}$-representation of CCR on a Hilbert space $\mathcal{H}$, then *there exists a covariant random field (Φ, T) from $\mathcal{V}$ to a probability space (Ω, Σ, μ) such that (U, V) is equivalent to (U_0, V_0) defined by Eqs. (7.8) and (7.9).*

PROOF. (a) It is clear that V_0 is unitary and that $V_0(\phi + \psi) = V_0(\phi) V_0(\psi)$ for all $\phi, \psi \in \mathcal{V}$. If $\phi_i \to \phi$ in $\mathcal{V}$ and $f \in \mathcal{K}$ then

$$\| V_0(\phi_i) f - V_0(\phi) f \|^2 = \| V_0(\phi_i - \phi) f - f \|^2$$

$$\leq \int |f|^2 |e^{i\Phi(\phi_i - \phi)} - 1|^2 \, d\mu. \tag{7.10}$$

The right-hand side of (7.10) goes to zero as $i \to \infty$ by the dominated convergence theorem. Hence V_0 is strongly continuous and therefore is a unitary representation of $\mathcal{V}$ on $\mathcal{K}$. Since μ_ψ is absolutely continuous relative to μ, by the Radon–Nikodym theorem there exist unique nonnegative functions $f_\psi \in L^1(\Omega, \Sigma, \mu)$ such that $\mu_\psi(A) = \int_A f_\mu \, d\mu$ for all $A \in \Sigma$ and $\psi \in \mathcal{V}$ or symbolically, $d\mu(T_\psi^{-1}\omega) = f_\psi(\omega) \, d\mu(\omega)$. Define the random functional $\Psi(\psi) = f_\psi^{1/2}(\omega)$. Then $U_0(\psi)$ is unitary since for all $f, g \in \mathcal{K}$ we have

$$\langle U_0(\psi) f, U_0(\psi) g \rangle = \int f_\psi(\omega) f(T_\psi^{-1}\omega) g^*(T_\psi^{-1}\omega) \, d\mu(\omega)$$

$$= \int f(T_\psi^{-1}\omega) g^*(T_\psi^{-1}\omega) \, d(T_\psi^{-1}\omega)$$

$$= \int f(\omega) g^*(\omega) \, d\mu(\omega) = \langle f, g \rangle.$$

We now show that $f_{\phi + \psi}(\omega) = f_\phi(\omega) f_\psi(T_\phi^{-1}\omega)$. Indeed, for any $A \in \Sigma$ we have

$$\int_A f_\phi(\omega) f_\psi(T_\phi^{-1}\omega) \, d\mu(\omega) = \int_A f_\psi(T_\phi^{-1}\omega) \, d\mu(T_\phi^{-1}\omega)$$

$$= \int_A d\mu(T_\psi^{-1} T_\phi^{-1}\omega) = \mu(T_{\phi+\psi} A) = \int_A f_{\phi + \psi}(\omega) \, d\mu(\omega).$$

The following is then obtained for any $\phi, \psi \in \mathcal{V}, f \in \mathcal{K}$.

$$[U_0(\phi + \psi) f](\omega) = [\Psi(\phi + \psi)](\omega)(\hat{T}_{\phi + \psi} f)(\omega)$$

$$= [\Psi(\phi)](\omega)[\Psi(\psi)](T_\phi^{-1}\omega) f(T_{\phi + \psi}^{-1}\omega)$$

$$= [\Psi(\phi)](\omega)\{ T_\phi [\Psi(\psi)(\omega) f(T_\psi^{-1}\omega)] \}$$

$$= [U_0(\phi) U_0(\psi) f](\omega).$$

Hence $U_0(\phi + \psi) = U_0(\phi) U_0(\psi)$. We now show that $\phi \to U_0(\phi)$ is strongly continuous. If $A \in \Sigma$ and $\psi_i \to \psi$, it follows from the continuity of T_ψ that

$$\int_A f_{\psi_i}(\omega) \, d\mu(\omega) = \mu(T_{\psi_i}A) = \int \chi_A(T_{\psi_i}\omega) \, d\mu(\omega)$$

$$\to \int \chi_A(T_\psi\omega) \, d\mu(\omega) = \mu(T_\psi A) = \int_A f_\psi(\omega) \, d\mu(\omega).$$

We then conclude that f_{ψ_i} converges to f_ψ in probability. For $f \in \mathcal{H}$ we have

$$\|U_0(\psi_i)f - U_0(\psi)f\|^2 = \|U_0(\psi_i - \psi)f - f\|^2$$

$$= \int |U_0(\psi_i - \psi)f - f|^2 \, d\mu$$

$$= \int |f_{\psi_i - \psi}(\omega)f(T_{\psi_i - \psi}^{-1}\omega)|^2 \, d\mu(\omega). \qquad (7.11)$$

Applying the dominated convergence theorem, the right-hand side of (7.11) approaches zero as $i \to \infty$ and hence U_0 is strongly continuous. It follows that U_0 is a unitary representation of $\mathcal{V}$ on $\mathcal{H}$. To show that the commutation relations hold we use the T-covariance of Φ to obtain

$$[U_0(\phi)V_0(\psi)f](\omega) = \Psi(\phi)[V_0(\psi)f](T_\phi^{-1}\omega)$$

$$= \Psi(\phi)\exp\{-i[\Phi(\psi)](T_\phi^{-1}\omega)\}f(T_\phi^{-1}\omega)$$

$$= \Psi(\phi)\exp\{-i[\Phi_\phi(\psi)](\omega)\}f(T_\phi^{-1}\omega)$$

$$= \Psi(\phi)e^{-i\Phi(\psi)(\omega)}e^{i\langle\phi,\psi\rangle}f(T_\phi^{-1}\omega)$$

$$= e^{i\langle\phi,\psi\rangle}[V_0(\psi)U_0(\phi)f](\omega).$$

Finally, since Σ is generated by $\{\Phi(\phi) : \phi \in \mathcal{V}\}$ it follows that $\mathrm{span}\{e^{i\Phi(\phi)} : \phi \in \mathcal{V}\}$ is dense in $\mathcal{H}$. Thus, since $U_0(\phi)\mathbf{1} = e^{i\Phi(\phi)}$, we see that $\mathbf{1}$ is a cyclic vector.

(b) We first show that (U, V) has an unique extension to a cyclic $\mathcal{V}$-representation of the CCR on $\mathcal{H}$. For $\phi \in \overline{\mathcal{V}}$, let ϕ_i be a sequence in $\mathcal{V}$ converging to ϕ. Now $U(\phi_i)$ is strongly Cauchy in $\mathcal{H}$ since for every $f \in \mathcal{H}$ we have

$$\lim_{i,j \to \infty} \|U(\phi_i)f - U(\phi_j)f\| = \lim_{i,j \to \infty} \|f - U(\phi_j - \phi_i)f\| = 0.$$

Defining $U(\phi)f = \lim U(\phi_i)f$, gives a well-defined linear operator which is bounded by the uniform boundedness theorem. Extend V to $\overline{\mathcal{V}}$ in a similar manner. By taking limits it is straightforward to show that (U, V) extended in this way gives a cyclic $\overline{\mathcal{V}}$-representation of the CCR on $\mathcal{H}$. Let $\{\psi_i\}$ be an orthonormal basis for $\overline{\mathcal{V}}$ where $\psi_i \in \mathcal{V}$, $i = 1, 2, \ldots$. Let $\{f_i\}$ be an orthonormal basis for $L^2(\mathbb{R}, dx)$ where f_i are in the Schwartz space $\mathcal{S}(\mathbb{R})$,

$i = 1, 2, \ldots$. Define the isomorphism $J: \bar{\mathcal{V}} \to L^2(\mathbb{R}, dx)$ by $J(\psi_i) = f_i$, $i = 1, 2 \ldots$. Then $(U \circ J^{-1}, V \circ J^{-1})$ is a cyclic $L^2(\mathbb{R}, dx)$-representation of the CCR on $\mathcal{K}$. In particular its restriction $(U \circ J^{-1}|\mathcal{S}(\mathbb{R}), V \circ J^{-1}|\mathcal{S}(\mathbb{R}))$ to $\mathcal{S}(\mathbb{R})$ gives a cyclic $\mathcal{S}(\mathbb{R})$-representation of the CCR on $\mathcal{K}$. Applying a theorem due to Gelfand–Vilenkin [87], we have the following three conclusions.

(1) There exists a unique Borel probability measure μ on the dual $\mathcal{S}'(\mathbb{R})$ of $\mathcal{S}(\mathbb{R})$ such that for every Borel set A and every $\phi \in \mathcal{S}(\mathbb{R})$, $\mu(A) = 0$ implies $\mu(A + f_\phi) = 0$, where $f_\phi \in \mathcal{S}'(\mathbb{R})$ is defined by $f_\phi(\psi) = \langle \psi, \phi \rangle$ for every $\psi \in \mathcal{S}(\mathbb{R})$.
(2) For every $\phi \in \mathcal{S}(\mathbb{R})$ there exists a functional F_ϕ on $\mathcal{S}'(\mathbb{R})$ such that (U_1, V_1) is a cyclic $\mathcal{S}(\mathbb{R})$-representation of the CCR on $L^2(\mathcal{S}'(\mathbb{R}), \mu)$ where $[V_1(\phi)F](f) = e^{-if(\phi)}F(f)$ and $[U_1(\phi)F](f) = F_\phi(f)F(f + f_\phi)$.
(3) There exists an isomorphism $M: \mathcal{K} \to L^2(\mathcal{S}'(\mathbb{R}), \mu)$ such that $MU \circ J^{-1}|\mathcal{S}(\mathbb{R})M^{-1} = U_1$ and $MV \circ J^{-1}|\mathcal{S}(\mathbb{R})M^{-1} = V_1$.

Since U and V are strongly continuous, it follows that U_1 and V_1 are strongly continuous (relative to the inner product topology on $\mathcal{S}(\mathbb{R})$). By an earlier argument, (U_1, V_1) has a unique cyclic extension (U_2, V_2) to a $L^2(\mathbb{R}, dx)$-representation of the CCR on $L^2(\mathcal{S}'(\mathbb{R}), \mu)$. It is straightforward to show that (U_2, V_2) satisfies the following conditions.

(1') For every Borel set A of $\mathcal{S}'(\mathbb{R})$ and every $\phi \in L^2(\mathbb{R}, dx)$, $\mu(A) = 0$ implies $\mu(A + f_\phi) = 0$, where $f_\phi \in \mathcal{S}'(\mathbb{R})$ is defined by $f_\phi(\psi) = \langle \psi, \phi \rangle$ for every $\psi \in \mathcal{S}(\mathbb{R})$.
(2') For every $\phi \in L^2(\mathbb{R}, dx)$ there exists a functional F_ϕ on $\mathcal{S}'(\mathbb{R})$ such that $[V_2(\phi)F](f) = e^{-if(\phi)}F(f)$ and $[U_2(\phi)F](f) = F_\phi(f)F(f + f_\phi)$.
(3') $MU \circ J^{-1}M^{-1} = U_2$ and $MV \circ J^{-1}M^{-1} = V_2$.

Define the random field $\Phi: \mathcal{V} \to (\mathcal{S}'(\mathbb{R}), \mu)$ by $[\Phi(\phi)](f) = f(J\phi)$ and the random functional $\Psi: \mathcal{V} \to (\mathcal{S}'(\mathbb{R}), \mu)$ by $[\Psi(\phi)](f) = F_{J\phi}(f)$. Furthermore, for every $\phi \in \mathcal{V}$, define the action $T_\phi: \mathcal{S}'(\mathbb{R}) \to \mathcal{S}'(\mathbb{R})$ by $T_\phi f = f + f_{J\phi}$. Finally, define the $\mathcal{V}$-representation of the CCR (U_0, V_0) by $U_0(\phi) = U_2(J\phi)$, $V_0(\phi) = V_2(J\phi)$. It can be shown that (Φ, T) and (U_0, V_0) satisfy the conditions of the theorem. Furthermore, (U, V) is equivalent to (U_0, V_0) since $U = M^{-1}U_0 M$ and $V = M^{-1}V_0 M$. □

There is another approach to representations of the CCR which is frequently used. Let (U, V) be a $\mathcal{V}$-representation of the Weyl form of the CCR on $\mathcal{K}$. For notational convenience we replace (U, V) by a single representation operator W over the complexification $\mathcal{V}_c$ of $\mathcal{V}$. To be precise, $\mathcal{V}_c$ is the set of ordered pairs $\{(\phi,\psi):\phi,\psi\in\mathcal{V}\}$. We define addition componentwise:

$$(\phi,\psi)+(\phi_1,\psi_1)=(\phi+\phi_1,\psi+\psi_1), \tag{7.12}$$

and if $\alpha+i\beta\in C$, $\alpha,\beta\in\mathbb{R}$, we define

$$(\alpha+i\beta)(\phi,\psi)=(\alpha\phi,\alpha\psi)+(-\beta\psi,\beta\phi). \tag{7.13}$$

It is straightforward to show that $\mathcal{V}_c$ is a complex linear space. We define an inner product on $\mathcal{V}_c$ by

$$\langle(\phi,\psi),(\phi_1,\psi_1)\rangle=\langle\phi,\phi_1\rangle+\langle\psi,\psi_1\rangle+i(\langle\psi,\phi_1\rangle-\langle\phi,\psi_1\rangle). \tag{7.14}$$

Then $\mathcal{V}_c$ becomes a complex inner product space. We use the notation $\phi+i\psi=(\phi,\psi)$ and now the natural operations of addition, scalar multiplication and inner product become applicable.

Now let (U,V) be a $\mathcal{V}$-representation of the CCR on $\mathcal{H}$. For $\phi+i\psi\in\mathcal{V}_c$ we define

$$W(\phi+i\psi)=e^{-i\langle\phi,\psi\rangle/2}U(\phi)V(\psi).$$

The following lemma, whose proof is left as a simple exercise, shows that W is equivalent to (U,V).

Lemma 7.4.

(i) $W(\phi)$ *is a unitary operator on* $\mathcal{H}$ *satisfying*

    (a) $\phi\mapsto W(\phi)$ *is continuous in the strong operator topology*,
    (b) $W(\phi)^*=W(-\phi)$ *for all* $\phi\in\mathcal{V}_c$,
    (c) $W(\phi)W(\psi)=W(\phi+\psi)e^{i\,\mathrm{Im}\langle\phi,\psi\rangle/2}$ *for all* $\phi,\psi\in\mathcal{V}_c$.

(ii) *Conversely, if* $W(\phi)$ *is a unitary operator on* $\mathcal{H}$ *satisfying* (a), (b), *and* (c) *and we define* $U(\phi)=W(\phi)$, $V(\phi)=W(i\phi)$ *for every* $\phi\in\mathcal{V}$, *then* (U,V) *is a* $\mathcal{V}$-*representation of the* CCR.

We call a map $\phi\mapsto W(\phi)$ satisfying (a), (b), and (c) a *complex* $\mathcal{V}$-representation of the CCR on $\mathcal{H}$. We say that W is *cyclic* with *cyclic vector* $h\in\mathcal{H}$ if $\overline{\mathrm{span}}\{W(\phi):\phi\in\mathcal{V}_c\}=\mathcal{H}$.

We now form a Banach *-algebra from $\mathcal{V}_c$. Let $\Delta(\mathcal{V}_c)$ be the set of maps $f:\mathcal{V}_c\to C$ such that $f(\phi)=0$ except for finitely many ϕ's in $\mathcal{V}_c$. Then $\Delta(\mathcal{V}_c)$ is a complex linear space under pointwise addition and scalar multiplication:

$$(f_1+f_2)(\phi)=f_1(\phi)+f_2(\phi),\qquad(\lambda f)(\phi)=\lambda f(\phi).$$

Multiplication of $f,g\in\Delta(\mathcal{V}_c)$ is defined by the convolution

$$(fg)(\phi)=\sum\left\{f(\psi)g(\phi-\psi)e^{i\,\mathrm{Im}\langle\psi,\phi\rangle/2}:\psi\in\mathcal{V}_c\right\}$$

An involution * is defined by $f^*(\phi) = \bar{f}(-\phi)$. We define a norm $\|\cdot\|$ on $\Delta(\mathcal{V}_c)$ by

$$\|f\| = \sum \{|f(\phi)| : \phi \in \mathcal{V}_c\}.$$

Then $\Delta(\mathcal{V}_c)$ becomes a normed *-algebra with identity $\chi_{\{0\}}$. We denote by $\Delta_1(\mathcal{V}_c)$ the Banach *-algebra obtained by completing $\Delta(\mathcal{V}_c)$ with respect to the norm $\|\cdot\|$. One can define another norm which is smaller than $\|\cdot\|$ relative to which $\Delta(\mathcal{V}_c)$ is completed to a C^*-algebra but that will not be necessary here.

Now let W be a cyclic complex $\mathcal{V}$-representation of the CCR on $\mathcal{H}$ with unit cyclic vector $h \in \mathcal{H}$. Define $w: \mathcal{V}_c \to C$ by $w(\phi) = \langle W(\phi)h, h \rangle$. We would now like to find the properties of w. First, since W is continuous we have

(1) w is continuous.

Applying (b) and (c) of Lemma 7.4 we conclude that $W(0) = I$ and hence

(2) $w(0) = 1$.

Finally, suppose that $\phi_i \in \mathcal{V}_c$ and $\lambda_i \in C$, $i = 1, \ldots, n$. Then

$$\sum_{j,k=1}^{n} \lambda_j \lambda_k^* w(\phi_j - \phi_k) e^{i \operatorname{Im}\langle \phi_j, \phi_k \rangle / 2}$$

$$= \sum_{j,k=1}^{n} \lambda_j \lambda_k^* \langle W(\phi_j - \phi_k)h, h \rangle e^{i \operatorname{Im}\langle \phi_j, \phi_k \rangle / 2}$$

$$= \sum_{j,k=1}^{n} \lambda_j \lambda_k^* \langle W(\phi_j) W(\phi_k)^* h, h \rangle$$

$$= \left\langle \left[\sum_j \lambda_j W(\phi_j) \right] \left[\sum_k \lambda_k W(\phi_k) \right]^* h, h \right\rangle.$$

Hence,

(3) $\displaystyle \sum_{j,k=1}^{n} \lambda_j \lambda_k^* w(\phi_j - \phi_k) e^{i \operatorname{Im}\langle \phi_j, \phi_k \rangle / 2} \geq 0.$

The next theorem gives the converse of these results.

Theorem 7.5. *If $w: \mathcal{V}_c \to C$ satisfies (1), (2), and (3), then there exists a unique (up to unitary equivalence) cyclic complex $\mathcal{V}$-representation W of the CCR with cyclic vector h such that $w(\phi) = \langle W(\phi)h, h \rangle$ for every $\phi \in \mathcal{V}_c$.*

PROOF. Any $f \in \Delta(\mathcal{V}_c)$ has a unique representation of the form $f = \sum_{j=1}^{n} \lambda_j \chi_{\{\phi_j\}}$, where $0 \neq \phi_j \in \mathcal{V}_c$ are distinct. Define $w: \Delta(\mathcal{V}_c) \to C$ by $w(f) = \sum_{j=1}^{n} \lambda_j w(\phi_j)$. Then w is a linear functional on $\Delta(\mathcal{V}_c)$ and, moreover, w is positive since

$$w\left[\left(\sum \lambda_j \chi_{\{\phi_j\}}\right)^* \left(\sum \lambda_j \chi_{\{\phi_j\}}\right)\right] = w\left[\left(\sum \lambda_k^* \chi_{\{-\phi_k\}}\right)\left(\sum \lambda_j \chi_{\{\phi_j\}}\right)\right]$$

$$= \sum_{j,k} \lambda_j \lambda_j^* w(\chi_{\{-\phi_k\}} \chi_{\{\phi_j\}})$$

$$= \sum_{j,k} \lambda_j \lambda_k^* w(\phi_j - \phi_k) e^{i \operatorname{Im}\langle \phi_j, \phi_k \rangle / 2} \geq 0,$$

where the last equality follows from

$$\chi_{\{-\phi_k\}} \chi_{\{\phi_j\}}(\phi) = \sum_\psi \chi_{\{-\phi_k\}}(\psi) \chi_{\{\phi_j\}}(\phi - \psi) e^{i \operatorname{Im}\langle \psi, \phi \rangle / 2}$$

$$= \chi_{\{\phi_j\}}(\phi + \phi_k) e^{i \operatorname{Im}\langle \phi, \phi_k \rangle / 2}$$

and hence

$$\chi_{\{-\phi_k\}} \chi_{\{\phi_j\}} = e^{i \operatorname{Im}\langle \phi_j, \phi_k \rangle / 2} \chi_{\{\phi_j - \phi_k\}}. \tag{7.15}$$

Also, w acting on the identity $\chi_{\{0\}}$ is

$$w(\chi_{\{0\}}) = w(0) = 1.$$

Applying (7.15) and Schwarz's inequality we have

$$|w(\chi_{\{\phi\}})|^2 \leq w(\chi^*_{\{\phi\}} \chi_{\{\phi\}}) = w(\chi_{\{0\}}) = 1$$

for all $\phi \in \mathcal{V}_c$. Hence if $f = \sum \lambda_i \chi_{\{\phi_i\}}$ we have

$$|w(f)| = \sum |\lambda_i| |w(\chi_{\{\phi_i\}})| \leq \sum |\lambda_i| = \sum_\phi |f(\phi)| = \|f\|.$$

Thus w is continuous on $\Delta(\mathcal{V}_c)$ and hence has a unique extension to a state on $\Delta_1(\mathcal{V}_c)$. Applying the GNS construction of Section 5.3 there exists a unique (to within equivalence) cyclic representation π of $\Delta_1(\mathcal{V}_c)$ with cyclic vector h such that $\langle \pi(f)h, h \rangle = w(f)$ for all $f \in \Delta_1(\mathcal{V}_c)$. Define $W(\phi) = \pi(\chi_{\{\phi\}})$ for every $\phi \in \mathcal{V}_c$. Equation (b) holds since

$$W(\phi)^* = \left[\pi(\chi_{\{\phi\}})\right]^* = \pi(\chi^*_{\{\phi\}}) = \pi(\chi_{\{-\phi\}}) = W(-\phi).$$

From (7.15) we obtain Eq. (c):

$$W(\phi)W(\psi) = \pi(\chi_{\{\phi\}})\pi(\chi_{\{\psi\}}) = \pi(\chi_{\{\phi\}}\chi_{\{\psi\}})$$

$$= \pi\left(e^{i \operatorname{Im}\langle \phi, \psi \rangle / 2} \chi_{\{\phi + \psi\}}\right) = e^{i \operatorname{Im}\langle \phi, \psi \rangle / 2} W(\phi + \psi).$$

Condition (a) follows from (1). Finally,

$$\langle W(\phi)h, h \rangle = \langle \pi(\chi_{\{\phi\}})h, h \rangle = w(\chi_{\{\phi\}}) = w(\phi). \qquad \square$$

7. Quantum Field Theory

As an application of Theorem 7.5 we outline a proof of the von Neumann uniqueness theorem. If $\mathcal{V}$ is one-dimensional we can assume that $\mathcal{V} = \mathbb{R}$. Then $\mathcal{V}_c = C$. Let W be an irreducible complex $\mathbb{R}$-representation of the CCR on $\mathcal{H}$. Then $W(\lambda)$ is a unitary operator on $\mathcal{H}, \lambda \in C$, satisfying:

(a') $\lambda \mapsto W(\lambda)$ is strongly continuous;
(b') $W(\lambda)^* = W(-\lambda)$ for every $\lambda \in C$;
(c') $W(\alpha)W(\beta) = e^{i\operatorname{Im}\alpha\bar\beta/2} W(\alpha + \beta)$.

Define the linear operator A on $\mathcal{H}$ by

$$\langle Ax, y \rangle = (2\pi)^{-1/2} \int \langle W(\lambda)x, y \rangle e^{-|\lambda|^2/4} d\lambda$$
$$= (2\pi)^{-1/2} \int \langle W(s,t)x, y \rangle e^{-(s^2+t^2)/4} ds\, dt, \qquad (7.16)$$

where $x, y \in \mathcal{H}$ and $\lambda = s + it$, $s, t \in \mathbb{R}$. Since $W(\lambda)$ is unitary,

$$|\langle Ax, y \rangle| \le (2\pi)^{-1/2} \int e^{-(s^2+t^2)/4} ds\, dt \|x\| \|y\| = \|x\| \|y\|$$

and hence A is bounded and $\|A\| \le 1$. Applying (b') gives

$$\langle x, Ay \rangle = \langle Ay, x \rangle^* = (2\pi)^{-1/2} \int \langle W(s,t)y, x \rangle^* e^{-(s^2+t^2)/4} ds\, dt$$
$$= (2\pi)^{-1/2} \int \langle W(-s,-t)x, y \rangle e^{-(s^2+t^2)/4} ds\, dt$$
$$= (2\pi)^{-1/2} \int \langle W(s,t)x, y \rangle e^{-(s^2+t^2)/4} ds\, dt = \langle Ax, y \rangle,$$

so A is self-adjoint. Using a straightforward (but tedious) calculation one can show that

$$AW(\lambda)A = Ae^{-|\lambda|^2/4} \qquad (7.17)$$

for every $\lambda \in C$. Moreover, one can show that $A \ne 0$. Letting $\lambda = 0$ in (7.17) we conclude that A is a nonzero projection. Hence there exists a unit vector $x \in \mathcal{H}$ such that $Ax = x$. Since W is irreducible, it follows that any nonzero vector in $\mathcal{H}$ is cyclic so, in particular, x is cyclic. Define $w: C \to C$ by $w(\lambda) = \langle W(\lambda)x, x \rangle$. Then applying (7.17),

$$w(\lambda) = \langle AW(\lambda)Ax, x \rangle = e^{-|\lambda|^2/4}.$$

Now let W_0 be the complex $\mathbb{R}$-representation of the CCR on $L^2(\mathbb{R})$ given by

$$W_0(\lambda)f(u) = W_0(s + it)f(u) = e^{ist/2} e^{itu} f(u - s). \qquad (7.18)$$

If $x_0 \in L^2(\mathbb{R})$ is the cyclic vector $x_0 = \pi^{-1/4} e^{-u^2/2}$, then

$$w_0(\lambda) = \langle W_0(\lambda) x_0, x_0 \rangle = \pi^{-1/2} \int e^{-ist/2} e^{itu} e^{-(u-s)^2/2} du$$

$$= e^{-(ist+s^2)/2} \pi^{-1/2} \int e^{(s+it)u - u^2} du$$

$$= e^{-(s^2+t^2)/4} \pi^{-1/2} \int e^{-[u-(s+it)/2]^2} du = e^{-|\lambda|^2/4}.$$

It follows from Theorem 7.5 that W and W_0 are unitarily equivalent. The generalization to any finite number of degrees of freedom is straightforward. However, this proof (and the result) cannot be generalized to an infinite number of degrees of freedom since in that case the measure

$$A \mapsto (2\pi)^{-1/2} \int_A e^{-(s^2+t^2)/4} \, ds \, dt$$

has no infinite-dimensional analog.

7.3. The Wightman Axioms

A. S. Wightman [277, 278] proposed his axioms for quantum field theory over 20 years ago. Since then his axioms have been studied intensively by many investigators. It is generally agreed by a majority of workers in this area that the Wightman axioms are a reasonable set of postulates for a quantum field theory. It can be shown (see Sections 7.4, 7.5) that there exist mathematical models for noninteracting quantum systems that satisfy Wightman's axioms. Unfortunately, mathematical models for quantum systems with nontrivial interactions satisfying the axioms have not yet been exhibited. However, as we shall see in Section 7.5, important progress has been made.

Quantum field theory describes relativistic quantum systems with an infinite number of degrees of freedom. The geometry of a relativistic system is given by *Minkowski space* M^4 where $M^4 = \mathbb{R}^4$ equipped with the indefinite inner product

$$x \cdot y = x_0 y_0 - x_1 y_1 - x_2 y_2 - x_3 y_3, \tag{7.19}$$

where $x = (x_0, x_1, x_2, x_3)$, $y = (y_0, y_1, y_2, y_3)$. We frequently write (7.19) in the more concise notation $x \cdot y = x_0 y_0 - \mathbf{x} \cdot \mathbf{y}$ where $\mathbf{x} = (x_1, x_2, x_3)$, $\mathbf{y} = (y_1, y_2, y_3)$. The x_0 component of a vector $x \in M^4$ represents the "time" coordinate ct, where c is the speed of light and the vector $\mathbf{x} \in \mathbb{R}^3$ represents the "space" coordinates. For this reason M^4 is frequently called *space–time*.

The *forward cone* V_+ is the set

$$V_+ = \{ x \in M^4 : x_0 > 0, \ x \cdot x = x_0^2 - \mathbf{x} \cdot \mathbf{x} > 0 \}.$$

179

The boundary ∂V_+ of V_+ is the set

$$\partial V_+ = \left\{ x \in M^4 : x_0 \geq 0,\ x \cdot x = x_0^2 - \mathbf{x} \cdot \mathbf{x} = 0 \right\}.$$

If a system is initially at the origin $0 \in M^4$ and travels forward in time at the speed of light, then its coordinate x at a later time t will satisfy $\mathbf{x} \cdot \mathbf{x} = c^2 t^2 = x_0^2$. Hence $x \in \partial V_+$. If a system is initially at the origin and travels forward in time at less than the speed of light, then its coordinate x at a later time t satisfies $\mathbf{x} \cdot \mathbf{x} < c^2 t^2 = x_0^2$ so $x \in V_+$. Since any physical system cannot have a speed faster than that of light, any system, initially at 0, moving forward in time must have coordinate $x \in \overline{V}_+$, the closure of V_+. Analogously, systems initially at 0 moving backward in time must have coordinate $x \in \overline{V}_-$ where V_- is the *backward cone*

$$V_- = \left\{ x \in M^4 : x_0 < 0,\ x \cdot x = x_0^2 - \mathbf{x} \cdot \mathbf{x} > 0 \right\}.$$

In this way V_+ represents the "future" and V_- represents the "past." Vectors in $\overline{V}_+ \cup \overline{V}_-$ are *timelike* and vectors in the complement $M^4 - (\overline{V}_+ \cup \overline{V}_-)$ (i.e., $x \cdot x < 0$) are called *spacelike*.

Two vectors $x, y \in M^4$ are *timelike separated* if $(x - y) \cdot (x - y) \geq 0$. Thus x, y are timelike separated if and only if $x - y \in \overline{V}_+ \cup \overline{V}_-$. Physically, timelike separated systems can interact since a signal traveling at a speed not greater than c can be transmitted from one to the other. Analogously, $x, y \in M^4$ are *spacelike separated* if $(x - y) \cdot (x - y) < 0$. Spacelike separated systems cannot interact since no physical communication is possible between them.

The symmetries of space–time are given by the *Lorentz group* $\mathcal{L}$ which consists of all linear transformations from M^4 onto M^4 which preserve the Minkowski inner product. That is, a linear transformation Λ of M^4 onto M^4 is in $\mathcal{L}$ if and only if $(\Lambda x) \cdot (\Lambda y) = x \cdot y$ for every $x, y \in M^4$. We would like to extend $\mathcal{L}$ to include space–time translations $x \mapsto x + a$, $a \in M^4$. This larger group is called the *Poincaré group* $\mathcal{P}$ or the *inhomogeneous Lorentz group*. Precisely, $\mathcal{P}$ is the set of all pairs (a, Λ), $a \in M^4$, $\Lambda \in \mathcal{L}$, where $(a, \Lambda) : M^4 \to M^4$ is the transformation defined by $(a, \Lambda)x = \Lambda x + a$, for every $x \in M^4$. Since

$$(a_1, \Lambda_1)(a_2, \Lambda_2)x = (a_1, \Lambda_1)(\Lambda_2 x + a_2)$$
$$= \Lambda_1 \Lambda_2 x + \Lambda_1 a_2 + a_1 = (a_1 + \Lambda_1 a_2, \Lambda_1 \Lambda_2)x,$$

the multiplication law for $\mathcal{P}$ is

$$(a_1, \Lambda_1)(a_2, \Lambda_2) = (a_1 + \Lambda_1 a_2, \Lambda_1 \Lambda_2). \tag{7.20}$$

It is straightforward to show that $\mathcal{P}$ is indeed a group and that $(a, \Lambda)^{-1}x = \Lambda^{-1}(x - a)$.

We are now ready to present Wightman's axioms. We first give some heuristic background and then state the axioms precisely. Our object is to give a mathematical description of a quantum phenomenon occurring in space–time M^4. We think of this phenomenon as a physical "field" in M^4. Ideally, each point $x \in M^4$ would give an observable $\Theta(x)$ representing the field at x. Such an idealization is impossible however, due to the imprecision of measurements and possible singularities of the physical field. [In fact, it can be proved [23] that such a $\Theta(x)$ with mild physical properties is impossible.] The best we can do is find the average field over small regions of M^4. This may be accomplished by "smearing" the field with a test function $\phi \in \mathcal{S}(\mathbb{R}^4)$. We have already encountered such situations in our study of random fields in Chapter 6. Thus for every $\phi \in \mathcal{S}(\mathbb{R}^4)$ we assume there exists an operator $\Theta(\phi)$ on a fixed Hilbert space $\mathcal{H}$. If ϕ is real valued, then $\Theta(\phi)$ is assumed to be symmetric. [Physically one would want $\Theta(\phi)$ to be self-adjoint, but we are trying to be very general here. Presumably, $\Theta(\phi)$ would have a self-adjoint extension if its domain is properly adjusted.] It is reasonable that $\phi \mapsto \Theta(\phi)$ is linear and weakly continuous. We say that $\phi, \psi \in \mathcal{S}(\mathbb{R}^4)$ are *spacelike separated* if their supports are spacelike separated; that is, $(x-y) \cdot (x-y) < 0$ for every $x \in \text{supp}\,\phi$ and $y \in \text{supp}\,\psi$. If ϕ, ψ are spacelike separated, then they correspond to regions of space–time which cannot interact. We then assume that $\Theta(\phi)$ and $\Theta(\psi)$ can be simultaneously measured and hence commute. This is called *locality* or *microscopic causality*.

If we make a symmetry transformation in M^4, then there should be a corresponding symmetry transformation of the states in $\mathcal{H}$. We therefore assume that there exists a unitary representation U of the Poincaré group $\mathcal{P}$ in $\mathcal{H}$. For $\phi \in \mathcal{S}(\mathbb{R}^4)$, $(a, \Lambda) \in \mathcal{P}$, $\phi \circ (a, \Lambda)^{-1}$ represents the test function ϕ after transformation by the symmetry (a, Λ). The field operator $\Theta(\phi)$ is transformed into the field operator $\Theta[\phi \circ (a, \Lambda)^{-1}]$. By the physical interpretation of $U(a, \Lambda)$ we should have

$$U(a, \Lambda)\Theta(\phi)U(a, \Lambda)^* = \Theta\left[\phi \circ (a, \Lambda)^{-1}\right]. \tag{7.21}$$

Equation (7.21) is called *Lorentz* or *relativistic covariance*. Physically there should be a distinguished unit vector $h \in \mathcal{H}$ which represents the unique vacuum state; that is, the state in which no particles are present. The state represented by h should be invariant under all symmetry transformations. We thus assume that $U(a, \Lambda)h = h$ for every $(a, \Lambda) \in \mathcal{P}$ and that h is unique (up to a multiple of modulus one).

There is one more condition to be considered. For $a_0 \in \mathbb{R}$, the map $a_0 \mapsto U((a_0, 0, 0, 0), I)$ is a strongly continuous one-parameter unitary group.

Hence by Stone's theorem [252], there exists a self-adjoint operator P_0 on $\mathcal{H}$ such that $U((a_0,0,0,0),I) = e^{ia_0 P_0}$. In a similar way, we have the self-adjoint operators P_1, P_2, P_3 corresponding to $a_1, a_2, a_3 \in \mathbb{R}$. Hence, for $a = (a_0, a_1, a_2, a_3) \in \mathbb{R}^4$ we have

$$U(a,I) = e^{ia \cdot P} = e^{i[a_0 P_0 - \mathbf{a} \cdot \mathbf{P}]},$$

where $P = (P_0, P_1, P_2, P_3)$ and $\mathbf{P} = (P_1, P_2, P_3)$. We interpret P as the energy-momentum operator (P_0 corresponds to energy and $\mathbf{P}$ corresponds to momentum). Again by Stone's theorem there exists a spectral measure $E(\cdot)$ on $\mathbb{R}^4$ such that

$$U(a,I) = \int_{\mathbb{R}^4} e^{ia \cdot x} E(dx).$$

Our last condition is that $\operatorname{supp} E(dx) \subseteq \overline{V}_+$. This is called the *spectrum condition* and amounts to the fact that the total energy of the system is nonnegative in a certain sense.

We now give a concise formulation of the Wightman axioms for a neutral scalar field.

(1) (Hilbert Space) There exists a Hilbert space $\mathcal{H}$ with a distinguished unit vector $h \in \mathcal{H}$ called the vacuum.

(2) (Fields and Temperedness) There exists a dense subspace $D \subseteq \mathcal{H}$ and for every $\phi \in \mathcal{S}(\mathbb{R}^4)$ an operator $\Theta(\phi)$ with domain D such that:
   (a) for every $f, g \in D, \phi \mapsto \langle \Theta(\phi)f, g \rangle$ is a tempered distribution;
   (b) if ϕ is real valued, then $\Theta(\phi)$ is symmetric, that is, $\langle \Theta(\phi)f, g \rangle = \langle f, \Theta(\phi)g \rangle$ for every $f, g \in D$;
   (c) $\Theta(\phi)$ leaves D invariant, that is, $f \in D$ implies that $\Theta(\phi)f \in D$;
   (d) $h \in D$ and $D = \operatorname{span}\{\Theta(\phi_1) \ldots \Theta(\phi_n)h : \phi_1, \ldots, \phi_n \in \mathcal{S}(\mathbb{R}^n)\}$.

(3) (Covariance) There exists a unitary representation of the Poincaré group $\mathcal{P}$, $(a, \Lambda) \mapsto U(a, \Lambda)$, such that:
   (a) $U(a, \Lambda)$ leaves D invariant;
   (b) $U(a, \Lambda)h = h$ for every $(a, \Lambda) \in \mathcal{P}$;
   (c) $U(a, \Lambda)\Theta(\phi)U(a, \Lambda)^* f = \Theta[\phi \circ (a, \Lambda)^{-1}]f$ for all $f \in D$.

(4) (Spectrum Condition) The joint spectrum of the infinitesimal generators of $U(a, I)$ lies in $\overline{V}_+$.

(5) (Locality) If ϕ and ψ are spacelike separated, then $[\Theta(\phi)\Theta(\psi) - \Theta(\psi)\Theta(\phi)]f = 0$ for all $f \in D$.

(6) (Uniqueness of Vacuum) The only vectors in $\mathcal{H}$ left invariant by all the $U(a, I), a \in \mathbb{R}^4$, are multiples of h.

We call a four-tuple (Θ, D, U, h) satisfying conditions (1)–(6) above a *Wightman quantum field*. It can be shown that 3(a) above follows from 2(d), 3(b), and 3(c). There is an equivalent formulation of the Wightman axioms which is physically less transparent but technically simpler. This formulation is phrased entirely in terms of tempered distributions and hence avoids the use of the cumbersome unbounded operators. For a fixed integer $n \geq 1$, define

$$\mathcal{W}_n(\phi_1,\ldots,\phi_n) = \langle \Theta(\phi_1) \cdots \Theta(\phi_n)h, h \rangle.$$

For notational convenience we define $\mathcal{W}_0 : C \to C$ by $\mathcal{W}_0(\lambda) = \lambda$. The $\mathcal{W}_n$'s are clearly multilinear in $\phi_1,\ldots,\phi_n$ and by 2(a) separately continuous. It follows from the nuclear theorem (Theorem 5.7) that there exist tempered distributions $W_n \in \mathcal{S}'(\mathbb{R}^{4n})$ such that

$$\mathcal{W}_n(\phi_1,\ldots,\phi_n) = W_n(\phi_1 \otimes \cdots \otimes \phi_n).$$

The $\mathcal{W}_n$'s are called *vacuum expectation values* and the W_n's are called *Wightman distributions*. The properties of the Wightman distributions are summarized in the following theorem.

Theorem 7.6. *The Wightman distributions W_n associated with a system obeying the Wightman axioms satisfy*:

W1. (*Positive Definiteness*) *For given $\phi_0 \in C$,*

$$\phi_1 \in \mathcal{S}(\mathbb{R}^4),\ldots,\phi_n \in \mathcal{S}(\mathbb{R}^{4n}),$$

$$\sum_{i,j=0}^{n} W_{i+j}(\phi_j^* \otimes \phi_i) \geq 0,$$

where ϕ^* is given by (5.16).

W2. (*Reality*) *For every $\phi \in \mathcal{S}(\mathbb{R}^{4n})$, $W_n(\phi^*) = \overline{W}_n(\phi)$.*

W3. (*Covariance*) *For $(a,\Lambda) \in \mathcal{P}$ and $\phi \in \mathcal{S}(\mathbb{R}^{4n})$ define*

$$\phi \circ (a,\Lambda)^{-1}(x_1,\ldots,x_n) = \phi\big((a,\Lambda)^{-1}x_1,\ldots,(a,\Lambda)^{-1}x_n\big),$$

$x_1,\ldots,x_n \in \mathbb{R}^4$. *Then $W_n[\phi \circ (a,\Lambda)^{-1}] = W_n(\phi)$.*

W4. (*Spectrum Condition*) *For any integer $n > 0$ there is a tempered distribution $M_n \in \mathcal{S}'(\mathbb{R}^{4n-4})$ satisfying:*
   (a) *for every $\phi \in \mathcal{S}(\mathbb{R}^{4n})$, $W_n(\phi) = M_n(\hat{\phi})$, where*

$$\hat{\phi}(y_1,\ldots,y_{n-1}) = \phi(0, y_1, y_2 + y_1,\ldots,y_{n-1} + y_{n-2} + \cdots + y_1);$$

   (b) $\operatorname{supp} \mathcal{F}M_n \subseteq \{(x_1,\ldots,x_{n-1}) : x_i \in \overline{V}_+, i = 1,\ldots,n-1\}$.

W5. (*Locality*) *If $\phi_j, \phi_{j+1} \in \mathcal{S}(\mathbb{R}^4)$ are spacelike separated, then*

$$W_n(\phi_1 \otimes \cdots \otimes \phi_j \otimes \phi_{j+1} \otimes \cdots \otimes \phi_n)$$

$$= W_n(\phi_1 \otimes \cdots \otimes \phi_{j+1} \otimes \phi_j \otimes \cdots \otimes \phi_n).$$

W6. (*Cluster Property*) *If* $a \in M^4$ *is spacelike and* $T_{j,a}: \mathbb{S}(\mathbb{R}^{4n}) \to \mathbb{S}(\mathbb{R}^{4n})$ *denotes the translation operator*

$$T_{j,a}\phi(x_1,\ldots,x_n) = \phi(x_1,\ldots,x_j,x_{j+1}-a,\ldots,x_n-a)$$

for $0 \le j \le n$, *then* $\lim_{\lambda\to\infty} T_{j,\lambda a} W_n = W_j \otimes W_{n-j}$.

PROOF. Since linear combinations of test functions of the form $\phi_1 \otimes \cdots \otimes \phi_n$ are dense in $\mathbb{S}(\mathbb{R}^{4n})$, the verification of the conditions for such test functions is sufficient.

(W2) For $\phi \in \mathbb{S}(\mathbb{R}^4)$ it follows from 2(b) that $\langle\Theta(\bar{\phi})h,h\rangle = \langle h,\Theta(\phi)h\rangle$. Indeed, let $\phi = \phi_1 + i\phi_2$ where ϕ_1 and ϕ_2 are the real and imaginary parts of ϕ, respectively. Then

$$\langle\Theta(\bar{\phi})h,h\rangle = \langle\Theta(\phi_1 - i\phi_2)h,h\rangle = \langle[\Theta(\phi_1) - i\Theta(\phi_2)]h,h\rangle$$
$$= \langle h,[\Theta(\phi_1) + i\Theta(\phi_2)]h\rangle = \langle h,\Theta(\phi)h\rangle.$$

Now suppose that $\phi \in \mathbb{S}(\mathbb{R}^{4n})$ and $\phi = \phi_1 \otimes \cdots \otimes \phi_n$, $\phi_i \in \mathbb{S}(\mathbb{R}^4)$, $i = 1,\ldots,n$. Then

$$W_n(\phi^*) = W_n(\bar{\phi}_n \otimes \cdots \otimes \bar{\phi}_1) = \mathcal{W}_n(\bar{\phi}_n,\ldots,\bar{\phi}_1)$$
$$= \langle\Theta(\bar{\phi}_n)\cdots\Theta(\bar{\phi}_1)h,h\rangle = \langle h,\Theta(\phi_1)\cdots\Theta(\phi_n)h\rangle$$
$$= \overline{\mathcal{W}}_n(\phi_1,\ldots,\phi_n) = \overline{W}_n(\phi_1 \otimes \cdots \otimes \phi_n) = \overline{W}_n(\phi).$$

(W3) If $\phi \in \mathbb{S}(\mathbb{R}^{4n})$ and $\phi = \phi_1 \otimes \cdots \otimes \phi_n$, $\phi_i \in \mathbb{S}(\mathbb{R}^4)$, then by 3(b) and 3(c) we have

$$W_n[\phi \circ (a,\Lambda)^{-1}] = W_n[\phi_1 \circ (a,\Lambda)^{-1} \otimes \cdots \otimes \phi_n \circ (a,\Lambda)^{-1}]$$
$$= \mathcal{W}_n[\phi_1 \circ (a,\Lambda)^{-1},\ldots,\phi_n \circ (a,\Lambda)^{-1}]$$
$$= \langle\Theta[\phi_1 \circ (a,\Lambda)^{-1}]\cdots\Theta[\phi_n \circ (a,\Lambda)^{-1}]h,h\rangle$$
$$= \langle U(a,\Lambda)\Theta(\phi_1)\cdots\Theta(\phi_n)U(a,\Lambda)^*h,h\rangle$$
$$= \langle\Theta(\phi_1)\cdots\Theta(\phi_n)h,h\rangle = \mathcal{W}_n(\phi_1,\ldots,\phi_n) = W_n(\phi).$$

(W5) If $\phi_j,\phi_{j+1} \in \mathbb{S}(\mathbb{R}^4)$ are spacelike separated, then applying 2(c) and (5) we have

$$W_n(\phi_1 \otimes \cdots \otimes \phi_n) = \langle\Theta(\phi_1)\cdots\Theta(\phi_n)h,h\rangle$$
$$= \langle\Theta(\phi_j)\Theta(\phi_{j+1})\cdots\Theta(\phi_n)h,\Theta(\bar{\phi}_{j-1})\cdots\Theta(\bar{\phi}_1)h\rangle$$
$$= \langle\Theta(\phi_{j+1})\Theta(\phi_j)\cdots\Theta(\phi_n)h,\Theta(\bar{\phi}_{j-1})\cdots\Theta(\bar{\phi}_1)h\rangle$$
$$= \mathcal{W}_n(\phi_1,\ldots,\phi_{j+1},\phi_j,\ldots,\phi_n) = W_n(\phi_1 \otimes \cdots \otimes \phi_{j+1} \otimes \phi_j \otimes \cdots \otimes \phi_n).$$

(W1) For $\phi_1,\ldots,\phi_n \in S(\mathbb{R}^4)$ define

$$F_n(\phi_1,\ldots,\phi_n) = \Theta(\phi_1)\cdots\Theta(\phi_n)h.$$

Then F_n is a multilinear vector-valued function which is continuous in each argument. By a slight generalization of the nuclear theorem, there exists a vector-valued tempered distribution Θ_n on $S(\mathbb{R}^{4n})$ such that

$$F_n(\phi_1,\ldots,\phi_n) = \Theta_n(\phi_1 \otimes \cdots \otimes \phi_n).$$

Moreover, Θ_n satisfies

$$\langle \Theta_n(\phi), \Theta_m(\psi) \rangle = W_{n+m}(\psi^* \otimes \phi).$$

Indeed, if $\phi = \phi_1 \otimes \cdots \otimes \phi_n$ and $\psi = \psi_1 \otimes \cdots \otimes \psi_n$, then we have

$$\begin{aligned}
\langle \Theta_n(\phi), \Theta_m(\psi) \rangle &= \langle \Theta(\phi_1)\cdots\Theta(\phi_n)h, \Theta(\psi_1)\cdots\Theta(\psi_m)h \rangle \\
&= \langle \Theta(\bar{\psi}_m)\cdots\Theta(\bar{\psi}_1)\Theta(\phi_1)\cdots\Theta(\phi_n)h, h \rangle \\
&= \mathcal{W}(\bar{\psi}_m,\ldots,\bar{\psi}_1,\phi_1,\ldots,\phi_n) \\
&= W_{n+m}(\bar{\psi}_m \otimes \cdots \otimes \bar{\psi}_1 \otimes \phi_1 \otimes \cdots \otimes \phi_n) \\
&= W_{n+m}(\psi^* \otimes \phi).
\end{aligned}$$

Hence, for $\phi_0 \in C, \phi_1 \in S(\mathbb{R}^4),\ldots,\phi_n \in S(\mathbb{R}^{4n})$ we have

$$\sum_{i,j=0}^n W_{i+j}(\phi_j^* \otimes \phi_i) = \sum_{i,j=0}^n \langle \Theta_i(\phi_i), \Theta_j(\phi_j) \rangle$$

$$= \left\| \sum_{i=0}^n \Theta_i(\phi_i) \right\|^2 \geq 0.$$

(W4) For $\phi \in S(\mathbb{R}^{4n})$ define $\check{\phi} \in S(\mathbb{R}^{4n-4})$ by

$$\check{\phi}(x_1,\ldots,x_n) = \phi(x_2 - x_1, x_3 - x_2,\ldots,x_n - x_{n-1}).$$

Define $M_n \in S(\mathbb{R}^{4n-4})$ as $M_n(\phi) = W_n(\check{\phi})$. If $\phi \in S(\mathbb{R}^{4n})$, then

$$\begin{aligned}
(\hat{\phi})^\vee(x_1,\ldots,x_n) &= \hat{\phi}(x_2 - x_1, x_3 - x_2,\ldots,x_n - x_{n-1}) \\
&= \phi(0, x_2 - x_1, x_3 - x_1,\ldots,x_n - x_1).
\end{aligned}$$

Hence,

$$\begin{aligned}
M_n(\hat{\phi}) &= W_n[\phi(0, x_2 - x_1, x_3 - x_1,\ldots,x_n - x_1)] \\
&= W_n[\phi \circ (x_1, I)^{-1}] = W_n(\phi).
\end{aligned}$$

The proofs of W4(b) and W6 are fairly technical and will be omitted. $\quad\square$

7. Quantum Field Theory

We have shown that corresponding to any Wightman quantum field there exists a sequence W_n of tempered distributions on $S(\mathbb{R}^{4n})$, $n = 0, 1, \ldots$, satisfying conditions W1–W6. We shall now prove the converse. That is, a Wightman quantum field can be reconstructed from a sequence of tempered distributions satisfying W1–W6. Moreover, we shall show that Wightman quantum fields are characterized by a certain class of states on the "field algebra."

Let $S_n = S(\mathbb{R}^{4n})$, $n = 0, 1, \ldots$, and let $S = \oplus S_n$ with the operations as defined in Section 5.3. We call S the *field algebra*. We showed in Section 5.3 that S is a *-algebra. For $\phi \in S_1, f \in S$ define

$$(\phi \circ f)_i = \int_{\mathbb{R}^4} \phi(a)[(a, I)f_i] \, da \tag{7.22}$$

for $i = 0, 1, \ldots$. It is not hard to show that $\phi \circ f \in S$ for every $\phi \in S_1, f \in S$. Let I_s be the smallest left ideal in S containing the set

$$\{\phi \circ f : f \in S, f_0 = 0, \phi \in S_1, \mathcal{F}\phi(x) = 0, x \in V_+\}.$$

The ideal I_s is related to the spectrum condition. We shall also need an ideal related to locality. We say that $\phi \in S_n$ and $\psi \in S_m$ are *spacelike separated* if $(x_i - y_j) \cdot (x_i - y_j) < 0$ for every $(x_1, \ldots, x_n) \in \operatorname{supp} \phi$ and $(y_1, \ldots, y_m) \in \operatorname{supp} \psi$. We say that $f, g \in S$ are *spacelike separated* if $f_0 = g_0 = 0$ and f_i, g_j are spacelike separated for every $i, j > 0$. We denote by I_c the smallest two-sided ideal in S containing all elements of the form $f \times g - g \times f$, where f, g are spacelike separated. Finally, we say that a state w on S is *invariant* if

$$w(f) = w\left(f \circ (a, \Lambda)^{-1}\right)$$

for every $(a, \Lambda) \in \mathcal{P}, f \in S$.

If w is a state on S and $\phi_n \in S_n$ we define

$$w(\phi_n) = w(0, \ldots, \phi_n, 0, \ldots).$$

Let w be a state on $S, \phi \in S_n$ and $0 \leq j \leq n$ an integer. If we define

$$\psi(x_1, \ldots, x_j) = w[\phi(x_1, \ldots, x_j, \cdots)] \tag{7.23}$$

then it is not hard to show that $\psi \in S_j$. We define $w_j \otimes w_{n-j} : S_n \to C$ by $w_j \otimes w_{n-j}(\phi) = w(\psi)$. Now let (Θ, D, U, h) be a Wightman quantum field. Define $w_h : S \to C$ by

$$w_h(\phi_0, \phi_1, \ldots) = \sum_{n=0}^{\infty} W_n(\phi_n). \tag{7.24}$$

Theorem 7.7. (a) *Let* (Θ, D, U, h) *be a Wightman quantum field. Then* w_h *is an invariant state on* $\mathcal{S}$ *which annihilates the ideals* I_s, I_c *and satisfies:*

$$\lim_{\lambda \to \infty} w_h(T_{j,\lambda a}\phi) = w_j \otimes w_{n-j}(\phi) \tag{7.25}$$

for every spacelike $a \in M^4$.

(b) *Conversely, if* w *is an invariant state on* $\mathcal{S}$ *which annihilates the ideals* I_s, I_c *and satisfies* (7.25), *then there exists a Wightman quantum field* (Θ, D, U, h) *which is unique to within an equivalence such that*

$$\langle \Theta(\phi)h, h \rangle = w(\phi)$$

for every $\phi \in \mathcal{S}_1$.

PROOF. (a) Clearly w_h is linear on $\mathcal{S}$ and satisfies $w_h(1) = 1$. To show that w_h is positive, suppose $f = (\phi_0, \ldots, \phi_n, 0, 0, \ldots) \in \mathcal{S}$. Then

$$f^* \times f = \left(|\phi_0|^2, \phi_0^* \phi_1 + \phi_1^* \phi_0, \ldots, \sum_{i+j=n} \phi_i^* \otimes \phi_j, \ldots \right)$$

It follows from W1 that

$$w_h(f^* \times f) = \sum_{i,j=0} W_{i+j}(\phi_i^* \otimes \phi_j) \geq 0.$$

The invariance of w_h follows from W3. We now show that w_h annihilates I_s. Let $\phi \in \mathcal{S}_1$ satisfy $\mathcal{F}\phi(x) = 0$ for $x \in V_+$. Then by the spectrum condition W4

$$\int \phi(a) U(a, I) \, da = \int \phi(a) \left[\int e^{ip \cdot a} E(dp) \right] da$$

$$= \int \left[\int e^{ip \cdot a} \phi(a) \, da \right] E(dp) = \int \mathcal{F}\phi(p) E(dp) = 0.$$

Now let $f = \phi_1 \otimes \cdots \otimes \phi_n \in \mathcal{S}_n$. Then by covariance

$$0 = \left\langle \int \phi(a) U(a, I) \, da \Theta(\phi_1) \cdots \Theta(\phi_n) h, h \right\rangle$$

$$= \left\langle \int \phi(a) \Theta(\phi_1 \circ (a, I)^{-1}) \cdots \Theta(\phi_n \circ (a, I)^{-1}) \, dah, h \right\rangle$$

$$= \int \phi(a) W_n((a, I)f) = W_n\left(\int \phi(a)(a, I)f \, da \right)$$

$$= W_n(\phi \circ f) = w_h(\phi \circ f).$$

Since $w_h(\phi \circ f) = 0$ for every $f \in \mathcal{S}$ and since the kernel of w_h is a left ideal,

$w_h(I_s) = 0$. To show that w_h annihilates I_c, let $\phi, \psi \in \mathcal{S}_1$ be spacelike separated and let $\phi_1, \phi_2 \in \mathcal{S}_1$. Then by W5 we have

$$w_h[\phi_1 \otimes (\phi \otimes \psi - \psi \otimes \phi) \otimes \phi_2] = W_4(\phi_1 \otimes \phi \otimes \psi \otimes \phi_2) - W_4(\phi_1 \otimes \psi \otimes \phi \otimes \phi_2) = 0.$$

The general case follows similarly. Hence $w_h(I_c) = 0$. Equation (7.25) follows from W6.

(b) Let w be a state on $\mathcal{S}$ satisfying the given conditions.

We proceed using the GNS construction. Let D be the vector space $D = \mathcal{S}/\mathcal{G}$ with inner product $\langle f, g \rangle = w(g^* \times f)$, let $\mathcal{H}$ be the Hilbert space completion of D and let $h = [1]$. Define $\Theta(\phi) = \pi(\phi)$, $\phi \in \mathcal{S}_1$, where π is the GNS construction. Conditions (1) and (2) for a Wightman quantum field hold. We now define a unitary representation U of $\mathcal{P}$ on $\mathcal{H}$. For $[f] \in D$, $(a, \Lambda) \in \mathcal{P}$, define $U(a, \Lambda)[f] = [f \circ (a, \Lambda)^{-1}]$. The transformation $U(a, \Lambda)$ is well-defined since w is invariant and (a, Λ) leaves the left ideal $\mathcal{G} \cap I_s \cap I_c$ invariant. It is clear that $U(a, \Lambda)$ is linear and leaves D invariant. Moreover, $U(a, \Lambda)$ is isometric on D since

$$\langle U(a, \Lambda)[f], U(a, \Lambda)[g] \rangle = w[f \circ (a, \Lambda)^{-1*} \times g \circ (a, \Lambda)]$$
$$= w[(f^* \times g) \circ (a, \Lambda)^{-1}] = w(f^* \times g) = \langle f, g \rangle.$$

Since D is dense in $\mathcal{H}$, $U(a, \Lambda)$ has a unique unitary extension to $\mathcal{H}$. To show that U is a unitary representation of $\mathcal{P}$ we have

$$U(a_1, \Lambda_1) U(a_2, \Lambda_2)[f] = U(a_1, \Lambda_1)[f \circ (a_2, \Lambda_2)^{-1}]$$
$$= [f \circ (a_2, \Lambda_2)^{-1}(a_1, \Lambda_1)^{-1}]$$
$$= [f \circ ((a_1, \Lambda_1)(a_2, \Lambda_2))^{-1}]$$
$$= U[(a_1, \Lambda_1)(a_2, \Lambda_2)][f].$$

The representation is continuous since w is continuous in $\mathcal{H}$. Moreover,

$$U(a, \Lambda) h = U(a, \Lambda)[1] = [1 \circ (a, \Lambda)^{-1}] = [1] = h$$

and hence conditions 3(a) and 3(b) hold. To show that 3(c) holds, we have for every $f \in D$

$$U(a, \Lambda) \Theta(\phi) U(a, \Lambda)^*[f] = U(a, \Lambda) \Theta(\phi)[f \circ (a, \Lambda)]$$
$$= U(a, \Lambda)[\phi \times f \circ (a, \Lambda)] = [\phi \times f \circ (a, \Lambda)] \circ (a, \Lambda)^{-1}$$
$$= [\phi \circ (a, \Lambda)^{-1} \times f] = \Theta(\phi \circ (a, \Lambda)^{-1})[f].$$

To show that (4) holds, let $\mathcal{F}\phi(x) = 0$ for $x \in V_+$. Then for every $g \in \mathcal{S}_1$ we have

$$\int \phi(a) U(a,I)[\,g\,]\, da = \int \phi(a)\big[\,g \circ (a,I)^{-1}\,\big]\, da$$
$$= \phi \circ g = 0$$

since $\phi \circ g \in I_s \subseteq \mathcal{G}$. Since $g \in S_1$ was arbitrary, it follows that $\int \phi(a) U(a,I)$ $da = 0$. Therefore, supp $\mathcal{F}^{-1} U(a,I) \subseteq V_+$, which is equivalent to the spectrum condition. For (5), if $\phi, \psi \in S_1$ are spacelike separated, then

$$\|\Theta(\phi)\Theta(\psi)[\,g\,] - \Theta(\psi)\Theta(\phi)[\,g\,]\|^2$$
$$= w(\,g \otimes (\psi \otimes \phi - \phi \otimes \psi) \otimes (\psi \otimes \phi - \phi \otimes \psi) \otimes g) = 0$$

since $\psi \otimes \phi - \phi \otimes \psi \in I_c$. Condition (6) follows from (7.25). $\qquad\square$

As a corollary we obtain the converse of Theorem 7.6.

Corollary 7.8. *If W_n, $n = 0, 1, \ldots$ is a sequence of tempered distributions on* $S(\mathbb{R}^{4n})$, $n = 0, 1, \ldots$, *satisfying* W1–W6, *then there exists a unique (to within equivalence) Wightman quantum field (Θ, D, U, h) such that*

$$W_n(\phi_1 \otimes \cdots \otimes \phi_n) = \langle \Theta(\phi_1) \cdots \Theta(\phi_n) h, h \rangle.$$

PROOF. Define $w_h : S \to C$ by (7.24). As in the proof of Theorem 7.7(a), w_h is an invariant state which annihilates the ideals I_s, I_c and satisfies (7.25). The result follows from Theorem 7.7(b). $\qquad\square$

7.4. Fock Space

In this section we illustrate the concepts of the previous two sections for the important example of Fock space. Let $\mathcal{H}$ be a complex Hilbert space whose unit vectors represent the pure states for some quantum system. For illustrative purposes, suppose the quantum system consists of a single particle p. We would now like to describe the quantum system consisting of n particles all identical to p; for example, a system composed of n electrons. This system is represented by the tensor product $\mathcal{H}^n = \mathcal{H}_1 \otimes \cdots \otimes \mathcal{H}_n$, where $\mathcal{H}_i = \mathcal{H}$, $i = 1, \ldots, n$. If $\phi_1, \phi_2 \in \mathcal{H}$ are unit vectors representing pure states of a particle, then one might think that $\phi_1 \otimes \phi_2 \in \mathcal{H}^2$ represents a system of two identical particles in which one particle is in state ϕ_1, and one is in state ϕ_2. But since the particles are indistinguishable, the state must not be altered upon an interchange of the two particles. Hence, we would obtain $\phi_2 \otimes \phi_1 = c \phi_1 \otimes \phi_2$ where $|c| = 1$. Since this is

189

7. Quantum Field Theory

impossible if ϕ_1 and ϕ_2 are linearly independent, $\phi_1 \otimes \phi_2$ is not the correct form for a state consisting of two identical particles. What is needed is a unit vector $f(\phi_1, \phi_2) \in \mathcal{H}^2$ satisfying $f(\phi_2, \phi_1) = cf(\phi_1, \phi_2)$, where $|c| = 1$. Examples of such states are

$$a(\phi_1 \otimes \phi_2 + \phi_2 \otimes \phi_1) \tag{7.26}$$

and

$$b(\phi_1 \otimes \phi_2 - \phi_2 \otimes \phi_1), \tag{7.27}$$

where a and b are normalization constants. It has been found that the pure states of all known particles satisfy either (7.26) or (7.27). This is a law of physics called the *Pauli symmetrization principle*. Particles whose states satisfy (7.26) are called *bosons* and particles whose states satisfy (7.27) are called *fermions*. For concreteness we shall only consider bosons here. An analogous theory holds for fermions.

Let $\mathcal{H}$ be the Hilbert space for a single boson and $\mathcal{H}^n$ the Hilbert space for n identical bosons. Let A be the group of permutations of the set $\{1, \ldots, n\}$ and for $a \in A$ let $U(a): \mathcal{H}^n \to \mathcal{H}^n$ be the linear operator defined by

$$U(a)\phi_1 \otimes \cdots \otimes \phi_n = \phi_{a(1)} \otimes \cdots \otimes \phi_{a(n)}. \tag{7.28}$$

It is not hard to show that $a \mapsto U(a)$ is a unitary representation of A on $\mathcal{H}^n$. Now define the linear operator

$$S_+^n = \frac{1}{n!} \sum_{a \in A} U(a) \tag{7.29}$$

on $\mathcal{H}^n$. It is easy to show that S_+^n is an orthogonal projection on $\mathcal{H}^n$ and that $S_+^n U(a) = U(a)S_+^n$ for all $a \in A$. The Pauli symmetrization principle says that a unit vector $\phi \in \mathcal{H}^n$ represents the state of n identical bosons if and only if $S_+^n \phi = \phi$. We write $\mathcal{H}_+^n = S_+^n \mathcal{H}^n$ and call $\mathcal{H}_+^n$ the *boson* (or *symmetrized*) subspace of $\mathcal{H}^n$. If $\phi_1, \ldots, \phi_n \in \mathcal{H}$ are unit vectors, then the unit vector

$$(\|S_+^n \phi_1 \otimes \cdots \otimes \phi_n\|)^{-1} S_+^n \phi_1 \otimes \cdots \otimes \phi_n$$

represents the state of a system of n identical bosons in which the ith boson is in state ϕ_i, $i = 1, \ldots, n$.

Now suppose we have a system of identical bosons, but the total number of bosons is unknown. Such situations are common is quantum field theory since interactions may occur in which particles are created or annihilated so the total number may change. If we define $\mathcal{H}_+^0 = C$, then this system is described by the direct sum Hilbert space

$$\mathcal{F}_+(\mathcal{H}) = \sum_{n=0}^{\infty} \mathcal{H}_+^n.$$

We call $\mathcal{F}_+(\mathcal{H})$ the *boson Fock space* (there is an analogous fermion Fock space which we shall not treat). The boson Fock space is a closed subspace of the *Fock space* $\mathcal{F}(\mathcal{H}) = \Sigma \mathcal{H}^n$ and if we define the orthogonal projection $S_+ = \Sigma S_+^n$, then $\mathcal{F}_+(\mathcal{H}) = S_+ \mathcal{F}(\mathcal{H})$. We identify $\mathcal{H}_+^n$ with the corresponding subspace of $\mathcal{F}_+(\mathcal{H})$ and write an element $\phi \in \mathcal{F}_+(\mathcal{H})$ as

$$\phi = \sum_{n=0}^{\infty} \phi^{(n)}, \tag{7.30}$$

where $\phi^{(n)} \in \mathcal{H}_+^n$ and $\Sigma \|\phi^{(n)}\|^2 < \infty$. If ϕ in (7.30) is a unit vector corresponding to a pure state, then $\|\phi^{(n)}\|^2$ gives the probability that there are n bosons in the state corresponding to ϕ. The vector $\phi_0 \in C$ given by $\phi_0 = 1$ is the *vacuum* vector representing the state with no bosons present.

If $\{u_k\}$ is an orthonormal basis for $\mathcal{H}$, then

$$\left\{ u_{k_1} \otimes \cdots \otimes u_{k_n} : \{k_1, \ldots, k_n\} \subseteq \{1, 2, \ldots\} \right\}$$

is an orthonormal basis for $\mathcal{H}^n$. Let n_i be the number of indices among the $k_1, \ldots, k_n$ in the vector $u_{k_1} \otimes \cdots \otimes u_{k_n}$ that equals i. Of course, $n_i = 0$ except for finitely many i. By direct computation we have

$$\| S_+^n u_{k_1} \otimes \cdots \otimes u_{k_n} \| = \frac{n_1! n_2! \cdots n_i! \cdots}{n!}. \tag{7.31}$$

The unit vector

$$u(n_1, n_2, \ldots) = \left(\frac{n!}{n_1! n_2! \cdots} \right)^{1/2} S_+^n u_{k_1} \otimes \cdots \otimes u_{k_n} \tag{7.32}$$

represents the state in which there are n bosons, n_i of which are in the state u_i. Define the linear operator N on $\mathcal{F}_+(\mathcal{H})$ as follows. The domain $D(N)$ is

$$D(N) = \left\{ \phi = \sum \phi^{(n)} \in \mathcal{F}_+(\mathcal{H}) : \sum \|n\phi^{(n)}\|^2 < \infty \right\}$$

and for $\phi \in D(N)$, $N\phi = \Sigma n\phi^{(n)}$. The operator N is self-adjoint and is called the *number of particles operator*.

For $\phi \in \mathcal{H}$, define the bounded linear operator $C(\phi) : \mathcal{H}^n \to \mathcal{H}^{n+1}$ by

$$C(\phi)\phi_1 \otimes \cdots \otimes \phi_n = \phi \otimes \phi_1 \otimes \cdots \otimes \phi_n.$$

Thus $\|C(\phi)\| = \|\phi\|$ and $C(\phi)$ corresponds to the creation of a particle in the state ϕ. The adjoint operator $C(\phi)^* : \mathcal{H}^{n+1} \to \mathcal{H}^n$ satisfies $C(\phi)^* = 0$ if $\phi \in \mathcal{H}^0$ and

$$C(\phi)^* \phi_1 \otimes \cdots \otimes \phi_n = \langle \phi, \phi_1 \rangle \phi_2 \otimes \cdots \otimes \phi_n$$

for $n \geq 1$. The operator $C(\phi)^*$ corresponds to the annihilation of a particle in the state ϕ. We define $C(\phi)$ and $C(\phi)^*$ on $\mathcal{F}(\mathcal{H})$ by

$$C(\phi)\left(\sum \phi^{(n)}\right) = \sum C(\phi)\phi^{(n)},$$

$$C(\phi)^*\left(\sum \phi^{(n)}\right) = \sum C(\phi)^*\phi^{(n)}.$$

Let D_0 be the dense subspace of $\mathcal{F}_+(\mathcal{H})$ defined by

$$D_0 = \left\{ \sum \phi^{(n)} \in \mathcal{F}_+(\mathcal{H}) : \phi^{(n)} = 0 \quad \text{except for finitely many} \quad n \right\}.$$

For $\phi \in \mathcal{H}$ define the following operators on D_0:

$$\mathcal{C}(\phi) = S_+ C(\phi)^* \sqrt{N} \; ; \tag{7.33}$$

$$\mathcal{C}^*(\phi) = \sqrt{N} \, S_+ C(\phi). \tag{7.34}$$

It can be shown that $\mathcal{C}(\phi)$ and $\mathcal{C}^*(\phi)$ are closed operators on D_0 and that $\mathcal{C}^*(\phi)$ is the adjoint of $\mathcal{C}(\phi)$. We call $\mathcal{C}(\phi)$ and $\mathcal{C}^*(\phi)$ *annihilation* and *creation* operators, respectively. It is not hard to show that $\mathcal{C}$ and $\mathcal{C}^*$ satisfy the following commutation relations on D_0:

$$[\mathcal{C}(\phi), \mathcal{C}(\psi)] = [\mathcal{C}^*(\phi), \mathcal{C}^*(\psi)] = 0,$$

$$[\mathcal{C}(\phi), \mathcal{C}^*(\psi)] = \langle \psi, \phi \rangle I, \tag{7.35}$$

where $[A, B] = AB - BA$. Moreover, it is easy to show that

$$u(n_1, n_2, \ldots) = (n_1! n_2! \cdots)^{-1/2} [\mathcal{C}^*(u_{k_1}) \mathcal{C}^*(u_{k_2}) \cdots] \phi_0, \tag{7.36}$$

where $u(n_1, n_2, \ldots)$ is defined by (7.32). Since vectors of this form are dense in $\mathcal{F}_+(\mathcal{H})$ this shows that the entire boson Fock space is generated by creating bosons from the vacuum ϕ_0.

We now define the operators $p(\phi)$, $q(\phi)$ on D_0 by

$$p(\phi) = \frac{1}{\sqrt{2} \, i} [\mathcal{C}(\phi) - \mathcal{C}^*(\phi)],$$

$$q(\phi) = \frac{1}{\sqrt{2}} [\mathcal{C}(\phi) + \mathcal{C}^*(\phi)].$$

Then $p(\phi)$ and $q(\phi)$ are essentially self-adjoint and satisfy the commutation relations

$$[q(\phi), q(\psi)] = i \operatorname{Im}\langle \psi, \phi \rangle,$$

$$[p(\phi), p(\psi)] = i \operatorname{Im}\langle \psi, \phi \rangle,$$

$$[q(\phi), p(\psi)] = -i \operatorname{Im}\langle \psi, \phi \rangle. \tag{7.37}$$

If we define $W(\phi) = e^{iq(\phi)}$, then W satisfies (a), (b), (c) of Lemma 7.4 so W is a complex representation of the CCR. Moreover, it is not hard to show that W is cyclic. In a similar way, $e^{ip(\phi)}$ is a cyclic complex representation of the CCR. An explicit computation gives

$$\langle W(\phi)\phi_0, \phi_0\rangle = e^{-(1/4)\|\phi\|^2}. \tag{7.38}$$

Now let us suppose that $\mathcal{H}$ is a real Hilbert space. Then the $q(\phi), \phi \in \mathcal{H}$, are essentially self-adjoint and commute since $\mathrm{Im}\langle\psi,\phi\rangle = 0$. In fact, it can be shown that the spectral measures of the $q(\phi), \phi \in \mathcal{H}$, commute so the spectral measures generate a commutative C^*-algebra $\mathcal{C}$. It follows from the Gelfand–Naimark theorem [194] that there is an isometric isomorphism h from $\mathcal{C}$ onto the set of continuous functions $C(\Omega)$ on a compact Hausdorff space Ω. The state $A \mapsto \langle A\phi_0, \phi_0\rangle$ on $\mathcal{C}$ is mapped under h to a positive linear functional l on $C(\Omega)$ such that $l(1) = 1$. Applying the Riesz representation theorem, there exists a probability measure μ on the Borel sets $\mathcal{B}(\Omega)$ such that $\langle A\phi_0, \phi_0\rangle = \int h(A) d\mu$ for every $A \in \mathcal{C}$. Under the map h, $q(\phi)$ becomes a random variable $\Phi_{\mathcal{F}}(\phi)$ on the probability space $(\Omega, \mathcal{B}(\Omega), \mu)$ for every $\phi \in \mathcal{H}$. Since $q(\cdot)$ is linear, so is $\Phi_{\mathcal{F}}(\cdot)$ and it follows that $\Phi_{\mathcal{F}}$ is a random field. Moreover, the characteristic functional of $\Phi_{\mathcal{F}}$ satisfies

$$E\left[e^{i\Phi_{\mathcal{F}}(\phi)}\right] = \int e^{i\Phi_{\mathcal{F}}(\phi)} d\mu = \langle e^{iq(\phi)}\phi_0, \phi_0\rangle = e^{-(1/4)\|\phi\|^2}.$$

Thus $\Phi_{\mathcal{F}}$ is the unit Gaussian random field on $\mathcal{H}$ (except for the unimportant normalization constant $\frac{1}{4}$). We thus see that the boson Fock space over $\mathcal{H}$ is equivalent to the unit Gaussian random field on $\mathcal{H}$. More precisely, applying Theorem 6.13 we have the following result.

Theorem 7.9. *If $\mathcal{H}$ is a real Hilbert space then there is an isometry U from $\mathcal{F}_+(\mathcal{H})$ onto $\Gamma(\mathcal{H})$ such that* (a) $U\phi_0 = 1$, (b) $Uq(\phi)U^{-1} = \Phi_u(\phi)$, (c) $U\mathcal{H}_+^n = \Gamma_n$.

7.5. Euclidean Random Fields

In this section we give E. Nelson's method for constructing quantum fields from a certain class of random fields [199, 200]. Let $H^{-1}(\mathbb{R}^n), n \geq 2$, be the Hilbert space consisting of all distributions $f \in \mathcal{S}'(\mathbb{R}^n)$ whose Fourier transforms $\hat{f}$ are functions and for which the norm

$$\|f\|_{-1}^2 = \int |\hat{f}(k)|^2 (k^2 + 1)^{-1} d^n k < \infty. \tag{7.39}$$

The Hilbert space $H^{-1}(\mathbb{R}^n)$ is an example of a Sobolev space. To see that $H^{-1}(\mathbb{R}^n)$ is a Hilbert space and to get an idea of where this space originates, we can write (7.39) in terms of the usual $L^2(\mathbb{R}^n)$ inner product

$$\langle f,g\rangle_{-1}=\langle \hat{f}(k), \frac{1}{k^2+1}\hat{g}(k)\rangle. \tag{7.40}$$

Since the Fourier transform of the Laplacian operator Δ goes over to multiplication by $-k^2$, taking the usual function inverse Fourier transform of (7.40) gives

$$\langle f,g\rangle_{-1}=\langle f,(-\Delta+1)^{-1}g\rangle.$$

Thus $H^{-1}(\mathbb{R}^n)$ is isomorphic to the Hilbert space of functions on $\mathbb{R}^n$ satisfying $\langle f,(-\Delta+1)^{-1}f\rangle<\infty$.

Equip $\mathbb{R}^n, n\geq 2$, with its usual inner product $x\cdot y=\sum_{i=1}^{n}x_iy_i$. The *Euclidean group* $\mathfrak{E}_n$ on $\mathbb{R}^n$ is the group of all nonsingular inhomogeneous linear transformations that preserve the inner product. By a *representation* of $\mathfrak{E}_n$ on a probability space (Ω,Σ,μ) we mean a group homomorphism $\beta\mapsto T_\beta$ of $\mathfrak{E}_n$ into the group of measure-preserving transformations on (Ω,Σ,μ) such that for all $u,v\in L^\infty(\Omega,\Sigma,\mu)$, $\beta\mapsto\int u(v\circ T_\beta)d\mu$ is measurable. A *covariant random field* on $H^{-1}(\mathbb{R}^n)$ is a random field $\Phi:H^{-1}(\mathbb{R}^n)\rightarrow R(\Omega,\Sigma,\mu)$ together with a representation T of $\mathfrak{E}_n$ on (Ω,Σ,μ) such that $\Phi(\phi)\circ T_\beta=\Phi(\phi\circ\beta)$ for every $\phi\in H^{-1}(\mathbb{R}^n)$ and $\beta\in\mathfrak{E}_n$. A covariant random field is *ergodic* if the only elements of $\Sigma/\mathcal{I}_\mu$ left invariant by T_β for every translation β are Ω and $\varnothing$.

Let $\Phi:H^{-1}(\mathbb{R}^n)\rightarrow R(\Omega,\Sigma,\mu)$ be a random field. For any open or closed set $A\subseteq\mathbb{R}^n$, let Σ_A be the σ-algebra generated by

$$\{\Phi(\phi):\operatorname{supp}\phi\subseteq A\}$$

and let E_A denote the conditional expectation $E(\cdot|\Sigma_A)$. We say that Φ is a *Markov* random field on $H^{-1}(\mathbb{R}^n)$ if for every closed set $A\subseteq\mathbb{R}^n$ and any Σ_{A^c}-measurable function f we have $E_A(f)=E_{\partial A}(f)$, where A^c denotes the complement of A and ∂A the boundary of A. Intuitively, the Markov condition says that if we wish to predict some aspect (namely, f) of the field's behavior inside A, then a knowledge Σ_{A^c} of the field outside A gives no more information than the knowledge $\Sigma_{\partial A}$ of the field on the boundary of A. This is a multi-dimensional generalization of the usual Markov condition for a stochastic process. For a stochastic process, $\mathbb{R}^n$ is replaced by $\mathbb{R}$ and is thought of as time. In this case, if $A=(-\infty,t]$, then the Markov condition states that the conditional expectation of future events given the past depends only on the present. More prosaically, the future is independent of the past given the present.

A random field Φ is *hermitian* if $\Phi(\bar{\phi})=\overline{\Phi}(\phi)$ for every ϕ. A *Euclidean random field* on $H^{-1}(\mathbb{R}^n)$ is a full, hermitian, covariant, ergodic, Markov random field on $H^{-1}(\mathbb{R}^n)$. Before we study the properties of Euclidean random fields, it is convenient to introduce some notation. Let (Φ, T) be a covariant random field. We denote by U_β the unitary representation of $\mathfrak{S}_n$ on $L^2(\Omega, \Sigma, \mu)$ given by $(U_\beta f)(\omega) = f(T_\beta^{-1} \omega)$. By a standard theorem in the theory of representations of locally compact groups [194], $\beta \mapsto U_\beta$ is strongly continuous. We frequently denote an element of $\mathbb{R}^n$ by (x, t), where $x \in \mathbb{R}^{n-1}, t \in \mathbb{R}$, and we let $\mathbb{R}_0^{n-1}$ denote the hyperplane $\{(x, 0): x \in \mathbb{R}^{n-1}\}$. We may think of t as the "time" coordinate. We let $\beta_t, t \in \mathbb{R}$, denote the translation $(x, s) \mapsto (x, s + t)$ and denote the corresponding unitary operator by $U_t = U_{\beta_t}$. Moreover, $\rho \in \mathfrak{S}_n$ will be the time reflection $\rho(x, t) = (x, -t)$. We denote the projection operator $E_{\mathbb{R}_0^{n-1}}$ on $L^2(\Omega, \Sigma, \mu)$ by E_0 and the closed subspace $L^2(\Omega, \Sigma_{\mathbb{R}_0^{n-1}}, \mu) = E_0 L^2(\Omega, \Sigma, \mu)$ of $L^2(\Omega, \Sigma, \mu)$ by $\mathcal{H}$. Finally, we use the shorthand notation $E_t \equiv E_{\beta_t}(\mathbb{R}_0^{n-1}) = U_t E_0 U_t^{-1}$ and for $B \in \mathcal{B}(\mathbb{R}), E_B = E_{\mathbb{R}^{n-1} \times B}$.

In classical probability theory, Markov processes are naturally associated with contraction semigroups [65, 68]. We now show that there is an analogous result for covariant Markov random fields.

Theorem 7.10. *Let (Φ, T) be a covariant Markov random field on $H^{-1}(\mathbb{R}^n)$ and let $P_t = E_0 U_t | \mathcal{H}$. Then $\{P_t : t \geq 0\}$ is a strongly continuous self-adjoint contraction semigroup and $P_{-t} = P_t$ for every $t \in \mathbb{R}$.*

PROOF. Since $t \mapsto U_t$ is strongly continuous so is $t \mapsto P_t$ and since U_t is unitary, we have

$$P_t^* = (E_0 U_t E_0)^* = E_0 U_t^{-1} E_0 = P_{-t}.$$

Suppose that $f \in H^{-1}(\mathbb{R}^n)$ and $\operatorname{supp} f \subseteq \mathbb{R}_0^{n-1}$. We now show that $f \circ \rho = f$. Since $\operatorname{supp} f \subseteq \{(x, 0): x \in \mathbb{R}^{n-1}\}$, it follows from Theorem 5.6 that there exist tempered distributions $g_j \in \mathcal{S}'(\mathbb{R}^{n-1}), 0 \leq j \leq m$, such that

$$f = \sum_{j=0}^m g_j \otimes \delta_0^{(j)}.$$

To find the Fourier transform of $\delta_0^{(j)}$ we have for any $\phi \in \mathcal{S}(\mathbb{R})$

$$\delta_0^{(j)\hat{}}(\phi) = \delta_0^{(j)}(\hat{\phi}) = \delta^{(j)}(2\pi)^{-1/2} \int e^{its} \phi(s) \, ds$$

$$= (2\pi)^{-1/2} \int (is)^{-j} \phi(s) \, ds.$$

Hence $\delta_0^{(j)\hat{}} = (2\pi)^{-1/2}(is)^j$. We thus have

$$\| g_j \otimes \delta_0^{(j)} \|_{-1}^2 = (2\pi)^{-1} \int \int | \hat{g}_j(k) |^2 |s|^{2j} (k^2 + s^2 + 1) d^{n-1} k \, ds. \quad (7.41)$$

The s integration in (7.41) is finite if and only if $j = 0$. It follows that

$f = g_0 \otimes \delta_0$ so clearly $f \circ \rho = f$. Hence, by covariance, we have

$$U_\rho \Phi(f)(\omega) = \Phi(f)\big(T_\rho^{-1}\omega\big) = \Phi(f) \circ T_\rho^{-1}(\omega)$$
$$= \Phi(f \circ \rho)(\omega) = \Phi(f)(\omega).$$

It follows that U_ρ leaves the range of E_0 pointwise invariant so that

$$U_\rho E_0 = E_0 U_\rho = E_0.$$

Hence,

$$P_{-t} = E_0 U_{-t} E_0 = E_0 U_\rho U_t E_0 = E_0 U_t E_0 = P_t.$$

We conclude that $P_t^* = P_t$ is self-adjoint for every $t \in \mathbb{R}$. Since $E_t = U_t E_0 U_{-t}$ we have

$$U_t E_s = U_t U_s E_0 U_{-s} = U_{t+s} E_0 U_{-(t+s)} U_t = E_{t+s} U_t.$$

In particular,

$$U_t E_{-t} = E_0 U_t, \qquad U_s E_0 = E_s U_s. \tag{7.42}$$

Moreover, by the Markov condition, if $t, s \geq 0$, we have

$$E_{-t} E_0 E_s = E_{-t} E_{(-\infty, 0]} E_s = E_{-t} E_s. \tag{7.43}$$

Applying (7.42) and (7.43) for $t, s \geq 0$, we have

$$P_t P_s = E_0 U_t E_0 U_s E_0 = E_0 (E_0 U_t) E_0 (U_s E_0) E_0$$
$$= E_0 U_t E_{-t} E_0 E_s U_s E_0 = E_0 (U_t E_{-t})(E_s U_s) E_0$$
$$= E_0 U_t U_s E_0 = P_{t+s}.$$

Since $P_0 = I$ and $\|P_t\| \leq 1$, $\{P_t : t \geq 0\}$ is a contraction semigroup. $\qquad \square$

Corollary 7.11. *If Φ and P_t are defined as in the previous theorem, then there exists a unique positive self-adjoint operator H on $\mathcal{H}$ such that $P_t = e^{-|t|H}, t \in \mathbb{R}$.*

PROOF. The proof follows from a standard result on self-adjoint contraction semigroups [131]. $\qquad \square$

Following Nelson [199] we now show how a Wightman quantum field can be constructed from a Euclidean field Φ on $H^{-1}(\mathbb{R}^n)$. Construct the Hilbert space $\mathcal{H}$ and the positive self-adjoint operator as in Theorem 7.10 and Corollary 7.11. We first construct the "time zero" quantum field Θ_0. If $f \in S(\mathbb{R}^{n-1})$ then $f \otimes \delta_0 \in H^{-1}(\mathbb{R}^n)$. Indeed, since $\hat{\delta}_0 = (2\pi)^{-1/2}$, we have

$$\|f \otimes \delta_0\|_{-1}^2 = (2\pi)^{-1} \int \int |\hat{f}(k)|^2 (k^2 + s^2 + 1)^{-1} d^{n-1}k \, ds$$

$$\leq \int \int |\hat{f}(k)|^2 (s^2 + 1)^{-1} d^{n-1}k \, ds \leq \|\hat{f}\|_\infty \int (s^2 + 1)^{-1} ds < \infty.$$

For $\phi \in \mathcal{S}(\mathbb{R}^{n-1})$, define $\Theta_0(\phi) = \Phi(\phi \otimes \delta_0)$. Since $\operatorname{supp} \phi \otimes \delta_0 \subseteq \mathbb{R}_0^{n-1}, \Theta_0(\phi)$ is a random variable which is measurable with respect to $\Sigma_{\mathbb{R}_0^{n-1}}$ and hence multiplication by the function $\Theta_0(\phi)$ is a linear operator (unbounded, in general) on some domain in $\mathcal{H} = L^2(\Omega, \Sigma_{\mathbb{R}_0^{n-1}}, \mu)$. In the sequel we consider $\Theta_0(\phi)$ to be such a linear operator. We define the "sharp time" quantum field $\Theta_t, t \in \mathbb{R}$, by $\Theta_t(\phi) = e^{itH} \Theta_0(\phi) e^{-itH}, \phi \in \mathcal{S}(\mathbb{R}^{n-1})$. Thus, for every $\phi \in \mathcal{S}(\mathbb{R}^{n-1}), t \in \mathbb{R}, \Theta_t(\phi)$ is a linear operator in $\mathcal{H}$ whose domain we shall define presently.

In order for $\Theta_t(\phi)$ to have a sufficiently large domain, Nelson introduces a mild regularity condition which we now discuss. For every integer k, let $\mathcal{H}^k$ be the completion of the domain $D(H^{k/2})$ in the norm $\|u\|_k = \|(I + H)^{k/2} u\|$ and let $\mathcal{H}^\infty = \cap \mathcal{H}^k, \mathcal{H}^{-\infty} = \cup \mathcal{H}^k$ [199]. All of these subspaces are dense in $\mathcal{H}$. By $\mathcal{L}(\mathcal{H}^k, \mathcal{H}^l)$ we mean the Banach space of bounded linear transformations from $\mathcal{H}^k$ to $\mathcal{H}^l$ equipped with the norms $\|\cdot\|_k, \|\cdot\|_l$, respectively. A Euclidean random field on $H^{-1}(\mathbb{R}^n)$ is *regular* if there exist integers k and l such that for every $\phi \in \mathcal{S}(\mathbb{R}^{n-1})$, $\Theta_0(\phi) \in \mathcal{L}(\mathcal{H}^k, \mathcal{H}^l)$ and $\phi \mapsto \Theta_0(\phi)$ is continuous. If Φ is regular, since e^{itH} leaves each $\mathcal{H}^k$ invariant and is unitary on it for every $t \in \mathbb{R}$, we have

$$\Theta_t(\phi) = e^{itH} \Theta_0(\phi) e^{-itH} \in \mathcal{L}(\mathcal{H}^k, \mathcal{H}^l)$$

for every $t \in \mathbb{R}$.

We are now ready to construct the quantum field Θ associated with Φ. For $\phi \in \mathcal{S}(\mathbb{R}^4)$, let $\phi_t(x) = \phi(x, t), x \in \mathbb{R}^3, t \in \mathbb{R}$, and define

$$\Theta(\phi) = \int_\mathbb{R} \Theta_t(\phi_t) \, dt = \int_\mathbb{R} e^{itH} \Theta_0(\phi_t) e^{-itH} \, dt.$$

It is not hard to show that $\Theta(\phi) \in \mathcal{L}(\mathcal{H}^\infty)$. The next theorem shows that Θ is a Wightman quantum field.

Theorem 7.12 (Nelson). *If Φ is a regular Euclidean random field on $H^{-1}(\mathbb{R}^4)$, then Θ is a Wightman quantum field with vacuum vector* **1**.

PROOF. We shall just give the idea of the proof; for details the reader is referred to [199]. First, the uniqueness of the vacuum follows almost immediately from ergodicity. For the remainder of the proof, form the expectation values

$$\mathcal{W}_n(\phi_1, \ldots, \phi_n) = \langle \Theta(\phi_1) \cdots \Theta(\phi_n) \mathbf{1}, \mathbf{1} \rangle$$

and the corresponding Wightman distributions given by

$$W_n(\phi_1 \otimes \cdots \otimes \phi_n) = \mathcal{W}_n(\phi_1, \ldots, \phi_n).$$

Positive definiteness (W1) and reality (W2) follow immediately. Relativistic covariance (W3) follows from the Euclidean covariance of Φ since in a

197

certain sense M^4 is just $\mathbb{R}^4$ with t replaced by it. The spectrum condition (W4) is a consequence of the positivity of H and relativistic covariance. The most difficult part of the proof is locality (W5). Intuitively, the idea is the following. If ϕ_j and ϕ_{j+1} are spacelike separated, then by relativistic covariance, we may assume that they have supports in the hyperplane $\mathbb{R}_0^3$. But $\Theta = \Phi$ on this hyperplane and the values of Φ commute since they are random variables. $\qquad \square$

B. Simon [250] has given a converse to Theorem 7.12. He has shown that a certain large class of Wightman quantum fields can be constructed from regular Euclidean random fields using Nelson's method.

Nelson's approach to the construction of quantum fields has many desirable features. Reducing the study of quantum fields to Euclidean random fields has the following advantages.

1. The values of a random field are random variables, so one does not have the cumbersome domain problems of unbounded operators. Moreover, the values of a random field always commute.
2. The Euclidean group is much easier to work with than the Poincaré group.
3. One can bring into play the powerful methods and results of probability theory. In particular, results analogous to those for Markov processes have already been widely exploited. Furthermore, using existence theorems of probability theory, the existence of random fields with specific properties are much easier to prove than the existence of quantum fields.

In Section 7.4 we showed that the unit Gaussian random field Φ over a real Hilbert space $\mathcal{H}$ is unitarily equivalent to the boson Fock space $\mathcal{F}_+(\mathcal{H})$ (in the sense of Theorem 7.9). Nelson has shown that the unit Gaussian random field over $H^{-1}(\mathbb{R}^4)$ is a regular Euclidean random field [200]. This not only shows that regular Euclidean random fields exist, but also that free quantum fields can be constructed from the unit Gaussian random field.

7.6. Notes and References

For a general discussion of the canonical commutation relations CCR, the canonical anticommutation relations and other aspects of algebraic quantum field theory we recommend [72]. Our approach to the CCR in Section 7.2 is similar to that in [6, 87]. For more details of the C^*-algebra framework applied to the CCR see [6–10, 13, 46, 61, 86, 127, 182, 231, 256, 257, 266]. The Wightman axioms are considered in detail in [23, 150, 161,

229, 230, 237, 250, 254, 255, 277–279]. The field algebra approach to quantum fields is due to Borchers [28, 29]. For a further discussion of Fock space we refer the reader to [14, 16, 21, 44, 62]. Our exposition on Euclidean random fields follows [197, 199, 200]. Much of the work of Nelson and others in this area has been strongly influenced by the pioneering investigations of Segal [238–245]. Other papers on random fields and quantum field theory include [3, 4, 34, 37, 38, 45, 80, 125, 126, 152, 153, 202, 203, 260, 281].

7.7. Exercises

1. For Q_1 and P_1 defined in (7.4), show that Q_1 is bounded and that P_1 has discrete point spectrum.

2. Verify Eq. (7.6).

3. Show that the complexification $\mathcal{V}_c$ of $\mathcal{V}$ defined by (7.12) and (7.13) is a complex linear space.

4. Show that $\mathcal{V}_c$ is a complex inner product space under the inner product (7.14).

5. Prove Lemma 7.4.

6. Prove that $\Delta(\mathcal{V}_c)$ is a normed *-algebra with identity $\chi_{\{0\}}$.

7. Prove Eq. (7.17).

8. If A is defined as in (7.16), prove that $A \neq 0$.

9. Show that W_0 given in (7.18) is a complex $\mathbb{R}$-representation of the CCR.

10. Show that $x_0 = \pi^{-1/4} e^{-u^2/2}$ is a cyclic vector for the representation W_0 of (7.18).

11. Prove that $\mathcal{L}$ and $\mathcal{P}$ are groups and that $(a, \Lambda)^{-1} x = \Lambda^{-1}(x - a)$, for every $x \in M^4$.

12. Show that 3(a) follows from 2(d), 3(b), and 3(c) in Section 7.3.

13. If $\phi \circ f$ is defined by (7.22), show that $\phi \circ f \in \mathcal{S}$.

14. If we define ψ by (7.23), show that $\psi \in \mathcal{S}_j$.

15. In the proof of Theorem 7.7(b) show that $U(a, \Lambda)$ is well defined.

16. Supply the details for the proof of Corollary 7.8.

17. For U defined by (7.28), show that $a \mapsto U(a)$ is a unitary representation of A on $\mathcal{H}^n$.

18. For S_+^n defined by (7.29) show that S_+^n is an orthogonal projection on $\mathcal{H}^n$ and that $S_+^n U(a) = U(a) S_+^n$ for every $a \in A$.

19. Prove Eq. (7.31).

20. Prove that the number of particles operator N is self-adjoint.

21. Prove that $\mathscr{C}(\phi), \mathscr{C}^*(\phi)$ defined by (7.33), (7.34) are closed operators and $\mathscr{C}^*(\phi)$ is the adjoint of $\mathscr{C}(\phi)$.

22. Verify Eq. (7.35).

23. Verify Eq. (7.36).

24. Verify Eq. (7.37).

25. If $W(\phi) = e^{iq(\phi)}$, show that W is a cyclic complex representation of the CCR.

26. Prove Eq. (7.38).

27. Prove that a random field Φ is hermitian if and only if $\Phi(\phi)$ is real whenever ϕ is real.

Bibliography

1. Accardi, L., On the non-commutative Markof property, *Funct. Anal. Appl.* 9, 1–12 (1975).
2. Accardi, L., Nonrelativistic quantum mechanics as a non-commutative Markof Process, *Adv. Math.* 20, 329–366 (1976).
3. Albevario, S., and R. Hoegh-Krohn, Uniqueness of the physical vacuum and the Wightman functions in the infinite limit for some non-polynomial interactions, *Commun. Math. Phys.* 30, 171–200 (1973).
4. Albevario, S., and R. Hoegh-Krohn, Homogeneous random fields and statistical mechanics, *J. Funct. Anal.* 19, 242–272 (1975).
5. Ali, S., and G. Emch, Fuzzy observables in quantum mechanics, *J. Math. Phys.* 15, 176–182 (1974).
6. Araki, H., On representations of the canonical commutation relations, *Commun. Math. Phys.* 20, 9–25 (1971).
7. Araki, H., Hamiltonian formalism and the canonical commutation relations in quantum field theory, *J. Math. Phys.* 1, 492–504 (1960).
8. Araki, H., *Local Quantum Theory*, New York: Benjamin, 1970.
9. Araki, H., and E. Woods, Representations of the canonical commutation relations describing a nonrelativistic infinite free Bose gas, *J. Math. Phys.* 4, 637–662 (1963).
10. Araki, H., and W. Wyss, Representations of canonical anticommutation relations, *Helv. Phys. Acta* 37, 136–159 (1964).
11. Atkinson, D., and M. Halpern, Non-usual topologies on space–time and high-energy scattering, *J. Math. Phys.* 8, 373–387 (1967).
12. Ballentine, L., The statistical interpretation of quantum mechanics, *Rev. Mod. Phys.* 42, 358–381 (1970).

Bibliography

13. Balsev, E., J. Manuceau, and A. Verbeure, Representations of anticommutation relations and Bogoliubov transformations, *Commun. Math. Phys.* 8, 315–326 (1968).
14. Barton, G., *Introduction to Advanced Field Theory*, New York: Wiley (Interscience), 1963.
15. Bell, J., On the problem of hidden variables in quantum mechanics, *Rev. Mod. Phys.* 38, 447–452 (1966).
16. Berezin, F., *The Method of Second Quantization*, New York: Academic Press, 1966.
17. Billingsley, P., *Convergence of Probability Measures*, New York: Wiley, 1968.
18. Birkhoff, G., *Lattice Theory*, Providence, Rhode Island: Amer. Math. Soc., 1948.
19. Birkhoff, G., Lattices in applied mathematics, in *Proceedings of the Symposia in Pure Mathematics, Vol. 2 (Lattice Theory)*, Providence, Rhode Island: Amer. Math. Soc., 1961.
20. Birkhoff, G., and J. von Neumann, The logic of quantum mechanics, *Ann. Math.* 37, 823–843 (1936).
21. Bjorken, J., and S. Drell, *Relativistic Quantum Fields*, New York: McGraw–Hill, 1965.
22. Bodiou, G., *Théorie dialectique des probabilités*, Paris: Gauthier–Villar, 1964.
23. Bogolubov, N., A. Logunov, and I. Todorov, *Introduction to Axiomatic Quantum Field Theory*, New York: Benjamin, 1975.
24. Bohm, D., *Quantum Theory*, Englewood Cliffs, New Jersey: Prentice–Hall, 1951.
25. Bohm, D., and J. Bub, A proposed solution of the measurement problem in quantum mechanics by hidden variables, *Rev. Mod. Phys.* 38, 453–469 (1966).
26. Bohm, D., and J. Bub, A refutation of the proof by Jauch and Piron that hidden variables can be excluded in quantum mechanics, *Rev. Mod. Phys.* 38, 470–475 (1966).
27. Bohr, N., Can quantum-mechanical description of reality be considered complete?, *Phys. Rev.* 48, 696–702 (1935).
28. Borchers, H., On structure of the algebra of field operators, *Nuovo Cimento* 24, 214–236 (1962).
29. Borchers, H., On the theory of local observables, in *Cargese Lectures in Theoretical Physics* (F. Lucat, ed.), New York: Gordon and Breach, 1967.
30. Born, M., Zur Quantenmechanik der Strossvorgange, *Z. Phys.* 37, 863–867 (1926).
31. Breiman, L., *Probability*, Reading, Massachusetts: Addison–Wesley, 1968.
32. Brown, R., A brief account of microscopical observations made in the months of June, July, and August, 1827, on the particles contained in the pollen of plants; and on the general existence of active molecules in organic and inorganic bodies, *Philos. Mag.* 4, 161–173 (1828).
33. Brown, R., Additional remarks on active molecules, *Philos. Mag.* 6, 161–166 (1829).
34. Canon, J., Continuous sample paths in quantum field theory, *Commun. Math. Phys.* 43, 215–233 (1974).

35. Cantoni, V., Generalized "transition probability", *Commun. Math. Phys.* 44, 125–128 (1975).
36. Chacon, R., and N. Friedman, Additive functionals, *Arch. Rat. Mech. Anal.* 18, 230–240 (1965).
37. Chaiken, J., Finite-particle representations and states of the canonical commutation relations, *Ann. Phys.* 42, 23–80 (1967).
38. Chaiken, J., Number operators for representations of the canonical commutation relations, *Commun. Math. Phys.* 8, 164–184 (1968).
39. Chung, K., *Markov Chains with Stationary Transition Probabilities*, Berlin and New York: Springer-Verlag, 1967.
40. Cockcroft, A., S. Gudder, and R. Hudson, A quantum mechanical functional central limit therorem, *J. Mult. Anal.* 7, 125–148 (1977).
41. Crockcroft, A., and R. Hudson, Quantum mechanical Wiener Process, *J. Mult. Anal.* 7, 107–124 (1977).
42. Cohen, L., Can quantum mechanics be formulated as a classical probability theory? *Philos. Science* 33, 317–322 (1966).
43. Comisar, G., Browian-motion model of nonrelativistic quantum mechanics, *Phys. Rev.* 138B, 1332–1337 (1965).
44. Cook, J., The mathematics of second quantization, *Trans. Amer. Math. Soc.* 74, 222–245 (1953).
45. Cotella, P., and O. Lanford III, Sample field behavior for the free Markov random field, in *Constructive Quantum Field Theory*, (G. Velo and A. Wightman, eds.), Berlin and New York: Springer-Verlag, 1973.
46. Courbage, M., S. Miracle-Sole, and D. Robinson, Normal states and representations of the canonical commutation relations, *Ann. Inst. Henri Poincáre* A14, 171–178 (1971).
47. Cushen, C., and R. Hudson, A quantum-mechanical central limit theorem, *J. Appl. Prob.* 8, 454–469 (1971).
48. Cycon, H., and K. E. Hellwig, Conditional expectations in generalized probability theory, *J. Math. Phys.* 18, 1154–1165 (1977).
49. Dahn, G., An attempt at an axiomatic formulation of quantum mechanics and more general theories, *Commun. Math. Phys.* 9, 192–211 (1968).
50. Dahn, G., Two equivalent criteria for modularity of the lattice of all physical decision effects, *Commun. Math. Phys.* 30, 69–78 (1973).
51. Davies, E. B., Quantum stochastic processes, *Commun. Math. Phys.* 15, 277–304 (1969).
52. Davies, E. B., Quantum stochastic processes, II, *Commun. Math. Phys.* 19, 83–105 (1970).
53. Davies, E. B., Quantum stochastic processes, III, *Commun. Math. Phys.* 22, 51–70 (1971).
54. Davies, E. B., *Quantum Theory of Open Systems*, New York: Academic Press, 1976.
55. Davies, E. B., and J. Lewis, An operational approach to quantum probability, *Commun. Math. Phys.* 17, 239–260 (1970).
56. de Korvin, A., Complete sets of expectations on a von Neumann algebra, *Quart. J. Math.* 22, 135–142 (1971).

57. de la Pena-Auerbach, L., A simple derivation of the Schroedinger equation from the theory of Markoff processes, *Phys. Lett.* 24A, 603–604 (1967).
58. de la Pena-Auerbach, L., and L. Garcia-Colin, Quantum-mechanical description of a Brownian particle, *J. Math. Phys.* 9, 668–674 (1968).
59. de la Pena-Auerbach, L., A new formulation of stochastic theory and quantum mechanics, *Phys. Lett.* 27A, 594–595 (1968).
60. de la Pena-Auerbach, L., New formulation of stochastic theory and quantum mechanics, *Rev. Mexicana Fisica*, 19, 133–145 (1970).
61. Dell'Antonio, G., S. Doplicher, and D. Ruelle, A theorem on canonical commutation and anticommutation relations, *Commun. Math. Phys.* 2, 223–230 (1966).
62. Dell'Antonio, G., and S. Doplicher, Total number of particles and Fock representation, *J. Math. Phys.* 8, 663–666 (1967).
63. Dirac, P., *The Principles of Quantum Mechanics*, Oxford: Clarendon, 1930.
64. Dixmier, J., *Les C* − algèbres et leurs representations*, Paris: Gauthier–Villais, 1969.
65. Doob, J., *Stochastic Processes*, New York: John Wiley, 1953.
66. Drewnowski, L., and W. Orlicz, Continuity and representation of orthogonally additive functionals, *Bull. Acad. Polon. Sci.* 27, 647–653 (1969).
67. Dunford, N., and J. Schwartz, *Linear Operators*, Parts 1 and 2, New York: Wiley (Interscience), 1958 and 1963.
68. Dykin, E., *Markov Processes*, Vols. I and II, Berlin and New York: Springer-Verlag, 1965.
69. Edwards, C. M., The operational approach to algebraic quantum theory I, *Commun. Math. Phys.* 16, 207–230 (1970).
70. Edwards, C. M., Classes of operations in quantum theory, *Commun. Math. Phys.* 20, 26–56 (1971).
71. Einstein, A., B. Podolsky, and N. Rosen, Can quantum-mechanical description of reality be considered complete?, *Phys. Rev.* 47, 777–780 (1935).
72. Emch, G., *Algebraic Methods in Statistical Mechanics and Quantum Field Theory*, New York: Wiley (Interscience), 1972.
73. Emch, G., and C. Piron, Symmetry in quantum theory, *J. Math. Phys.* 4, 469–473 (1963).
74. Favella, L., Brownian motions and quantum mechanics, *Ann. Inst. Henri Poincaré* 7, 77–94 (1967).
75. Finch, P., On the structure of quantum logic, *J. Sym. Logic* 34, 275–282 (1969).
76. Fine, T., *Theories of Probability: An Examination of Foundations*, New York: Academic Press, 1973.
77. Finkelstein, D., The logic of quantum physics, *Trans. N. Y. Acad. Sci.* 25, 621–637 (1963).
78. Finklestein, D., J. Jauch, S. Schiminovich, and D. Speiser, Foundations of quaternion quantum mechanics, *J. Math. Phys.* 3, 207–220 (1963).
79. Foulis, D., and C. Randall, Operational statistics I. Basic concepts, *J. Math. Phys.* 13, 1167–1675 (1972).

80. Gallavotti, G., and G. Jona-Lasinio, Limit theorems for multidimensional Markov processes, *Commun. Math. Phys.* 41, 301–307 (1975).
81. Gamow, G., *Thirty Years That Shook Physics*, Garden City, New York: Doubleday, 1966.
82. Garczynski, W., Stochastic approach to quantum mechanics and to quantum field theory, *Acta Phys. Austriaca* 6, 501–517 (1969).
83. Garczynski, W., On quantum stochastic processes, *Bull. Acad. Polon. Sci.* 17, 775–779 (1969).
84. Garczynski, W., Quantum stochastic processes and the Feyman path integral for a single spinless particle, *Rep. Math. Phys.* 4, 21–46 (1973).
85. Garczynski, W., and J. Peisert, Some examples of quantum Markovian processes, *Acta. Phys. Pol.* B3, 459–473 (1972).
86. Garding, L., and A. Wightman, Representations of the commutation relations, *Proc. Nat. Acad. Sci.* 40, 622–626 (1954).
87. Gelfand, I., and N. Vilenkin, *Generalized Functions*, New York: Academic Press, 1964.
88. Gelfand, I., and A. Yaglom, Integration in function spaces and its application to quantum physics, *J. Math. Phys.* 1, 48–69 (1960).
89. Gilson, J., On stochastic theories of quantum mechanics, *Proc. Cambridge Philos. Soc.* 64, 1061–1070 (1968).
90. Gleason, A., Measures on closed subspaces of a Hilbert space, *J. Rat. Mech. Anal.* 6, 885–893 (1975).
91. Greechie, R., Orthomodular lattices admitting no states, *J. Comb. Theory* 10, 119–132 (1971).
92. Gudder, S., Spectral methods for a generalized probability theory, *Trans. Amer. Math. Soc.* 119, 420–422 (1965).
93. Gudder, S., Uniqueness and existence properties of bounded observables, *Pacific J. Math.* 19, 81–93, 588–589 (1966).
94. Gudder, S., Hilbert space, independence and generalized probability, *J. Math. Anal. Appl.* 20, 48–61 (1967).
95. Gudder, S., Coordinate and momentum observables in axiomatic quantum mechanics, *J. Math. Phys.* 8, 1848–1858 (1967).
96. Gudder, S., Dispersion-free states and the existence of hidden variables, *Proc. Amer. Math. Soc.* 19, 319–324 (1968).
97. Gudder, S., Joint distribution of observables, *J. Math. Mech.* 18, 296–302 (1968).
98. Gudder, S., Elementary length topologies in physics, *SIAM J. Appl. Math.* 16, 1011–1019 (1968).
99. Gudder, S., Quantum probability spaces, *Proc. Amer. Math. Soc.* 21, 296–302 (1969).
100. Gudder, S., On the quantum logic approach to quantum mechanics, *Commun. Math. Phys.* 12, 1–15 (1969).
101. Gudder, S., Axiomatic quantum mechanics and generalized probability theory, in *Probabilistic Methods in Applied Mathematics* (A. Bharucha-Reid, ed.), New York: Academic Press, 1970.

102. Gudder, S., A superposition principle in physics, *J. Math. Phys.* 11, 1037–1040 (1970).
103. Gudder, S., On hidden-variable theories, *J. Math. Phys.* 11, 431–436 (1970).
104. Gudder, S., Quantum logics, physical space, position observables and symmetry, *Rep. Math. Phys.* 3, 193–202 (1972).
105. Gudder, S., State automorphisms in axiomatic quantum mechanics, *Intern. J. Theor. Phys.* 7, 205–211 (1973).
106. Gudder, S., Generalized measure theory, *Found. Phys.* 3, 399–411 (1973).
107. Gudder, S., Convex structures and operational quantum mechanics, *Commun. Math. Phys.* 29, 249–264 (1973).
108. Gudder, S., Convexity and mixtures, *SIAM Rev.* 19, 221–240 (1977).
109. Gudder, S., Gaussian random fields, *Found. Phys.* 8, 295–302 (1978).
110. Gudder, S., and S. Boyce, A comparison of the Mackey and Segal models for quantum mechanics, *Intern. J. Theor. Phys.* 3, 7–12 (1970).
111. Gudder, S., and W. Cornett, The mixture of quantum states, *J. Math. Phys.* 15, 842–850 (1974).
112. Gudder, S., and R. Hudson, A noncommutative probability theory, *Trans. Amer. Math. Soc.* 245, 1–41 (1978).
113. Gudder, S., and J. P. Marchand, Non-commutative probability on von Neumann algebras, *J. Math. Phys.* 13, 799–806 (1972).
114. Gudder, S., and J. P. Marchand, Conditional expectations on von Neumann algebras, *Rep. Math. Phys.* 12, 317–329 (1977).
115. Gudder, S., and C. Piron, Observables and the field in quantum mechanics, *J. Math. Phys.* 12, 1583–1588 (1971).
116. Gudder, S., and D. Strawther, Orthogonally additive and orthogonally monotone functions on vector spaces, *Pacific J. Math.* 58, 427–436 (1975).
117. Gudder, S., and D. Strawther, A converse of Pythagoras' theorem, *Amer. Math. Monthly* 84, 551–553 (1977).
118. Guenin, M., Axiomatic foundations of quantum theories, *J. Math. Phys.* 7, 271–282 (1966).
119. Gunson, J., On the algebraic structure of quantum mechanics, *Commun. Math. Phys.* 6, 262–285 (1967).
120. Guz, W., Markovian processes in classical and quantum statistical mechanics, *Rep. Math. Phys.* 7, 205–214 (1975).
121. Haag, R., and D. Kastler, An algebraic approach to quantum field theory, *J. Math. Phys.* 5, 848–861 (1964).
122. Haag, R., and B. Schroer, Postulates of quantum field theory, *J. Math. Phys.* 3, 248–256 (1962).
123. Halmos, P., *Measure Theory*, Princeton, New Jersey: Van Nostrand Reinhold, 1950.
124. Heelen, P., Quantum and classical logic: their respective roles, *Synthèse* 21, 1–33 (1970).
125. Hegerfeldt, G., From Euclidean to relativistic fields and on the notion of Markoff fields, *Commun. Math. Phys.* 35, 155–171 (1974).

126. Hegerfeldt, G., Probability measures on distribution spaces and quantum field theoretical models, *Rep. Math. Phys.* 7, 403–409 (1975).
127. Hegerfeldt, G., and O. Melsheimer, The form of representations of the canonical commutation relations for Bose fields and connection with finitely many degrees of freedom, *Commun. Math. Phys.* 12, 304–323 (1969).
128. Heisenberg, W., *The Physical Principles of Quantum Theory*, (C. Eckhart and F. Hoyt, trans.) New York: Dover, 1930.
129. Hellwig, K., and K. Kraus, Pure operations and measurements, *Commun. Math. Phys.* 11, 214–220 (1969).
130. Hellwig, K., and K. Kraus, Operations and measurements II, *Commun. Math. Phys.* 16, 142–147 (1970).
131. Hille, E., and R. Phillips, *Functional Analysis and Semigroups*, Providence, Rhode Island: Coll. Publ. Amer. Math. Soc., 1957.
132. Holevo, A., An analogue of the theory of statistical decisions in non-commutative probability theory, *Trans. Moscow Math. Soc.* 26, 113–149 (1972).
133. Holevo, A., Statistical decision theory for quantum systems, *J. Mult. Anal.* 3, 337–394 (1973).
134. Holland, S., The current interest in orthomodular lattices, in *Trends in Lattice Theory* (J. Abbott, ed.), Princeton, New Jersey: Van Nostrand Reinhold, 1970.
135. Horvath, J., *Topological Vector Spaces and Distributions*, Reading, Massachusetts: Addison–Wesley, 1966.
136. Ingarden, R., Quantum information theory, *Rep. Math. Phys.* 10, 43–72 (1976).
137. Jammer, M., *The Philosophy of Quantum Mechanics*, New York: Wiley, 1974.
138. Jauch, J., Systems of observables in quantum mechanics, *Helv. Phys. Acta* 33, 711–726 (1960).
139. Jauch, J., The problem of measurement in quantum mechanics, *Helv. Phys. Acta* 34, 293–316 (1961).
140. Jauch, J., *Foundations of Quantum Mechanics*, Reading, Massachusetts: Addison–Wesley, 1968.
141. Jauch, J., and B. Misra, Supersymmetry and essential observables, *Helv. Phys. Acta* 34, 699–709 (1961).
142. Jauch, J., and C. Piron, Can hidden variables be excluded from quantum mechanics?, *Helv. Phys. Acta* 36, 827–837 (1965).
143. Jauch, J., and C. Piron, On the structure of quantal proposition systems, *Helv. Phys. Acta* 42, 842–848 (1969).
144. Jordan, T., and E. Sudarshan, Dynamical mappings of density operators in quantum mechanics, *J. Math. Phys.* 2, 772–775 (1961).
145. Jordan, T., M. Pinsky, and E. Sundarshan, Dynamical mappings of density operators in quantum mechanics II, *J. Math. Phys.* 3, 772–775 (1961).
146. Jordan, P., J. von Neumann, and E. Wigner, On an algebraic generalization of the quantum mechanical formalism, *Ann. Math.* 33, 29–64 (1934).
147. Jost, P., *The General Theory of Quantized Fields*, Providence, Rhode Island: Amer. Math. Soc., 1965.

Bibliography

148. Kac, M., On a characterization of the normal distribution, *Amer. J. Math.* 61, 726–728 (1939).
149. Kadison, R., Isometries of operator algebras, *Ann. Math.* 54, 325–338 (1951).
150. Kadison, R., The energy–momentum spectrum of quantum fields, *Commun. Math. Phys.* 4, 258–260 (1967).
151. Kano, Y., Probability distribution functions relating to blackbody radiation, *J. Phys. Soc. Japan*, 19, 1555–1560 (1964).
152. Karwowski, W., On Borchers class of Markov fields, *Proc. Cambridge Philos. Soc.* 76, 457–463 (1974).
153. Karwowski, W., A class of quantum fields constructed from Markoff fields, *Rep. Math. Phys.* 7, 411–416 (1975).
154. Kastler, D., A C^*–algebra approach to field theory, in *Analysis in Function Spaces* (T. Martin and I. Segal, eds.), Cambridge, Massachusetts: MIT Press, 1964.
155. Kastler, D., The C^*–algebra of a free Boson field, *Commun. Math. Phys.* 1, 14–48 (1965).
156. Kastler, D., Topics in the algebraic approach to field theory, in *Cargese Lectures in Theoretical Physics* (F. Lurcat, ed.), New York: Gordon and Breach, 1967.
157. Kershaw, D., Theory of hidden variables, *Phys. Rev.* 136B, 1850–1856 (1964).
158. Klauder, J., and E. Sudarshan, *Fundamentals of Quantum Optics*, New York: Benjamin, 1968.
159. Kochen, S., and E. Specker, The problem of hidden variables in quantum mechanics, *J. Math. Mech.* 17, 59–87 (1967).
160. Kolmogorov, A., *Foundations of the Theory of Probability*, New York: Chelsea, 1956.
161. Kraus, K., An algebraic spectrum condition, *Commun. Math. Phys.* 16, 138–141 (1970).
162. Kraus, K., General state changes in quantum theory, *Ann. Phys.* 64, 311–335 (1971).
163. Kuo, H., *Gaussian Measures in Banach Spaces*, Berlin and New York: Springer–Verlag, 1975.
164. Lamperti, J., *Probability*, New York: Benjamin, 1966.
165. Lewis, J., and L. Thomas, On the existence of a class of stationary quantum stochastic processes. *Ann. Inst. Henri Poincaré* 22A, 241–248 (1975).
166. Loève, M., *Probability Theory*, Princeton, New Jersey: Van Nostrand Reinhold, 1963.
167. Loomis, L., On the representation of σ–complete Boolean algebras, *Bull. Amer. Math. Soc.* 53, 757–760 (1947).
168. Louton, T., A theorem on simultaneous observability, *Pacific J. Math.* 59, 147–159 (1975).
169. Lowdenslader, D., On postulates for general quantum mechanics, *Proc. Amer. Math. Soc.* 8, 88–91 (1957).
170. Ludwig, G., Attempt at an axiomatic foundation of quantum mechanics and more general theories II, *Commun. Math. Phys.* 4, 331–348 (1967).

171. Ludwig, G., Attempt at an axiomatic foundation of quantum mechanics and more general theories III, *Commun. Math. Phys.* 9, 1–12 (1968).
172. Mackey, G., Quantum mechanics and Hilbert space, *Amer. Math. Monthly* 64, 45–57 (1957).
173. Mackey, G., *The Mathematical Foundations of Quantum Mechanics*, New York: Benjamin, 1963.
174. Mackey, G., *Induced Representations*, New York: Benjamin, 1968.
175. MacLaren, M., Atomic orthocomplemented lattices, *Pacific J. Math.* 14, 597–612 (1964).
176. MacLaren, M., Notes on axioms for quantum mechanics, *AEC Research and Development Report* ANL – 7065 (1965).
177. Maczynski, M., Hilbert space formalism of quantum mechanics without the Hilbert space axiom, *Rep. Math. Phys.* 3, 209–219 (1972).
178. Maczynski, M., The field of real numbers in axiomatic quantum mechanics, *J. Math. Phys.* 14, 1469–1471 (1973).
179. Maczynski, M., Functional properties of quantum logics, *Intern. J. Theor. Phys.* 11, 149–156 (1974).
180. Maczynski, M., Orthomodularity and lattice characterization of Hilbert spaces, *Bull. Acad. Pol. Sci.* 24, 481–484 (1976).
181. Maeda, F., and S. Maeda, *Theory of Symmetric Lattices*, Berlin and New York: Springer–Verlag, 1970.
182. Manuceau, J., C^* – algèbre de relations de commutation, *Ann. Inst. Henri Poincaré* 8, 139–161 (1968).
183. Marchand, J. P., Relative coarse-graining, *Found. Phys.* 7, 35–49 (1977).
184. Marchand, J. P., and W. Wyss, Statistical inference and entropy, *J. Stat. Phys.* 16, 349–355 (1977).
185. McKean, H., *Stochastic Integrals*, New York: Academic Press, 1969.
186. Messiah, A., *Quantum Mechanics*, Amsterdam: North Holland, 1961.
187. Mielnik, B., Geometry of quantum states, *Commun. Math. Phys.* 9, 55–80 (1968).
188. Mielnik, B., Theory of filters, *Commun. Math. Phys.* 15, 1–46 (1969).
189. Mielnik, B., Generalized quantum mechanics, *Commun. Math. Phys.* 37, 221–256 (1974).
190. Moyal, J., Quantum mechanics as a statistical theory, *Proc. Cambridge Philos. Soc.* 45, 99–124 (1947).
191. Mullikin, H., and S. Gudder, Measure theoretic convergences of observables and operators, *J. Math. Phys.* 14, 234–242 (1973).
192. Nakamura, M., and T. Turumaru, Expectations in an operator algebra, *Tohoku Math. J.* 6, 182–188 (1954).
193. Nakamura, M., and H. Umegaki, On von Neumann's theory of measurement in quantum statistics, *Math. Japan.* 7, 151–157 (1962).
194. Naimark, M., *Normed Algebras*, Gröningen, The Netherlands: Noordhoff, 1972.
195. Namioka, I., Partially ordered linear topological spaces, *Mem. Amer. Math. Soc.* 24 (1957).

196. Nelson, E., Derivation of the Schrödinger equation from Newtonian mechanics, *Phys. Rev.* 150, 1079–1085 (1966).

197. Nelson, E., Probability theory and Euclidean field theory, in *Constructive Quantum Field Theory* (G. Velo and O. Wightman, eds.), Berlin and New York: Springer–Verlag, 1973.

198. Nelson, E., *Dynamical Theories of Brownian Motion*, Princeton, New Jersey: Princeton Univ. Press, 1967.

199. Nelson, E., Construction of quantum fields from Markoff Fields, *J. Funct. Anal.* 12, 97–112 (1973).

200. Nelson, E., The free Markov field, *J. Funct. Anal.* 12, 211–227 (1973).

201. Neubrunn, T., A note on quantum probability spaces, *Proc. Amer. Math. Soc.* 25, 672–675 (1970).

202. Newman, C., The construction of stationary two-dimensional Markoff fields with an application to quantum field theory, *J. Funct. Anal.* 14, 44–61 (1973).

203. Noda, A., Gaussian random fields with projective invariances, *Nagoya Math. J.* 59, 65–76 (1975).

204. Ochs, W., On the foundations of quantal proposition systems, *Z. Naturforsch.* 27, 893–900 (1972).

205. Ochs, W., On the covering law in quantal proposition systems, *Commun. Math. Phys.* 25, 245–252 (1972).

206. Ochs, W., On the strong law of large numbers in quantum probability theory, *Univ. of Munich, Dept. of Physics* (preprint).

207. Ord, W., On some causes of Brownian movements, *J. Royal Micr. Soc.* 2, 656–662 (1879).

208. Packel, E., Hilbert space operators and quantum mechanics, *Amer. Math. Monthly* 81, 863–873 (1974).

209. Park, J., and H. Margenau, Simultaneous measurability in quantum theory, *Inter. J. Theor. Phys.* 1, 211–283 (1968).

210. Perrin, J., Brownian movement and molecular reality, *Ann. Chimie Phys.* 8 (1909).

211. Piron, C., Axiomatique quantique, *Helv. Phys. Acta* 37, 439–468 (1964).

212. Piron, C., *Foundations of Quantum Mechanics*, New York: Benjamin, 1976.

213. Plymen, R., A modification of Piron's axioms, *Helv. Phys. Acta* 41, 69–74 (1968).

214. Plymen, R., C*-algebras and Mackey's axioms, *Commun. Math. Phys.* 8, 132–146 (1968).

215. Pool, J., Baer *-semigroups and the logic of quantum mechanics, *Commun. Math. Phys.* 9, 118–141 (1968).

216. Pool, J., Semi-modularity and the logic of quantum mechanics, *Commun. Math. Phys.* 9, 212–228 (1968).

217. Powers, R., Self-adjoint algebras of unbounded operators, *Commun. Math. Phys.* 21, 85–124 (1971).

218. Prohorov, Yu., Convergence of random processes and limit theorems in probability theory, *Teor. Veroj. Prim.* 1, 177–238 (1956).

219. Prugovecki, E., *Quantum Mechanics in Hilbert Space*, New York: Academic Press, 1971.

220. Putnam, H., Is logic empirical?, *Boston Studies in the Philosophy of Science* 5, 216–241 (1969).

221. Ramsey, A., A theorem on two commuting observables, *J. Math. Mech.* 15, 227–234 (1966).

222. Randall, C., and D. Foulis, Operational statistics II, manuals of operations and their logics, *J. Math. Phys.* 14, 1472–1480 (1972).

223. Rao, M., Local functionals and generalized random fields, *Bull. Amer. Math. Soc.* 74, 288–293 (1968).

224. Rao, M., Local functionals and generalized random fields with independent values, *Teor. Veroj. Prim.* 16, 466–483 (1971).

225. Reed, M., and B. Simon, *Methods of Modern Mathematical Physics, Vol. 1, Functional Analysis*, New York: Academic Press, 1972.

226. Rickart, C., *General Theory of Banach Algebras*, Princeton, New Jersey: Van Nostrand Reinhold, 1960.

227. Riesz, F., and B. Sz. Nagy, *Functional Analysis*, New York: Ungar, 1955.

228. Roberts, J., and G. Roepstorff, Some basic concepts of algebraic quantum theory, *Commun. Math. Phys.* 11, 321–338 (1969).

229. Robinson, D., Algebraic aspects of relativistic quantum field, in *Axiomatic Field Theory* (M. Chretien and S. Deser, eds.), New York: Gordon and Breach, 1966.

230. Roman, P., *Introduction to Quantum Field Theory*, New York: John Wiley, 1969.

231. Ruelle, D., Quantum statistical mechanics and canonical commutation relations, in *Cargese Lectures in Theoretical Physics* (F. Lucat, ed.), New York: Gordon and Breach, 1967.

232. Rüttiman, G., Jauch–Piron states, *J. Math. Phys.* 18, 189–193 (1977).

233. Rylov, Y., Quantum mechanics as a theory of relativistic Brownian motion, *Ann. Phys.* 27, 1–11 (1971).

234. Sakai, S., *C*-Algebras and W*-Algebras*, Berlin and New York: Springer-Verlag, 1971.

235. Schrödinger, E., *Collected Papers on Wave Mechanics* (J. Shearer and W. Deans, trans.), London: Blackie and Sons, 1928.

236. Schwartz, L., *Théorie des distributions*, Paris: Hermann, 1951.

237. Schweber, S., *An Introduction to Relativistic Quantum Field Theory*, Evanston, Illinois: Row and Peterson, 1961.

238. Segal, I., Postulates for general quantum mechanics, *Ann. Math.* 48, 930–948 (1947).

239. Segal, I., A non-commutative extension of abstract integration, *Ann. Math.* 57, 401–457 (1953).

240. Segal, I., Abstract probability spaces and a theorem of Kolmogoroff, *Amer. J. Math.* 76, 721–732 (1954).

241. Segal, I., Tensor algebras over Hilbert space I, *Trans. Amer. Math. Soc.* 81, 106–134 (1956).

242. Segal, I., Tensor algebras over Hilbert space II, *Ann. Math.* 63, 160–175 (1956).

Bibliography

243. Segal, I., Foundations of the theory of infinite-dimensional dynamical systems II, *Canad. J. Math.* 13, 1–18 (1961).
244. Segal, I., Mathematical characterization of the physical vacuum for a linear Bose Einstein field, *Ill. J. Math.* 6, 500–523 (1962).
245. Segal, I., *Mathematical Problems of Relativistic Physics*, Providence, Rhode Island: Amer. Math. Soc., 1963.
246. She, C., and H. Heffner, Simultaneous measurement of non-commuting observables, *Phys. Rev.* 152, 1103–1110 (1966).
247. Sherman, S., Non-negative observables are squares, *Proc. Amer. Math. Soc.* 2, 31–33 (1951).
248. Sherman, S., On Segal's postulates for general quantum mechanics, *Ann. Math.* 64, 593–601 (1956).
249. Sikorski, R., On the inducing of homomorphisms by mappings, *Fund. Math.* 36, 7–22 (1949).
250. Simon, B., *The $P(\phi)_2$ Euclidean Quantum Field Theory*, Princeton, New Jersey: Princeton Univ. Press, 1974.
251. Srinivas, M., Foundations of quantum probability theory, *J. Math. Phys.* 16, 1672–1685 (1973).
252. Stone, M., On one-parameter unitary groups in Hilbert Space, *Ann. Math.* 33, 643–648 (1932).
253. Stone, M., Postulates for the barycentric calculus, *Ann. Math.* 29, 25–30 (1949).
254. Streater, R., Outline of axiomatic relativistic quantum field theory, *Rep. Prog. Phys.* 38, 771–846 (1975).
255. Streater, R., and A. Wightman, *PCT, Spin and Statistics and All That*, New York: Benjamin, 1964.
256. Streit, L., Test function spaces for direct product representations of the canonical commutation relations, *Commun. Math. Phys.* 4, 22–31 (1967).
257. Streit, L., Generalized free fields as cyclic representations of the canonical commutation relations, *Acta. Phys. Austriaca* 32, 107–112 (1970).
258. Sundaresan, K., Orthogonality and nonlinear functionals on Banach Spaces, *Proc. Amer. Math. Soc.* 34, 187–190 (1972).
259. Suppes, P., The probabilistic argument for a nonclassical logic for quantum mechanics, *Philos. Sci.* 33, 14–21 (1966).
260. Symanzik, K., Euclidean quantum field theory I. Equations for a scalar model, *J. Math. Phys.* 7, 510–525 (1966).
261. Uhlmann, A., The "transition probability" in the state space of a *-algebra, *Rep. Math. Phys.* 9, 273–279 (1976).
262. Umegaki, H., Conditional expectation in an operator Algebra I, *Tohoku Math. J.* 6, 177–181 (1954).
263. Umegaki, H., Conditional expectation in an operator algebra IV, *Kodai. Math. Sem. Rep.* 14, 59–85 (1962).
264. Urbanik, K., Generalized stochastic processes with independent values, *Fourth Berkeley Symposium* 2, 569–580 (1961).
265. Urbanik, K., Joint distribution of observables in quantum mechanics, *Studia Math.* 21, 117–133 (1961).

266. Van Daele, A., and A. Verbeure, Unitary equivalence of Fock representation of the Weyl algebra, *Commun. Math. Phys.* 20, 268–278 (1971).

267. Varadarajan, V., Probability in physics and a theorem on simultaneous observability, *Commun. Pure Appl. Math.* 15, 189–217 (1962).

268. Varadarajan, V., *Geometry of Quantum Theory*, Vols. 1 and 2, 1968 and 1970. Princeton, New Jersey: Van Nostrand Reinhold.

269. von Neumann, J., On an algebraic generalization of the quantum mechanical formalism I, *Mat. Sb.* 11, 429–559 (1936). [Also in *Collected Works*, Vol. 3, New York: Pergamon, 1961.]

270. von Neumann, J., *Mathematical Foundations of Quantum Mechanics*, Princeton, New Jersey: Princeton Univ. Press, 1955.

271. von Neumann, J., and O. Morgenstern, *Theory of Games and Economic Behavior*, Princeton, New Jersey: Princeton Univ. Press, 1944.

272. Watanabe, S., Evaluation and selection of variables in pattern recognition, in *Computer and Information Sciences*, Vol. 2, (J. Tow, ed.), New York: Academic Press, 1967.

273. Watanabe, S., Modified concepts of logic, probability and information based on generalized continuous characteristic function, *Information and Control* 2, 1–21 (1969).

274. Watanabe, S., *Pattern Recognition as Information Compression, Frontiers of Pattern Recognition*, New York: Academic Press, 1972.

275. Weisskopf, V., *Physics in the Twentieth Century*, Cambridge, Massachusetts: MIT Press, 1972.

276. Wiener, N., Differential space, *J. Math. Phys. Math. Inst. Tech.* 2, 131–174 (1923).

277. Wightman, A., Quantum field theory in terms of vacuum expectation values, *Phys. Rev.* 110, 860–866 (1956).

278. Wightman, A., Quelques problèmes mathématiques de la théorie quantique relativiste, in *Les problèmes mathématiques de la théorie quantique des champs*, Paris: CNRS, 1959.

279. Wightman, A., What is the point of the so-called "axiomatic field theory"?, *Phys. Today* 22, 53–58 (1969).

280. Wigner, E., On the quantum correction for thermodynamic equilibrium, *Phys. Rev.* 40 (2), 749–759 (1932).

281. Wilde, J., The free Fermion field as a Markov field, *J. Funct. Anal.* 15, 12–21 (1974).

282. Woyczynski, W., Additive operators, *Bull. Acad. Polon. Sci.* 17, 447–451 (1969).

283. Yeh, J., *Stochastic Processes and the Wiener Integral*, New York: Marcel Dekker, 1973.

284. Zierler, N., Axioms for non-relativistic quantum mechanics, *Pacific J. Math.* 11, 1151–1169 (1961).

285. Zierler, N., Order properties of bounded observables, *Proc. Amer. Math. Soc.* 14, 346–351 (1963).

286. Zierler, N., and M. Schlessinger, Boolean embeddings of orthomodular sets and quantum logic, *Duke J. Math.* 32, 251–262 (1965).

Index

A CATALOG OF SELECTED
DOVER BOOKS
IN SCIENCE AND MATHEMATICS

Physics

OPTICAL RESONANCE AND TWO-LEVEL ATOMS, L. Allen and J. H. Eberly. Clear, comprehensive introduction to basic principles behind all quantum optical resonance phenomena. 53 illustrations. Preface. Index. 256pp. 5⅜ x 8½. 0-486-65533-4

QUANTUM THEORY, David Bohm. This advanced undergraduate-level text presents the quantum theory in terms of qualitative and imaginative concepts, followed by specific applications worked out in mathematical detail. Preface. Index. 655pp. 5⅜ x 8½. 0-486-65969-0

ATOMIC PHYSICS (8th EDITION), Max Born. Nobel laureate's lucid treatment of kinetic theory of gases, elementary particles, nuclear atom, wave-corpuscles, atomic structure and spectral lines, much more. Over 40 appendices, bibliography. 495pp. 5⅜ x 8½. 0-486-65984-4

A SOPHISTICATE'S PRIMER OF RELATIVITY, P. W. Bridgman. Geared toward readers already acquainted with special relativity, this book transcends the view of theory as a working tool to answer natural questions: What is a frame of reference? What is a "law of nature"? What is the role of the "observer"? Extensive treatment, written in terms accessible to those without a scientific background. 1983 ed. xlviii+172pp. 5⅜ x 8½. 0-486-42549-5

AN INTRODUCTION TO HAMILTONIAN OPTICS, H. A. Buchdahl. Detailed account of the Hamiltonian treatment of aberration theory in geometrical optics. Many classes of optical systems defined in terms of the symmetries they possess. Problems with detailed solutions. 1970 edition. xv + 360pp. 5⅜ x 8½. 0-486-67597-1

PRIMER OF QUANTUM MECHANICS, Marvin Chester. Introductory text examines the classical quantum bead on a track: its state and representations; operator eigenvalues; harmonic oscillator and bound bead in a symmetric force field; and bead in a spherical shell. Other topics include spin, matrices, and the structure of quantum mechanics; the simplest atom; indistinguishable particles; and stationary-state perturbation theory. 1992 ed. xiv+314pp. 6⅛ x 9¼. 0-486-42878-8

LECTURES ON QUANTUM MECHANICS, Paul A. M. Dirac. Four concise, brilliant lectures on mathematical methods in quantum mechanics from Nobel Prize-winning quantum pioneer build on idea of visualizing quantum theory through the use of classical mechanics. 96pp. 5⅜ x 8½. 0-486-41713-1

THIRTY YEARS THAT SHOOK PHYSICS: THE STORY OF QUANTUM THEORY, George Gamow. Lucid, accessible introduction to influential theory of energy and matter. Careful explanations of Dirac's anti-particles, Bohr's model of the atom, much more. 12 plates. Numerous drawings. 240pp. 5⅜ x 8½. 0-486-24895-X

ELECTRONIC STRUCTURE AND THE PROPERTIES OF SOLIDS: THE PHYSICS OF THE CHEMICAL BOND, Walter A. Harrison. Innovative text offers basic understanding of the electronic structure of covalent and ionic solids, simple metals, transition metals and their compounds. Problems. 1980 edition. 582pp. 6⅛ x 9¼. 0-486-66021-4

HYDRODYNAMIC AND HYDROMAGNETIC STABILITY, S. Chandrasekhar. Lucid examination of the Rayleigh-Benard problem; clear coverage of the theory of instabilities causing convection. 704pp. 5⅝ x 8¼. 0-486-64071-X

INVESTIGATIONS ON THE THEORY OF THE BROWNIAN MOVEMENT, Albert Einstein. Five papers (1905–8) investigating dynamics of Brownian motion and evolving elementary theory. Notes by R. Fürth. 122pp. 5⅜ x 8½. 0-486-60304-0

THE PHYSICS OF WAVES, William C. Elmore and Mark A. Heald. Unique overview of classical wave theory. Acoustics, optics, electromagnetic radiation, more. Ideal as classroom text or for self-study. Problems. 477pp. 5⅜ x 8½. 0-486-64926-1

GRAVITY, George Gamow. Distinguished physicist and teacher takes reader-friendly look at three scientists whose work unlocked many of the mysteries behind the laws of physics: Galileo, Newton, and Einstein. Most of the book focuses on Newton's ideas, with a concluding chapter on post-Einsteinian speculations concerning the relationship between gravity and other physical phenomena. 160pp. 5⅜ x 8½. 0-486-42563-0

PHYSICAL PRINCIPLES OF THE QUANTUM THEORY, Werner Heisenberg. Nobel Laureate discusses quantum theory, uncertainty, wave mechanics, work of Dirac, Schroedinger, Compton, Wilson, Einstein, etc. 184pp. 5⅜ x 8½. 0-486-60113-7

ATOMIC SPECTRA AND ATOMIC STRUCTURE, Gerhard Herzberg. One of best introductions; especially for specialist in other fields. Treatment is physical rather than mathematical. 80 illustrations. 257pp. 5⅜ x 8½. 0-486-60115-3

AN INTRODUCTION TO STATISTICAL THERMODYNAMICS, Terrell L. Hill. Excellent basic text offers wide-ranging coverage of quantum statistical mechanics, systems of interacting molecules, quantum statistics, more. 523pp. 5⅜ x 8½. 0-486-65242-4

THEORETICAL PHYSICS, Georg Joos, with Ira M. Freeman. Classic overview covers essential math, mechanics, electromagnetic theory, thermodynamics, quantum mechanics, nuclear physics, other topics. First paperback edition. xxiii + 885pp. 5⅜ x 8½. 0-486-65227-0

PROBLEMS AND SOLUTIONS IN QUANTUM CHEMISTRY AND PHYSICS, Charles S. Johnson, Jr. and Lee G. Pedersen. Unusually varied problems, detailed solutions in coverage of quantum mechanics, wave mechanics, angular momentum, molecular spectroscopy, more. 280 problems plus 139 supplementary exercises. 430pp. 6½ x 9¼. 0-486-65236-X

THEORETICAL SOLID STATE PHYSICS, Vol. 1: Perfect Lattices in Equilibrium; Vol. II: Non-Equilibrium and Disorder, William Jones and Norman H. March. Monumental reference work covers fundamental theory of equilibrium properties of perfect crystalline solids, non-equilibrium properties, defects and disordered systems. Appendices. Problems. Preface. Diagrams. Index. Bibliography. Total of 1,301pp. 5⅜ x 8½. Two volumes. Vol. I: 0-486-65015-4 Vol. II: 0-486-65016-2

WHAT IS RELATIVITY? L. D. Landau and G. B. Rumer. Written by a Nobel Prize physicist and his distinguished colleague, this compelling book explains the special theory of relativity to readers with no scientific background, using such familiar objects as trains, rulers, and clocks. 1960 ed. vi+72pp. 5⅜ x 8½. 0-486-42806-0

Mathematics

FUNCTIONAL ANALYSIS (Second Corrected Edition), George Bachman and Lawrence Narici. Excellent treatment of subject geared toward students with background in linear algebra, advanced calculus, physics and engineering. Text covers introduction to inner-product spaces, normed, metric spaces, and topological spaces; complete orthonormal sets, the Hahn-Banach Theorem and its consequences, and many other related subjects. 1966 ed. 544pp. 6⅛ x 9¼. 0-486-40251-7

ASYMPTOTIC EXPANSIONS OF INTEGRALS, Norman Bleistein & Richard A. Handelsman. Best introduction to important field with applications in a variety of scientific disciplines. New preface. Problems. Diagrams. Tables. Bibliography. Index. 448pp. 5⅜ x 8½. 0-486-65082-0

VECTOR AND TENSOR ANALYSIS WITH APPLICATIONS, A. I. Borisenko and I. E. Tarapov. Concise introduction. Worked-out problems, solutions, exercises. 257pp. 5⅝ x 8¼. 0-486-63833-2

AN INTRODUCTION TO ORDINARY DIFFERENTIAL EQUATIONS, Earl A. Coddington. A thorough and systematic first course in elementary differential equations for undergraduates in mathematics and science, with many exercises and problems (with answers). Index. 304pp. 5⅜ x 8½. 0-486-65942-9

FOURIER SERIES AND ORTHOGONAL FUNCTIONS, Harry F. Davis. An incisive text combining theory and practical example to introduce Fourier series, orthogonal functions and applications of the Fourier method to boundary-value problems. 570 exercises. Answers and notes. 416pp. 5⅜ x 8½. 0-486-65973-9

COMPUTABILITY AND UNSOLVABILITY, Martin Davis. Classic graduate-level introduction to theory of computability, usually referred to as theory of recurrent functions. New preface and appendix. 288pp. 5⅜ x 8½. 0-486-61471-9

ASYMPTOTIC METHODS IN ANALYSIS, N. G. de Bruijn. An inexpensive, comprehensive guide to asymptotic methods–the pioneering work that teaches by explaining worked examples in detail. Index. 224pp. 5⅜ x 8½ 0-486-64221-6

APPLIED COMPLEX VARIABLES, John W. Dettman. Step-by-step coverage of fundamentals of analytic function theory–plus lucid exposition of five important applications: Potential Theory; Ordinary Differential Equations; Fourier Transforms; Laplace Transforms; Asymptotic Expansions. 66 figures. Exercises at chapter ends. 512pp. 5⅜ x 8½. 0-486-64670-X

INTRODUCTION TO LINEAR ALGEBRA AND DIFFERENTIAL EQUATIONS, John W. Dettman. Excellent text covers complex numbers, determinants, orthonormal bases, Laplace transforms, much more. Exercises with solutions. Undergraduate level. 416pp. 5⅜ x 8½. 0-486-65191-6

RIEMANN'S ZETA FUNCTION, H. M. Edwards. Superb, high-level study of landmark 1859 publication entitled "On the Number of Primes Less Than a Given Magnitude" traces developments in mathematical theory that it inspired. xiv+315pp. 5⅜ x 8½. 0-486-41740-9

CALCULUS OF VARIATIONS WITH APPLICATIONS, George M. Ewing. Applications-oriented introduction to variational theory develops insight and promotes understanding of specialized books, research papers. Suitable for advanced undergraduate/graduate students as primary, supplementary text. 352pp. 5⅜ x 8½.
0-486-64856-7

COMPLEX VARIABLES, Francis J. Flanigan. Unusual approach, delaying complex algebra till harmonic functions have been analyzed from real variable viewpoint. Includes problems with answers. 364pp. 5⅜ x 8½. 0-486-61388-7

AN INTRODUCTION TO THE CALCULUS OF VARIATIONS, Charles Fox. Graduate-level text covers variations of an integral, isoperimetrical problems, least action, special relativity, approximations, more. References. 279pp. 5⅜ x 8½.
0-486-65499-0

COUNTEREXAMPLES IN ANALYSIS, Bernard R. Gelbaum and John M. H. Olmsted. These counterexamples deal mostly with the part of analysis known as "real variables." The first half covers the real number system, and the second half encompasses higher dimensions. 1962 edition. xxiv+198pp. 5⅜ x 8½. 0-486-42875-3

CATASTROPHE THEORY FOR SCIENTISTS AND ENGINEERS, Robert Gilmore. Advanced-level treatment describes mathematics of theory grounded in the work of Poincaré, R. Thom, other mathematicians. Also important applications to problems in mathematics, physics, chemistry and engineering. 1981 edition. References. 28 tables. 397 black-and-white illustrations. xvii + 666pp. 6⅛ x 9¼.
0-486-67539-4

INTRODUCTION TO DIFFERENCE EQUATIONS, Samuel Goldberg. Exceptionally clear exposition of important discipline with applications to sociology, psychology, economics. Many illustrative examples; over 250 problems. 260pp. 5⅜ x 8½.
0-486-65084-7

NUMERICAL METHODS FOR SCIENTISTS AND ENGINEERS, Richard Hamming. Classic text stresses frequency approach in coverage of algorithms, polynomial approximation, Fourier approximation, exponential approximation, other topics. Revised and enlarged 2nd edition. 721pp. 5⅜ x 8½. 0-486-65241-6

INTRODUCTION TO NUMERICAL ANALYSIS (2nd Edition), F. B. Hildebrand. Classic, fundamental treatment covers computation, approximation, interpolation, numerical differentiation and integration, other topics. 150 new problems. 669pp. 5⅜ x 8½. 0-486-65363-3

THREE PEARLS OF NUMBER THEORY, A. Y. Khinchin. Three compelling puzzles require proof of a basic law governing the world of numbers. Challenges concern van der Waerden's theorem, the Landau-Schnirelmann hypothesis and Mann's theorem, and a solution to Waring's problem. Solutions included. 64pp. 5⅜ x 8½.
0-486-40026-3

THE PHILOSOPHY OF MATHEMATICS: AN INTRODUCTORY ESSAY, Stephan Körner. Surveys the views of Plato, Aristotle, Leibniz & Kant concerning propositions and theories of applied and pure mathematics. Introduction. Two appendices. Index. 198pp. 5⅜ x 8½. 0-486-25048-2

INTRODUCTORY REAL ANALYSIS, A.N. Kolmogorov, S. V. Fomin. Translated by Richard A. Silverman. Self-contained, evenly paced introduction to real and functional analysis. Some 350 problems. 403pp. 5⅜ x 8½. 0-486-61226-0

APPLIED ANALYSIS, Cornelius Lanczos. Classic work on analysis and design of finite processes for approximating solution of analytical problems. Algebraic equations, matrices, harmonic analysis, quadrature methods, much more. 559pp. 5⅜ x 8½. 0-486-65656-X

AN INTRODUCTION TO ALGEBRAIC STRUCTURES, Joseph Landin. Superb self-contained text covers "abstract algebra": sets and numbers, theory of groups, theory of rings, much more. Numerous well-chosen examples, exercises. 247pp. 5⅜ x 8½. 0-486-65940-2

QUALITATIVE THEORY OF DIFFERENTIAL EQUATIONS, V. V. Nemytskii and V.V. Stepanov. Classic graduate-level text by two prominent Soviet mathematicians covers classical differential equations as well as topological dynamics and ergodic theory. Bibliographies. 523pp. 5⅜ x 8½. 0-486-65954-2

THEORY OF MATRICES, Sam Perlis. Outstanding text covering rank, nonsingularity and inverses in connection with the development of canonical matrices under the relation of equivalence, and without the intervention of determinants. Includes exercises. 237pp. 5⅜ x 8½. 0-486-66810-X

INTRODUCTION TO ANALYSIS, Maxwell Rosenlicht. Unusually clear, accessible coverage of set theory, real number system, metric spaces, continuous functions, Riemann integration, multiple integrals, more. Wide range of problems. Undergraduate level. Bibliography. 254pp. 5⅜ x 8½. 0-486-65038-3

MODERN NONLINEAR EQUATIONS, Thomas L. Saaty. Emphasizes practical solution of problems; covers seven types of equations. ". . . a welcome contribution to the existing literature...."–*Math Reviews.* 490pp. 5⅜ x 8½. 0-486-64232-1

MATRICES AND LINEAR ALGEBRA, Hans Schneider and George Phillip Barker. Basic textbook covers theory of matrices and its applications to systems of linear equations and related topics such as determinants, eigenvalues and differential equations. Numerous exercises. 432pp. 5⅜ x 8½. 0-486-66014-1

LINEAR ALGEBRA, Georgi E. Shilov. Determinants, linear spaces, matrix algebras, similar topics. For advanced undergraduates, graduates. Silverman translation. 387pp. 5⅜ x 8½. 0-486-63518-X

ELEMENTS OF REAL ANALYSIS, David A. Sprecher. Classic text covers fundamental concepts, real number system, point sets, functions of a real variable, Fourier series, much more. Over 500 exercises. 352pp. 5⅜ x 8½. 0-486-65385-4

SET THEORY AND LOGIC, Robert R. Stoll. Lucid introduction to unified theory of mathematical concepts. Set theory and logic seen as tools for conceptual understanding of real number system. 496pp. 5⅜ x 8¼. 0-486-63829-4

CATALOG OF DOVER BOOKS

TENSOR CALCULUS, J.L. Synge and A. Schild. Widely used introductory text covers spaces and tensors, basic operations in Riemannian space, non-Riemannian spaces, etc. 324pp. 5⅜ x 8¼. 0-486-63612-7

ORDINARY DIFFERENTIAL EQUATIONS, Morris Tenenbaum and Harry Pollard. Exhaustive survey of ordinary differential equations for undergraduates in mathematics, engineering, science. Thorough analysis of theorems. Diagrams. Bibliography. Index. 818pp. 5⅜ x 8½. 0-486-64940-7

INTEGRAL EQUATIONS, F. G. Tricomi. Authoritative, well-written treatment of extremely useful mathematical tool with wide applications. Volterra Equations, Fredholm Equations, much more. Advanced undergraduate to graduate level. Exercises. Bibliography. 238pp. 5⅜ x 8½. 0-486-64828-1

FOURIER SERIES, Georgi P. Tolstov. Translated by Richard A. Silverman. A valuable addition to the literature on the subject, moving clearly from subject to subject and theorem to theorem. 107 problems, answers. 336pp. 5⅜ x 8½. 0-486-63317-9

INTRODUCTION TO MATHEMATICAL THINKING, Friedrich Waismann. Examinations of arithmetic, geometry, and theory of integers; rational and natural numbers; complete induction; limit and point of accumulation; remarkable curves; complex and hypercomplex numbers, more. 1959 ed. 27 figures. xii+260pp. 5⅜ x 8½. 0-486-63317-9

POPULAR LECTURES ON MATHEMATICAL LOGIC, Hao Wang. Noted logician's lucid treatment of historical developments, set theory, model theory, recursion theory and constructivism, proof theory, more. 3 appendixes. Bibliography. 1981 edition. ix + 283pp. 5⅜ x 8½. 0-486-67632-3

CALCULUS OF VARIATIONS, Robert Weinstock. Basic introduction covering isoperimetric problems, theory of elasticity, quantum mechanics, electrostatics, etc. Exercises throughout. 326pp. 5⅜ x 8½. 0-486-63069-2

THE CONTINUUM: A CRITICAL EXAMINATION OF THE FOUNDATION OF ANALYSIS, Hermann Weyl. Classic of 20th-century foundational research deals with the conceptual problem posed by the continuum. 156pp. 5⅜ x 8½. 0-486-67982-9

CHALLENGING MATHEMATICAL PROBLEMS WITH ELEMENTARY SOLUTIONS, A. M. Yaglom and I. M. Yaglom. Over 170 challenging problems on probability theory, combinatorial analysis, points and lines, topology, convex polygons, many other topics. Solutions. Total of 445pp. 5⅜ x 8½. Two-vol. set.
Vol. I: 0-486-65536-9 Vol. II: 0-486-65537-7